T0142317

Software Sustainability

Coral Calero • Mª Ángeles Moraga • Mario Piattini
Editors

Software Sustainability

Editors
Coral Calero
Alarcos Research Group, Institute of Technologies and Information Systems
University of Castilla-La Mancha (UCLM)
Ciudad Real, Spain

Mª Ángeles Moraga
Alarcos Research Group, Institute of Technologies and Information Systems
University of Castilla-La Mancha (UCLM)
Ciudad Real, Spain

Mario Piattini
Alarcos Research Group, Institute of Technologies and Information Systems
University of Castilla-La Mancha (UCLM)
Ciudad Real, Spain

ISBN 978-3-030-69972-7 ISBN 978-3-030-69970-3 (eBook)
https://doi.org/10.1007/978-3-030-69970-3

This Springer imprint is published by the registered company Springer Nature Switzerland AG.
The registered company address is: Gewerbestrasse 11, 6330 Cham, Switzerland

The editors want to dedicate this book to the Green Team from the Alarcos research group for their great work. We also want to dedicate it to all the readers who are interested in the sustainability of software.

To Olivia and Moisés. Because you deserve the best world to live!!

Coral Calero

To my sons, Carlos and Javier, and my niece, Elsa, who make my life more sustainable.

Mª Ángeles Moraga

To Beatriz, Catherine, and Sienna

Mario Piattini

Preface

Overview

The preservation of the environment has become one of the most urgent concerns of today's society. People have become aware of the need to cut down on energy consumption and to reduce our carbon footprint. This means that sustainability has arisen as a key aspect in several domains, guiding the development of the world's future. At an international level, there are many initiatives aiming to address these issues, and the main research and development programs include sizeable amounts of funding for projects seeking to achieve environmentally sound technologies. Also, at a governmental level there are efforts to align societal development with the goals of sustainability. The Paris Agreement is a good example of how countries (as representatives of their citizens) are involved in an effort to combat global climate change so as to ensure the best quality of life. According to the UN Climate Change website,[1] "... the Paris Agreement's central aim is to strengthen the global response to the threat of climate change." "It also aims to strengthen countries' ability to deal with the impacts of climate change," and "support them in their efforts...to make finance flows consistent with a pathway towards low greenhouse gas emissions and climate-resilient development." "To reach these ambitious goals, appropriate mobilisation and provision of financial resources, a new technology framework and enhanced capacity-building is to be put in place, thus supporting action by developing countries and the most vulnerable countries, in line with their own national objectives" (UN Climate Change Secretariat 2015).

But sustainability is not just a matter of CO_2 emissions, it also depends on other aspects. For example, it is necessary to develop software that takes into account not only its own energy efficiency but also aspects related to the amount of resources needed or the longevity of the software. It is important to integrate sustainability into

[1] https://unfccc.int/process-and-meetings/the-paris-agreement/the-paris-agreement/key-aspects-of-the-paris-agreement

the core business processes devoted to producing software or services, ensuring the continuity of the software industry and implementing appropriate risk-management programs and policies. And last, but not least, software workers must be taken into consideration as part of software sustainability: ethics, rights, protection, and training are among the necessary initiatives to support them.

There are therefore three key aspects of sustainability: the environment, society, and the actions necessary to ensure economic sustainability.

In fact, the United Nations' Brundtland Report defines sustainable development as the ability to "meet the needs of the present without compromising the ability of future generations to satisfy their own needs."[2]

Given the great relevance that software has today and the fact that it seems it will be even greater in the future, it is of utmost importance to consider sustainability as a key feature. The focus of this book is therefore on software sustainability, examined in terms of how software can be developed while taking into consideration environmental, social, and economic dimensions so as to meet the needs of the present without compromising the future.

Software sustainability has three dimensions (Fig. 1):

- Environmental sustainability: how software product development, maintenance, and use affect energy consumption and the consumption of other natural resources. Environmental sustainability is directly related to a software product characteristic. This dimension is also known as Green Software.
- Human sustainability: how software creation and use affect the sociological and psychological aspects of the software development community and its members. This encompasses topics such as labour rights, psychological health, social support, social equity, and liveability.
- Economic sustainability: how software development and use protect stakeholders' investments, ensure benefits, reduce risks, and maintain assets.

A topic that has attracted a lot of attention in the last year is Green IT and Green Software. As seen in Fig. 1, Green Software can be divided into Green IN Software (when the environmental issues are related to software itself) and Green BY Software (when software is used as a tool to support sustainability goals in any domain). As our focus is specifically on how software must be produced so as to be sustainable, the book will be focused on the Green IN Software part.

The aim of this book is therefore to present the latest advances related to software sustainability, the scope being those pieces of work developed within the environmental (Green Software), human, or economic dimensions of software sustainability, by way of a contribution on our part to raising the profile of software sustainability. To that end, we have brought together the findings on this matter of the main researchers in the field.

[2]United Nations World Commission on Environment and Development, "Report of the World Commission on Environment and Development: Our Common Future." At United Nations Conference on Environment and Development. 1987.

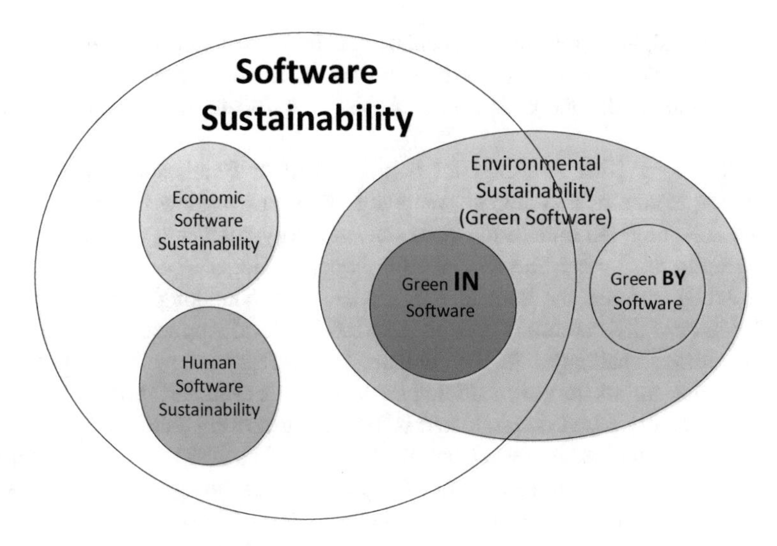

Fig. 1 Software sustainability dimensions

Organization

The book is composed of 16 chapters structured as follows.

Chapter 1, written by the editors, introduces the main general concepts related to software sustainability and defines its dimensions.

Then a set of chapters deal with the environmental (green) software dimension.

In Chap. 2, Achim Guldner, Eva Kern, Sandro Kreten, and Stefan Naumann describe a set of criteria and a label for sustainable software products, the German "Blue Angel," energy-efficient programming techniques which seek to reduce the consumption and provide a measurement method for the energy consumption of software.

In Chap. 3, Javier Mancebo, Ignacio Rodríguez, Mª Ángeles Moraga, Félix García, and Coral Calero present GSMP, a Green Software Measurement Process that integrates all the activities that must be carried out to measure and analyze the energy consumption of any given software.

In Chap. 4, Javier Mancebo, Coral Calero, and Félix García present FEETINGS (Framework for Energy Efficiency Testing to Improve eNvironmental Goals of the Software), a complete framework which aims to provide: (1) a solution to the lack of a unique and agreed terminology; (2) a process to evaluate the energy efficiency of the software; and (3) a technological environment that supports the process.

Chapter 5, by Daniel Feitosa, Luís Cruz, Rui Abreu, João Paulo Fernandes, Marco Couto, and João Saraiva, aims to demonstrate how patterns can help to build energy-efficient software.

In Chap. 6, written by Wellington Oliveira, Hugo Matalonga, Gustavo Pinto, Fernando Castor, and João Paulo Fernandes, the authors advocate that developers should leverage software diversity to make software systems more energy efficient.

Their main insight is that non-specialists can build software that consumes less energy in the development stage by alternating, at development time, between readily available, diversely designed pieces of software implemented by third parties.

Chapter 7, by Hina Anwar, Iffat Fatima, Dietmar Pfahl, and Usman Qamar, presents the results of a systematic mapping study to overview the state-of-the-art tools for detecting and refactoring code smells/energy bugs, and tools for detecting and migrating third-party libraries in Android applications.

Chapter 8, written by Samuel Chinenyeze and Xiaodong Liu, presents the knowledge base and reasoning around the Mobile Cloud Applications domain, and some identified challenges in the domain, and offers some directions and key considerations for an improved implementation and evaluation process in real-life scenarios—mainly based on renowned software engineering techniques.

In Chap. 9, Jutta Eckstein and Claudia Melo provide a new lens through which to understand the Agile Manifesto under the premise that the agile approach aims to fulfill its promise for sustainability, and they provide various case studies of companies attempting to use agile development to contribute to sustainability.

In Chap. 10, J. David Patón-Romero, Teresa Baldasarre, Moisés Rodríguez, and Mario Piattini present GMGIT ("Governance and Management Framework for Green IT") which establishes the characteristics and elements of governance and management that organizations should consider when implementing, assessing, and improving Green IT.

Turning to the human dimension, Chap. 11, by Alok Mishra and Deepti Mishra, looks at how sustainability can be included in various courses of the Software Engineering (SE) curriculum by considering ACM/IEEE guidelines, carrying out a review of the literature in this field, and also examining various viewpoints so that SE students can acquire knowledge of sustainable software engineering. It also includes an assessment of key competences in sustainability for proposed units in the SE curriculum.

Chapter 12, by Asif Imran and Tevfik Kosar, presents a detailed and inclusive study covering human factors (leadership, communication, etc.), related challenges, and approaches to software sustainability. This chapter groups the existing research efforts based on the above aspects. How these aspects affect software sustainability is studied via a survey of software practitioners.

Chapter 13, by Vijanti Ramautar, Sietse Overbeek, and Sergio España, aims to outline outsourcing approaches for facilitating human progress by conducting a semi-systematic literature review. The authors identify three outsourcing approaches that consider corporate social responsibility: impact sourcing, ethical outsourcing, and FairTrade Software. The ultimate aim is to understand the effect of these approaches on marginalized people, and the benefits and challenges for client organizations.

In Chap. 14, Eoin Martino Grua, Martina De Sanctis, Ivano Malavolta, Mark Hoogendoorn, and Patricia Lago present a reference architecture for enabling AI-based personalization and self-adaptation of mobile apps for e-Health. The proposed reference architecture is instantiated in the context of a mobile fitness

Fig. 2 Keyword cloud (created with https://worditout.com)

application and is exemplified through a series of typical usage scenarios extracted from industrial collaborations.

In Chap. 15, Mª Ángeles Moraga, por Ignacio García-Rodriguez de Guzmán, Félix García, and Coral Calero present "The importance of software sustainability in the CSRs of software companies." The authors analyze the Corporate Social Responsibility information of ten software companies and elaborate a list of specific actions related to software sustainability that the software company should include. They also suggest an initial set of actions which could be taken into account to improve the CSR of a company.

And finally, in Chap. 16, Bendra Ojameruaye and Rami Bahsoon propose an economics-driven architectural evaluation method which expands upon CBAM (Cost Benefits Analysis Method) and integrates principles of modern portfolio theory into the task of controlling risks when linking sustainability requirements to architectural design decisions.

We have created a keyword cloud (see Fig. 2) where the terms used most frequently in this book are written in a larger font to highlight the areas of special focus in the book.

Target Readership

The target readership for this book is assumed to have previous knowledge of information systems and software engineering, but we envisage CIOs (Chief Information Officers), CEOs (Chief Executive Officers), CSOs (Chief Sustainability Officers), CSRs (Chief Responsibility Officers), software developers, software managers, auditors, business owners, and quality professionals. It is also intended for masters' and bachelors' students studying Software Systems and Information

Systems, Computer Science, and Computer Engineering, and software researchers who want to inform themselves about the state of the art regarding software sustainability.

Ciudad Real, Spain
December 2020

Coral Calero
Mª Ángeles Moraga
Mario Piattini

Acknowledgments

We would like to express our gratitude to all those individuals and parties who helped us to produce this volume. In the first place, we would like to thank all the contributing authors and reviewers who helped to improve the final version. Special thanks to Springer-Verlag and Ralf Gerstner for believing in us once again and for giving us the opportunity to publish this work. We would also like to say how grateful we are to Natalia Pinilla of Universidad de Castilla-La Mancha for her support during the production of this book.

Finally, we wish to acknowledge the support of the SOS project (No. SBPLY/17/180501/000364), funded by the Department of Education, Culture and Sport of the Directorate General of Universities, Research and Innovation of the Regional Government of the Autonomous Region of Castilla-La Mancha—Junta de Communidades de Castilla-La Mancha (JCCM) and of the BIZDEVOPS-Global project (RTI2018-098309-B-C31), financed by the Spanish Ministry of Economy, Industry and Competitiveness and European FEDER funds.

Contents

Contributors

Rui Abreu Faculty of Engineering, University of Porto & INESC-ID, Porto, Portugal

Hina Anwar Institute of Computer Science, University of Tartu, Tartu, Estonia

Maria Teresa Baldassarre Department of Informatics, University of Bari "Aldo Moro", Bari, Italy

Rami Bashoon University of Birmingham, Birmingham, UK

Coral Calero Alarcos Research Group, Institute of Technologies and Information Systems, University of Castilla-La Mancha (UCLM), Ciudad Real, Spain

Fernando Castor Federal University of Pernambuco, Pernambuco, Brasil

Samuel Jaachimma Chinenyeze Edinburgh Napier University, Edinburgh, Scotland, UK

Marco Couto HASLab/INESC TEC and University of Minho, Braga, Portugal

Luís Cruz Delft University of Technology, Delft, The Netherlands

Jutta Eckstein Independent, Braunschweig, Germany

Sergio España Department of Information and Computing Sciences, Utrecht University, Utrecht, The Netherlands

Iffat Fatima College of Electrical and Mechanical Engineering, National University of Sciences and Technology, Islamabad, Pakistan

Daniel Feitosa University of Groningen, Groningen, The Netherlands

João Paulo Fernandes CISUC and University of Coimbra, Coimbra, Portugal

Félix García Alarcos Research Group, Institute of Technologies and Information Systems, University of Castilla-La Mancha (UCLM), Ciudad Real, Spain

Eoin Martino Grua Vrije Universiteit Amsterdam, Amsterdam, The Netherlands

Achim Guldner University of Applied Sciences Trier, Trier, Germany

Ignacio García-Rodríguez de Guzmán Alarcos Research Group, Institute of Technologies and Information Systems, University of Castilla-La Mancha (UCLM), Ciudad Real, Spain

Mark Hoogendoorn Vrije Universiteit Amsterdam, Amsterdam, The Netherlands

Asif Imran University at Buffalo, Buffalo, NY, USA

Eva Kern Leuphana University Lueneburg, Lueneburg, Germany
University of Applied Sciences Trier, Trier, Germany

Tevfik Kosar University at Buffalo, Buffalo, NY, USA

Sandro Kreten University of Applied Sciences Trier, Trier, Germany

Patricia Lago Vrije Universiteit Amsterdam, Amsterdam, The Netherlands
Chalmers University of Technology, Gothenburg, Sweden

Xiaodong Liu Driven Software Engineering Research Group, Edinburgh Napier University, Edinburgh, Scotland, UK

Ivano Malavolta Vrije Universiteit Amsterdam, Amsterdam, The Netherlands

Javier Mancebo Alarcos Research Group, Institute of Technologies and Information Systems, University of Castilla-La Mancha (UCLM), Ciudad Real, Spain

Hugo Matalonga Minho University, Minho, Portugal

Claudia de O. Melo International Agency (United Nations), Vienna, Austria

Alok Mishra Molde University College, Molde, Norway
Department of Software Engineering, Atilim University, Ankara, Turkey

Deepti Mishra Department of Computer Science, Norwegian University of Science and Technology, Gjøvik, Norway

Mª Ángeles Moraga Alarcos Research Group, Institute of Technologies and Information Systems, University of Castilla-La Mancha (UCLM), Ciudad Real, Spain

Stefan Naumann University of Applied Sciences Trier, Trier, Germany

Bendra Ojameruaye University of Birmingham, Birmingham, UK

Wellington Oliveira Federal University of Pernambuco, Pernambuco, Brasil

Sietse Overbeek Department of Information and Computing Sciences, Utrecht University, Utrecht, The Netherlands

J. David Patón-Romero University of Castilla-La Mancha, Ciudad Real, Spain
University of Bari "Aldo Moro", Bari, Italy
AQCLab, Ciudad Real, Spain

Dietmar Pfahl Institute of Computer Science, University of Tartu, Tartu, Estonia

Mario Piattini Alarcos Research Group, Institute of Technologies and Information Systems, University of Castilla-La Mancha (UCLM), Ciudad Real, Spain

Gustavo Pinto Federal University of Pará, Pará, Brasil

Usman Qamar College of Electrical and Mechanical Engineering, National University of Sciences and Technology, Islamabad, Pakistan

Vijanti Ramautar Department of Information and Computing Sciences, Utrecht University, Utrecht, The Netherlands

Moisés Rodríguez AQCLab, Ciudad Real, Spain

Martina De Sanctis Gran Sasso Science Institute, L'Aquila, Italy

João Saraiva HASLab/INESC TEC and University of Minho, Braga, Portugal

List of Abbreviations

ACL	Access Control List
ADB	Android Debug Bridge
AP	Additionally Performs
AS	Architectural Strategies
ASRs	Architecturally Significant Requirements
ATAM	Trade-off Analysis Method
BDD	Behavior-Driven Development
CBAM	Cost Benefits Analysis Method
CD	Continuous Delivery
CEO	Chief Executive Officer
CFO	Chief Financial Officer
CI	Continuous Integration
CIO	Chief Information Officer
CMD	Command-line Interface
CPS	Cyber Physical Systems
CSO	Chief Sustainability Officer
CSR	Corporate Social Responsibility
CTO	Chief Technology Officer
DAE	Data Aggregator and Evaluator
DBMSs	Database Management Systems
DCT	Dynamic Concurrency Throttling
DUT	Device Under Test
DVFS	Dynamic Voltage and Frequency Scaling
EC	Energy Consumption
ECG	Economy for the Common Good
EDS	Emergency Deployment System
EET	Energy Efficiency Tester
ETDC	End-Tagged Dense Code
FAAS	Function as a Service

FEETINGS	Framework for Energy Efficiency Testing to Improve eNvironmental Goals of the Software
FTSF	Fair Trade Software Foundation
GB	Green-BY Software
GH	Green Hardware
GHG	Greenhouse Gases
GI	Green-IN Software
GMGIT	Governance and Management Framework for Green IT
GoF	Gang of Four
GORE	Goal Oriented Requirement Engineering
GQM	Goal/Question/Metric
GS3M	Generic Sustainable Software Star Model
GSMO	Green Software Measurement Ontology
GSMP	Green Software Measurement Process
GSSE	Green and Sustainable Software Engineering
GUI	Graphical User Interface
HCI	Human Computer Interaction
HDF	Hierarchy Data Format
HID	Human Interface Device
IAAS	Infrastructure as a Service
ICE	Immigration and Customs Enforcement
ICT	Information and Communication Technologies
IDE	Integrated Development Environment
IoT	Internet of Things
IQR	Interquartile Range Method
IS	Information Systems
ISSP	Impact Sourcing Service Providers
IT	Information Technology
JCF	Java Collections Framework
JIT	Just-in-Time
LCA	Life Cycle Analysis
LoC	Lines of Code
LP	Laboratory Package
MADN	Median Absolute Deviations from the Median
MAPE	Monitor–Analyze–Plan–Execute
MCA	Mobile Cloud Applications
MCC	Mobile Cloud Computing
MDE	Model-Driven Engineering
MDGs	Millennium Development Goals
MSaPS	Mobile-enabled Self-adaptive Personalized Systems
MSRs	Machine-Specific Registers
OATH	Open Authentication
OCR	Optical Character Recognition
OECD	Organisation for Economic Co-operation and Development

OO	Object-Oriented
OTP	One-Time Password
P	Power
PHR	Personal Health Record
PM	Power Meter
PP	Primarily Performs
PSM	Practical Software Measurement
QA	Quality Attribute
RA	Reference Architecture
RAPL	Running Average Power Limit
SAAM	Software Architecture Analysis Method
SAAS	Software as a Service
SBMs	Sustainable Business Models
SDGs	Sustainable Development Goals
SDLC	Software Development Life Cycle
SE	Software Engineering
SE4S	Software Engineering for Sustainability
SMO	Software Measurement Ontology
SOA-PE	SOA for planning and execution
SOAs	Service-Oriented Architectures
SoS	Systems of Systems
SPLE	Software Product Line Engineering
SS	Software Sustainability
SSC	Sustainability Steering Committee
SSE	Sustainable Software Engineering
SSM	Soft Systems Methodology
STM	Software Transactional Memory
SUT	System Under Test
SUV	Sport Utility Vehicles
SVG	Scalable Vector Graphics
SWOT	Strengths, Weaknesses, Opportunities, and Threats
TLOC	Total Lines of Code
TPL	Third-party Library
UML	Unified Modeling Language
UN	United Nations
URSSI	US Research Software Sustainability Institute
VMs	Virtual Machines
WCED's	World Commission on Environment and Development
WG	Workload Generator
WWF	World Wide Fund for Nature

Chapter 1
Introduction to Software Sustainability

Coral Calero, Mª Ángeles Moraga, and Mario Piattini

Abstract Sustainability is gaining importance worldwide, reinforced by several initiatives that have highlighted the importance of reducing energy consumption and carbon footprint. Although these initiatives highlight ICTs as a key technology in achieving these goals, we must be aware that ICTs can also have a negative impact on the environment.

The main objective of this chapter is to provide an overview of the software sustainability concept and its dimensions (human, environmental, and economic), as well as the research efforts related to this area.

On the one hand, a review of the literature to define all the concepts related to software sustainability has been carried out. On the other, a bibliometric analysis is used to identify the main forum employed in the area for publishing the works and the percentage of papers related to each of the software sustainability dimensions.

Several definitions for the different sustainability levels are presented. As a result of the bibliometric analysis, it can be highlighted that the majority of the papers are published in conferences and are focused on the environmental dimension, whereas the number of books as well as the number of book chapters focused on software sustainability remains low.

Regarding the software sustainability dimensions, most of the works are on the environmental dimensions, highlighting the need for more research focused on the human and economic dimensions.

C. Calero · M. Á. Moraga (✉) · M. Piattini
Alarcos Research Group, Institute of Technologies and Information Systems, University of Castilla-La Mancha (UCLM), Ciudad Real, Spain
e-mail: Coral.Calero@uclm.es; MariaAngeles.Moraga@uclm.es; Mario.Piattini@uclm.es

C. Calero et al. (eds.), *Software Sustainability*,
https://doi.org/10.1007/978-3-030-69970-3_1

1.1 Introduction

Sustainability is gaining importance worldwide, reinforced by several initiatives that have received widespread media coverage such as Earth Hour,[1] a worldwide grass-roots movement uniting people to protect the planet, organized by the WWF (World Wide Fund for Nature). Other organizations such as the United Nations (UN) also highlight the importance of reducing energy consumption and our carbon footprint, including this issue among their Millennium Development Goals (MDGs[2]).

Although these initiatives highlight ICTs (Information and Communication Technologies) as a key technology in achieving these goals, we must be aware that ICTs can also have a negative impact on the environment. In fact, as noted by [1], when pursuing strategic sustainability, the impact of technology is simultaneously important from two different points of view. While technology helps organizations to tackle environmental issues (using videoconferences, reducing or eliminating materials, introducing more efficient processes, etc.), it is often responsible for major environmental degradation (e.g., in the amounts of energy consumed by the engineering processes used to manufacture products). The former concept is called "sustainability by IT," and the second "sustainability in IT."

The main difference between sustainability in IT and sustainability by IT is related to the role played by the specific IT. As indicated by [2], the difference lies in whether one considers IT as a producer, handling the emissions produced by the IT gadgets themselves, or as an enabler, facilitating the reduction of emissions across all areas of an enterprise.

This dual aspect of technology means that organizations also face two challenges: they need to have more sustainable processes, and they must also produce products that contribute to a more sustainable society.

It is therefore essential to control the use of ICTs, in order to reduce as far as possible their negative impact on sustainability. In this book we will focus specifically on software technology, because software is more complex to sell, service, and support than hardware; also, dollar for dollar, software generates more downstream economic activity than does hardware.

In order to gain an overview of the research efforts related to this area, a bibliometric study was carried out, at the beginning of November 2018. The dataset used in the study was obtained from the computer science category of Scopus between 2000 and 2018, was written in English, and resulted in the attainment of a total of 542 papers [3]. Of these, just 2 were books and 21 were book chapters. To update the analysis, and check if this low number of books and book chapters had changed, a new bibliometric study covering up to the end of 2019 was carried out, obtaining 151 new contributions. The results obtained from a total of 693 papers, listing the forum of publication, are shown in Table 1.1. As can be seen, the majority of the papers were published at conferences. Nonetheless, the number of books as

[1] http://www.earthhour.org/

[2] http://www.un.org/millenniumgoals/

Table 1.1 Update of the main forums used in the area

No.	Forum of publication	Publication	Percentage
1	Conference paper	417	62.9
2	Article	148	22.3
3	Conference review	48	7.2
4	Book chapter	28	4.2
5	Review	14	2.1
6	Editorial	3	0.5
7	Book	2	0.3
8	Note	1	0.2
9	Short survey	1	0.2
10	Undefined	1	0.2
	Total	663	100

well as the number of book chapters remains low. Consequently, we believe in the importance to researchers and software developers alike of works such as the present book, which collate the results of studies on software sustainability.

From the point of view of business, sustainability has also become an increasingly important consideration. A business that fails to make sustainable development one of its top priorities could receive considerable public criticism and subsequently lose market legitimacy [4]. All of this can be summarized under the concept of "strategic sustainability," as introduced by [5]. Most consumers claim that they will pay more for a green product [6]. In 2010, the ISO 26000 standard [7] for Corporate Social Responsibility (CSR) was published, providing executives with the necessary directions and measures for demonstrating social responsibility. In this standard, businesses are required to take a precautionary approach to protecting the environment; the aim is to promote greater environmental responsibility through business practices and to encourage the adoption of environmentally friendly information technologies. CSR involves companies in the voluntary integration of social and environmental concerns in their business operations, as well as in relationships with their partners [8]. Expectations of corporations are now higher than ever before. Investors and other stakeholders nowadays consider companies in terms of the "triple bottom line," reflecting financial performance, environmental practices, and corporate social responsibility (CSR). The present-day dominant conception of CSR implies that firms voluntarily integrate social and environmental concerns into their operations and their interactions with stakeholders [9]. In Chap. 15, the authors analyze whether software companies take account of software sustainability in their CSR.

In general, the initiatives that foster respect for the environment by means of ICT, IT, software, etc. are called "sustainability in IT," "Green ICT/IT/Software," etc. A problem that arises is that, as in any new discipline, there is as yet no clear map of concepts and definitions [1].

In the next section we will try to clarify the differences, similarities, and relationships between all these concepts.

1.2 Sustainability

The aim of this section is to give a general definition of the word "sustainability," without linking it to any particular context. To do so, we will first summarize the main existing definitions of sustainability.

Sustainability is a widely used term and refers to the capacity of something to last a long time. Some more precise definitions are as follows:

- Collins Dictionary [10] defines sustainability as "the ability to be maintained at a steady level without exhausting natural resources or causing severe ecological damage."
- A similar definition of "sustainable" can be found in the Merriam-Webster Dictionary: "of, relating to, or being a method of harvesting or using a resource so that the resource is not depleted or permanently damaged" [11].
- According to [12], a sustainable world is broadly defined as "one in which humans can survive without jeopardizing the continued survival of future generations of humans in a healthy environment."
- In [13], the authors affirm that "sustainability can be discussed with reference to a concrete system (ecological system, a specific software system, etc.), therefore, global sustainability implies the capacity for endurance given the functioning of all these systems in concert."
- "Sustainability is the capacity to endure and, for humans, the potential for long-term maintenance" [14].
- From another perspective, sustainability can be viewed as "one more central quality attribute in a row with the standard quality attributes of correctness, efficiency, and so forth" [14]. These same authors also define the term sustainable development as that which "includes the aspect to develop a sustainable product, as well as the aspect to develop a product using a sustainable development process."
- The Brundtland Report from the United Nations (UN) defines sustainable development as the ability to "meet the needs of the present without compromising the ability of future generations to satisfy their own needs" [15]. According to the UN, sustainable development needs to satisfy the requirements of three dimensions, which are society, the economy, and the environment.
- In [16], the author identifies the same dimensions of sustainable development as listed in the aforementioned UN report: economic development, social development, and environmental protection:
  - "Environmental sustainability ensures that the environment is able to replenish itself at a faster rate than it is destroyed by human actions. For instance, the use of recycled material for IT Hardware production helps to conserve natural resources.
  - Social development is concerned with creating a sustainable society which includes social justice or reducing poverty and, in general, with all actions that promote social equity and ethical consumerism.

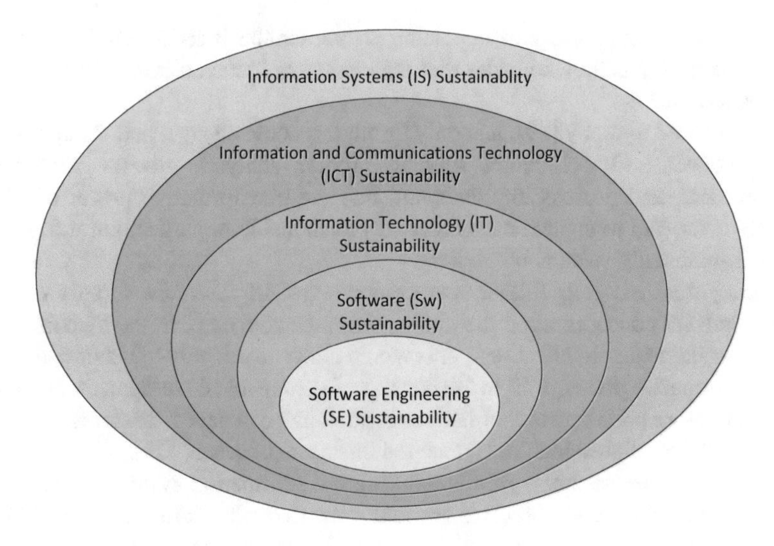

Fig. 1.1 Sustainability levels

– The economic pillar ensures that our economic growth maintains a healthy balance with our ecosystem; it integrates environmental and social concerns into business."

Of all the above definitions, the one most widely used is that established by the Brundtland Report of the United Nations (UN) [15].

If we take a close look at the variety of definitions offered, we can observe that there are two fundamental pillars underpinning the idea of sustainability: "The capacity of something to last a long time" and "the resources used."

Another aspect related to sustainability and found in the literature is the topic to which it is applied: Information Systems, ICT, Software, etc. (Figure 1.1 summarizes the different levels of sustainability according to the topic.)

In the following sections we will present definitions for each of the levels shown in Fig. 1.1. We have worked mainly with papers published in the area of software, software engineering and information systems, in line with the focus of this book.

1.2.1 IS Sustainability

It should be noted that, in general, authors do not differentiate between "IS Sustainability" and "Sustainable IS" (the same applies to the other levels), so in this book we take these concepts as being equivalent.

As articulated in the SIGGreen Statement, "the Information Systems discipline can have a central role in creating an ecologically sustainable society because of the field's five decades of experience in designing, building, deploying, evaluating, managing and studying information systems to resolve complex problems" [17].

In [18] the authors recommend placing greater emphasis on IS sustainability over IT sustainability, as they consider that the exclusive focus on information technologies is too narrow.

As remarked upon by [19], it is only through process change, and the application of process-centered techniques, such as process analysis, process performance measurement, and process improvement, that the transformative power of IS can be fully leveraged to create environmentally sustainable organizations and, in turn, an environmentally sustainable society.

Taking this one step further, we contend that IS researchers must consider process-related concepts when theorizing about the role of IT in the transformation towards sustainable organizations. This would allow us not only to better understand the transformative power of IS in the context of sustainable development, but also to proceed to more prescriptive, normative research that can have a direct impact on the implementation of sustainable, IT-enabled business processes [19].

Although there are some groups working on information systems and environmental friendliness, it is difficult to encounter the IS Sustainability concept. Most of the work being done is instead about Green IS. In [6] the authors consider that sustainability in IS must take account of such aspects as efficiency systems, forecasting, reporting and awareness, energy-efficient home computing, and behavior modification. Finally, the book [20] focuses on "Green Business Process Management," consolidating the global state-of-the-art knowledge about how business processes can be managed and improved in the light of sustainability objectives.

1.2.2 ICT/IT Sustainability

In [21] the authors remark that sustainable ICT can develop solutions that offer benefits both internally and across the enterprise

- By aligning all ICT processes and practices with the core principles of sustainability, which are to reduce, to reuse, and to recycle and
- By finding innovative ways to use ICT in business processes to deliver sustainability benefits across the enterprise and beyond.

The Ericsson Report [22] points to the reduction or elimination of materials and increased efficiency as the two main ways of aligning ICT with sustainability. Following the definition provided by [2], IT sustainability is seen as a shorthand for "global environmental sustainability," a characteristic of the Earth's future, in which certain essential processes persist for a period comparable to human lives.

1.2.3 Software Sustainability

There are several areas in which software sustainability should be applied: software systems, software products, web applications, data centers, etc. Various projects are currently being developed regarding the first of these areas, but most of this work concerns data centers—since the energy consumption of data centers is significantly higher than that of commercial office spaces [23].

As noted in [24], the main way to achieve sustainable software is by improving its power consumption. Whereas hardware has been constantly improved to be energy efficient, software has not. The software development lifecycle and related development tools and methodologies rarely, if ever, consider energy efficiency as an objective [25]. Energy efficiency has never been a key requirement in the development of software-intensive technologies, and so as a result there is a very large potential for efficiency improvements [26].

As remarked upon by [27], software plays a major role in sustainability, both as part of the problem and as part of the solution. The behavior of the software has a significant influence on whether the energy-saving features built into the platform are effective or not [28].

In [13] it is stated that "The term Sustainable Software can be interpreted in two ways: (1) the software code being sustainable, agnostic of purpose, or (2) the software purpose being to support sustainability goals. Therefore, in our context, sustainable software is energy-efficient, minimizes the environmental impact of the processes it supports, and has a positive impact on social and/or economic sustainability. These impacts can occur direct (energy), indirect (mitigated by service) or as rebound effect" [29].

According to [30], sustainable software is "software, whose impacts on economy, society, human beings, and environment that result from development, deployment, and usage of the software are minimal and/or which have a positive effect on sustainable development."

These authors subsequently use the same ideas for the concept of green and sustainable software, defining it as "software, whose direct and indirect negative impacts on economy, society, human beings, and environment that result from development, deployment, and usage of the software are minimal and/or which has a positive effect on sustainable development" [31]. They consider that direct impacts are related to resources and energy consumption during the production and use of the software, while indirect impacts are effects from the software product usage, together with other processes and long-term systemic effects.

One of the most complete definitions is the one proposed by [32], which considers that green and sustainable software is software whose

- "direct and indirect consumption of natural resources, which arise out of deployment and utilization, are monitored, continuously measured, evaluated and optimized already in the development process;
- appropriation and utilization aftermath can be continuously evaluated and optimized;

- development and production processes cyclically evaluate and minimize their direct and indirect consumption of natural resources and energy."

According to [3], software sustainability is about the capacity of software to last a long time by using only the resources that are strictly needed.

Another related term is "sustainable computing." It is used to convey the political concept of sustainability in the field of computer systems, including material components (hardware) as well as informational ones (software); it includes development as well as consumption processes [33].

Some of the literature contains some definitions of "sustainable" (or "sustainability"), while other scholarship refers to the term "green" (or "greenability").

This phenomenon is especially noteworthy in the case of software, because various authors such as [32] and [31] use both terms synonymously. We believe that this approach is faulty and that it ought to be avoided, since we are talking about two different concepts, as will be seen in due course.

What does seem true, however, is that software sustainability is a very important research topic whose significance has been growing in the last few years.

1.2.4 Software Engineering Sustainability

Within the context of software engineering, not many proposals have so far tackled the concept of sustainability [34], although in a recent updating of this work the authors observed that the number of proposals has increased considerably over the last few years [13]. This serves to demonstrate that there is an ever-growing concern to address sustainability in the context of software engineering.

Sustainability should generally be considered from the very first stages of software development. That is not always feasible, however, since it is not easy to change the way in which developers work.

There are many definitions of "sustainable software engineering." We will now go on to present some of these (see Table 1.2).

1.3 Dimensions of Software Sustainability

As detected in several definitions, sustainability is generally considered from three dimensions (the social, the economic, and the environmental) as provided by the UN [15]. There are some proposals, as discussed in [1], that add to these three characteristics which are, for instance, individual or technical. However, we consider that software sustainability is the same as sustainability software. For that reason, from our point of view, software sustainability has three dimensions that correspond to those of sustainability as proposed in the Brundtland Report. Therefore, taking into

Table 1.2 Sustainable software engineering

Reference	Term	Definition
[35]	Sustainable software engineering	Sustainable software engineering aims to create reliable, long-lasting software that meets the needs of users while reducing environmental impacts, their goal is to create better software so we will not have to compromise future generations' opportunities.
[36]	Sustainable software engineering	Sustainable software engineering aims to create reliable, long-lasting software that meets the needs of users, while reducing negative impact on the economy, society, and the environment.
[7]	Sustainable software engineering	Sustainable software engineering is the art of defining and developing software products in a way so that the negative and positive impacts on sustainability that result and/or are expected to result from the software product over its whole lifecycle are continuously assessed, documented, and optimized.
[37]	Sustainable software engineering	Sustainable software engineering is the development that balances rapid releases and long-term sustainability, whereas sustainability is meant as the ability to react rapidly to any change in the business or technical environment.
[38]	Green and sustainable software engineering	Green and sustainable software engineering is the art of developing green and sustainable software with a green and sustainable software engineering process. Therefore, it is the art of defining and developing software products in a way so that the negative and positive impacts on sustainable development that result and/or are expected to result from the software product over its whole life cycle are continuously assessed, documented, and used for a further optimization of the software product.
[39]	Green and sustainable software engineering	The objective of green and sustainable software engineering is the enhancement of software engineering, which targets: 1. the direct and indirect consumption of natural resources and energy as well as 2. the aftermath that are caused by software systems during their entire life cycle, the goal being to monitor, continuously measure, evaluate and optimize these facts.
[40]	Software engineering for sustainability	The aim of software engineering for sustainability (SE4S) is to make use of methods and tools in order to achieve this notion of sustainable software.

account the three types of resources required by software processes—human resources (people involved in carrying out the software processes), economic resources (needed to finance the software processes), and energy resources (all the resources that the software consumes during its life)—we can identify the three dimensions of software sustainability [1] (Fig. 1.2) as follows:

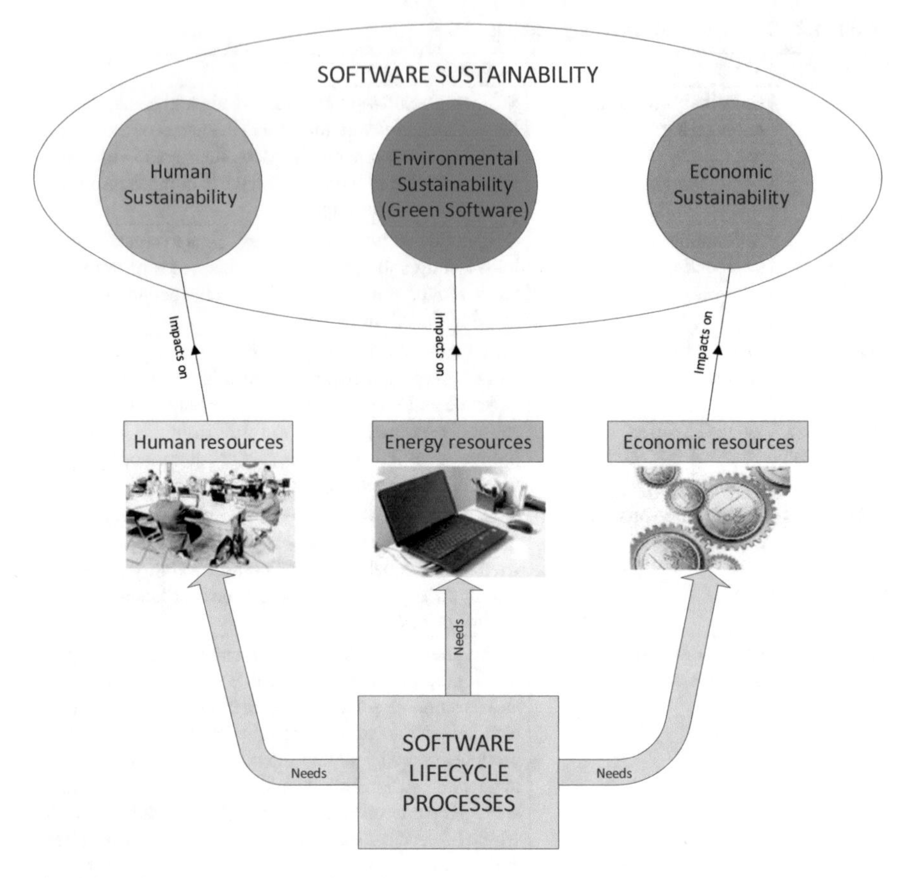

Fig. 1.2 Software sustainability dimensions

- Human sustainability: how software development and maintenance affect the sociological and psychological aspects of the software development community and its members. This encompasses topics such as labour rights, psychological health, social support, social equity, and liveability.
- Economic sustainability: how the software lifecycle processes protect stakeholders' investments, ensure benefits, reduce risks, and maintain assets.
- Environmental sustainability: how software product development, maintenance, and use affect energy consumption and the usage of other resources. Environmental sustainability is directly related to a software product characteristic that we call "Green Software."

As mentioned in the first section, we have extended a previous bibliometric study done on green and sustainable software [3] covering the period from 2000 to 2019 and obtaining a dataset of 663 papers. Analysing the dataset, we can see (Fig. 1.3) that the environmental dimension is the one most analyzed, followed by the social and the economic dimensions. However, we should be aware that only computer

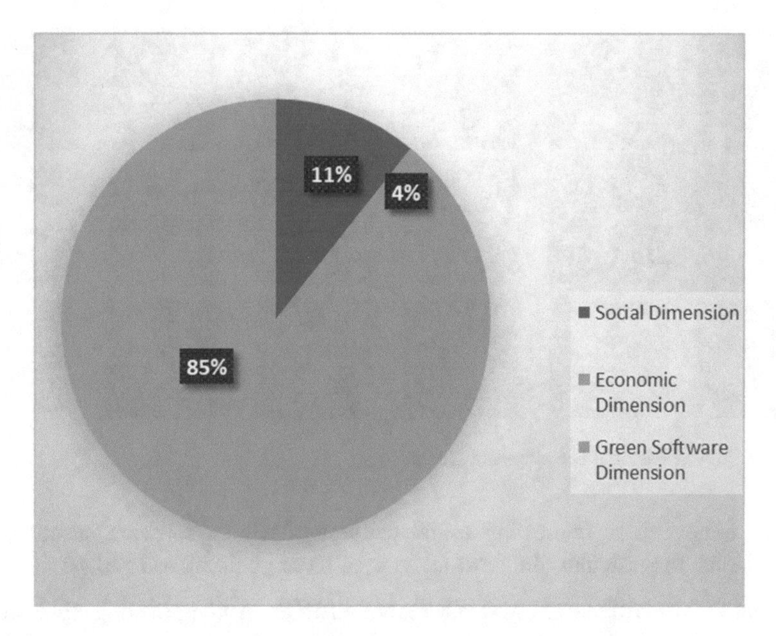

Fig. 1.3 Distributions of software sustainability dimensions in computer science literature

science papers were reviewed and that if we were to analyze economic or business sources the proportion would probably be different.

In this book 60% of the chapters are focused on the green software dimension whereas the social and economic dimensions each represent 20% respectively.

Comparing the percentages of contributions on each sustainability dimension of the bibliometric study and the number of chapters on each dimension in this book, we can observe that they are quite similar, although we have made a special effort to increase the number of chapters related to the human and economic dimensions. This is because we consider that in the future the research lines should be focused on filling the gap that currently exists in these dimensions.

1.3.1 Sustainability Dimensions and the UN's SDGs

The 2030 Agenda for Sustainable Development, adopted in 2015 by all United Nations Member States, provides a shared blueprint for peace and prosperity for people and the planet, now and into the future. At its heart are the 17 Sustainable Development Goals (SDGs) (see Fig. 1.4), which are an urgent call for action by all countries—developed and developing—in a global partnership.[3]

[3]https://sdgs.un.org/goals

Fig. 1.4 UN Sustainable Development Goals

We believe it is interesting to relate these SDGs to software sustainability dimensions. In particular, the most relevant of these goals are as follows:

- ***Goal 7***: AFFORDABLE AND CLEAN ENERGY. Among the objectives of this goal one can find the improvement of efficiency, which can be achieved through sustainable software, as pursued by the environmental dimension.
- ***Goal 8***: DECENT WORK AND ECONOMIC GROWTH. The creation of quality jobs and job opportunities, and decent working conditions, are among the sub-objectives of this goal, which is related to both the human and the economic dimensions of software sustainability.
- ***Goal 9***: INDUSTRY, INNOVATION, AND INFRASTRUCTURE. In this goal the need for inclusive and sustainable industrial development is mentioned, as is the need for technological solutions to ensure environmentally sound industrialization, plus the technological progress required to achieve environmental objectives, such as energy efficiency. All these objectives are related to the environmental dimension of software sustainability.
- ***Goal 12***: RESPONSIBLE CONSUMPTION AND PRODUCTION. This goal promotes, among other things, resource and energy efficiency, green and decent jobs, and a better quality of life for all, increasing net welfare gains from economic activities by reducing resource use, degradation, and pollution throughout the whole lifecycle, while improving the quality of life. It is thus related to all three dimensions.
- ***Goal 13***: CLIMATE ACTION. This is related to greenhouse gas emissions from human activities—in our case, the emissions produced during the software lifecycle. This has to do with the environmental dimension.
- ***Goal 17***: PARTNERSHIPS. These inclusive partnerships are built upon principles and values, a shared vision, and common goals that place people and the planet at the center. They also have to do with long-term investments in the information and communications technologies sector. This goal is thus related to the economic and the environmental dimensions.

Table 1.3 Relationship between Software Sustainability dimensions and the UN Sustainable Development Goals

	SDG 7	SDG 8	SDG 9	SDG 12	SDG 13	SGD 17
Human Software Sustainability		X		X		
Economic Software Sustainability		X		X		X
Environmental Software Sustainability	X		X	X	X	X

Table 1.3 shows the relationships between software sustainability dimensions and the UN sustainable development goals.

1.4 Conclusions

To date, the topic of sustainability has been of importance in several fields. However, this aspect has only been addressed more recently in the field of software, with software sustainability gaining increasing importance in the last decade. This is reflected in the growing number of papers on software sustainability that can be found in the literature and in the calls for it in both European and international initiatives.

We highlight the fact that the three dimensions that make up software sustainability are not being equally studied. As previously mentioned, the environmental dimension is the most studied of them all. In fact, according to our bibliometric analysis, there were no publications relating to the economic and social dimensions at all until 2011. Now that the three sustainability dimensions are related to some of the 17 Sustainable Development Goals proposed in the 2030 Agenda for Sustainable Development, they should henceforth be taken into consideration in any attempt to improve the global environment.

This book is a compilation of chapters related to the three dimensions, which we hope will serve as a reference for researchers and developers who are concerned with software sustainability. Moreover, we aim to raise awareness not only among software developers (software industries, development departments, etc.) but also among end-users, who hold in their hands the responsibility of choosing and demanding software that is more respectful of the environment.

References

1. Calero C, Piattini M (2017) Puzzling out Software Sustainability. Sustain Comput Informatics Syst 16:117–124. https://doi.org/10.1016/j.suscom.2017.10.011
2. Unhelkar B (2011) Green IT strategies and applications. Using environmental intelligence. CRC, Boca Raton, FL

3. Calero C, Mancebo J, García F, Moraga MÁ, Berná JAG, Fernández-Alemán JL, Toval A (2020) 5Ws of green and sustainable software. Tsinghua Sci Technol 25(3):401–414. https://doi.org/10.26599/TST.2019.9010006
4. Du W, Pan SL, Zuo M (2013) How to balance sustainability and profitability in technology organizations: an ambidextrous perspective. IEEE Trans Eng Manag 60(2):366–385. https://doi.org/10.1109/TEM.2012.2206113
5. Sroufe R, Sarkis J (2007) Strategic sustainability: the state of the art in corporate environmental management systems. Greenleaf, Sheffield
6. Cazier J, Hopkins B (2011) Doing the right thing for the environment just got easier with a little help from information
7. ISO/IEC (2010) ISO26000 Guidance on social responsibility
8. European Commission (2000) Green Book
9. Branco MC, Rodrigues LL (2006) Corporate social responsibility and resource-based perspectives. J Bus Ethics 69(2):111–132. https://doi.org/10.1007/s10551-006-9071-z
10. Collins (2020) Collins dictionary
11. Merriam-Webster (2020) Dictionary by Merriam-Webster
12. Brown BJ, Hanson ME, Liverman DM, Merideth RW (1987) Global sustainability: toward definition. Environ Manag 11(6):713–719. https://doi.org/10.1007/BF01867238
13. Penzenstadler B, Raturi A, Richardson D, Calero C, Femmer H, Franch X (2014) Systematic mapping study on software engineering for sustainability (SE4S). In: Proceedings of the 18th International Conference on Evaluation and Assessment in Software Engineering, New York, NY
14. Penzenstadler B, Fleischmann A (2011) Teach sustainability in software engineering? In: Proceedings of the 2011 24th IEEE-CS Conference on Software Engineering Education and Training, USA, pp 454–458
15. United Nations World Commission on Environment and Development (1987) Report of the World Commission on Environment and Development: our common future. In: United Nations conference on environment and development
16. Adams W (2006) The future of sustainability. Re-thinking environment and development in the twenty-first century: technical report. IUCN
17. Hasan H, Molla A, Cooper V (2012) Towards a green IS taxonomy, p 25
18. Watson R, Boudreau M-C, Chen A (2010) Information systems and environmentally sustainable development: energy informatics and new directions for the IS community. MIS Q 34 (1):23–38
19. Seidel S, vom Brocke J (2010) Call for action: investigating the role of business process management in green IS. p 132–133
20. vom Brocke J, Seidel S, Recker J (2012) Green business process management: towards the sustainable enterprise. Springer, Berlin, p XII, 263 p
21. Donnellan B, Sheridan C, Curry E (2011) A capability maturity framework for sustainable information and communication technology. IT Prof 13(1):33–40. https://doi.org/10.1109/MITP.2011.2
22. Ericsson (2013) Ericsson energy, carbon report. On the impact of the networked society. EAB-13:036469 Uen. Ericsson AB
23. Koomey J (2011) Growth in data center electricity use 2005 to 2010. Analytics, Oakland, CA
24. Calero C, Bertoa MF, Moraga MÁ (2013) A systematic literature review for software sustainability measures. In: Proceedings of the 2nd International Workshop on Green and Sustainable Software, pp 46–53
25. Capra E, Francalanci C, Slaughter SA (2012) Is software “green”? Application development environments and energy efficiency in open source applications. Inf Softw Technol 54 (1):60–71. https://doi.org/10.1016/j.infsof.2011.07.005
26. The Climate Group (2008) SMART 2020: enabling the low carbon economy in the information age. The Global eSustainability Initiative, Brussels

27. Easterbrook S (2010) Climate change: a grand software challenge. In: FoSER 2010, Santa Fe, New Mexico, USA, November 7–8. ACM 978-1-4503-0427-6/10/11, p 99–103
28. Steigerwald, B. and Agrawal, A. 2011. Developing green software.
29. Hilty L, Arnfalk P, Erdmann L, Goodman J, Lehmann M, Wäger P (2006) The relevance of information and communication technologies for environmental sustainability – a prospective simulation study. Environ Model Softw
30. Dick M, Drangmeister J, Kern E, Naumann S (2013) Green software engineering with agile methods. In: Proceedings of the 2nd International Workshop on Green and Sustainable Software, pp 78–85
31. Naumann S, Dick M, Kern E, Johann T (2011) The GREENSOFT Model: a reference model for green and sustainable software and its engineering. Sustain Comput Informatics Syst 1 (4):294–304. https://doi.org/10.1016/j.suscom.2011.06.004
32. Johann T, Dick M, Kern E, Naumann S (2011) Sustainable development, sustainable software, and sustainable software engineering: an integrated approach, pp 34–39
33. Mocigemba D (2006) Sustainable computing. Poiesis Prax 4(3):163–184. https://doi.org/10.1007/s10202-005-0018-8
34. Penzenstadler, B., Raturi, A., Richardson, D., Calero, C., Femmer, H. and Franch, X. 2014. Sustainability in software engineering: a systematic literature review for building up a knowledge base.
35. Amsel N, Ibrahim Z, Malik A, Tomlinson B (2011) Toward sustainable software engineering (NIER Track). In: Proceedings of the 33rd International Conference on Software Engineering, New York, NY, pp 976–979
36. Manteuffeal C, Loakeimidis S (2012) A systematic mapping study on sustainable software engineering: a research preview, pp 35–40
37. Tate K (2006) Sustainable software development: an agile perspective. Addison-Wesley, Upper Saddle River, NJ
38. Dick N, Naumann S (2010) Enhancing software engineering processes towards sustainable software product design. In: Greve K, Cremers AB (eds) EnviroInfo 2010: Integration of environmental information in Europe. Shaker, Aachen, pp 706–715
39. Kern E, Dick M, Naumann S, Guldner A, Johann T (2013) Green software and green software engineering – definitions, measurements, and quality aspects
40. IDC (2009) Aid to recovery: the economic impact of IT, software, and the Microsoft ecosystem on the global economy

Chapter 2
Criteria for Sustainable Software Products: Analyzing Software, Informing Users, and Politics

Achim Guldner, Eva Kern, Sandro Kreten, and Stefan Naumann

Abstract The energy consumption of information and communication technology is still increasing and comprises components such as data centers, the network, end devices, and also the software running on these components. Following the motto "What you can't measure you can't manage," it is helpful and reasonable to develop and validate criteria for software products. In our chapter we describe some of these criteria, and also introduce a label for sustainable software products, the German "Blue Angel." We also introduce some energy-efficient programming techniques in order to reduce consumption even during the development phase, and a measurement method for the energy consumption of software. We conclude with some implications of our results and an outlook.

2.1 Introduction

The environmental impacts of information and communication technology (ICT) regarding energy consumption and therefore greenhouse gas effects are still growing. Despite impacts from miniaturization (smartphones instead of desktop computers) or economic causes such as the 2008 financial crisis, the energy consumption is increasing with the growing number of devices, network traffic, computation power, and usage time. It is expected that by 2030 the overall energy consumption of ICT will exceed 20% of the worldwide total [1]. Therefore, it is necessary to find and define criteria that allow for structuring, measuring, organizing, and also forecasting this energy consumption. These criteria can help users, administrators, developers, and the whole ICT sector as well as decision makers in politics and society.

The hardware aspects of ICT have been the subject of research regarding energy for several years. Several labels, such as Energy Star and "TCO Certified,"

A. Guldner (✉) · E. Kern · S. Kreten · S. Naumann
University of Applied Sciences Trier, Environmental Campus Birkenfeld, Germany
e-mail: a.guldner@umwelt-campus.de; e.kern@umwelt-campus.de;
s.kreten@umwelt-campus.de; s.naumann@umwelt-campus.de

C. Calero et al. (eds.), *Software Sustainability*,
https://doi.org/10.1007/978-3-030-69970-3_2

communicate the resource and energy efficiency of the hardware products. For software, however, the situation is more complicated. Since there is a large variety of software products regarding usage options, software architecture, and of course purpose, it is necessary to take a deeper look at how especially software can be analyzed and structured regarding energy consumption. Other publications show that it is worth also taking software into account, since, for example, different software with similar functionality can differ in several ways in their energy consumption [2, 3]. In this chapter, we describe general criteria for measurements and tips for programmers, especially regarding distributed container software such as Docker,[1] which is in widespread use in data centers to support virtualization.

The chapter is organized as follows: In an overview of related work we give some definitions, discuss the meaning of sustainable software, and take a thorough look at the challenges of measuring software energy consumption and energy-efficient programming. Then, we present criteria for sustainable software products and describe a measurement method in detail. We then present aspects of energy-efficient development and deployment. The chapter closes with an in-depth discussion of the results and a conclusion with outlook.

2.2 Related Work

In general, the research field on "sustainable software" and its engineering is relatively new. Here, we focus on criteria for sustainable software, measurements of software sustainability, and programming guidelines for sustainable software. The following section provides an overview on the current state of research.

2.2.1 Sustainable Software

The environmental impacts and sustainability of software are discussed using different terms, depending on the context. This chapter is based on the definition by Dick et al. [4]: "Sustainable Software is software, whose impacts on economy, society, human beings, and environment that result from development, deployment, and usage of the software are minimal and/or which have a positive effect on sustainable development" [4, 5]. We use the terms "sustainable software," "green software," and "energy- and resource-efficient software" interchangeably. The latter terms point out the focus on consuming fewer natural resources, i.e., environmental and resource protection.

[1]https://www.docker.com/ (March 16, 2020).

Penzenstadler [6] places a similar emphasis on this consideration, focusing on energy efficiency while talking about "sustainable software." According to Calero et al. [7], sustainability of software can be understood as a non-functional requirement, i.e., the aspects need to be addressed at the latest in the design phase of a software development process. Ahmad et al. [8] point out that developers should aim at long-living systems. Further definitions of green and sustainable software are presented and discussed in [9–15].

Overall, it can be stated that, depending on the literature and research focus, different categorizations of criteria for sustainable software products are conceivable. Table 2.1 provides an overview of these categorizations which can be found in the literature.

In order to develop sustainable software, Penzenstadler [23] proposes a software development that considers the different dimensions of sustainability in corresponding application domains: development process, maintenance process, system production, and system usage going into aspects of system, function, and time. Along with the three dimensions according to the Brundtland Report [18]—environment, economy, and society—the following technical dimension is mentioned: "From a point of view of (software) systems engineering, there is another dimension that has to be considered. Technical sustainability has the central objective of long-time usage of systems and their adequate evolution with changing surrounding conditions and respective requirements" [24–26]. Penzenstadler et al. [25] summarize the definition in the question "How can software be created so that it can easily adapt to future change?" Additionally, a "human" [27] or an "individual" [25] dimension is defined in some literature.

In order to reach a sustainable software product, "software engineering for sustainability" is required [6]. With the Karlskrona Manifesto for Sustainability Design, Becker et al. [28] present principles for doing so. Additionally, Betz et al. [29] point out that it is important to also include the underlying business processes when integrating sustainability in the context of software systems. Table 2.2 presents an overview of literature presenting criteria and characteristics of sustainable software products.

2.2.2 *Measurement of Software Sustainability*

Once we understand what sustainable software is, the task of measuring seems rather difficult, because of the complexity of the topic. Assessing the sustainability or rather the environmental impact of a software product, several criteria were introduced, as described above. We will now take a look at the two criteria commonly used for measuring sustainability, *hardware usage* and *energy consumption*, to subsume proposed criteria such as efficiency, energy efficiency, power awareness, carbon footprint, pollution, and energy savings.

When commencing the task of measuring the hardware consumption and energy efficiency of a software product, one can find several different approaches. All of

Table 2.1 Categories of characteristics of sustainable software

Categories	Designation	Description
In or by [5, 6, 16]	Green in software	Activities on environmentally friendly ICT itself.
	Green by software	The question how ICT can contribute to sustainable development.
Relationship [11, 17]	Common criteria	Result from the known and standardized quality characteristics for software.
	Direct criteria	Criteria relating to first-order effects.
	Indirect criteria	Criteria concerning effects indirectly caused by ICTs (e.g., energy savings through process optimization) and on effects that have an indirect effect in the long term.
Effects [5, 17]	First order	"First-order effects" are environmental influences caused by production and usage of ICT, e.g., energy consumption during ICT use.
	Second order	"Second-order effects" are caused, for example, by dematerialization and produced substitution
	Third order	Longer-term environmental and social impacts are described as "third-order effects." So-called rebound effects are taken into account, which may reverse the savings of other energy efficiency measures to the opposite.
Sustainability aspects [18]	Social	Social aspects refer to society, but also to individuals and their participation in the community or enabling it.
	Ecological	Ecological aspects consider effects on the environment.
	Economic	Economic aspects aim at protecting economic resources.
Life cycle [5, 19]	Development	In the development phase, effects on a sustainable development, directly and indirectly through activities during the course of the software development.
	Distribution	With regard to a sustainable software product, the form of dissemination is relevant. Examples: resource consumption by printing a user manual, choice of media, and file size when downloading the product.
	Usage	The use of software is primarily concerned with the ecological aspects. Effects that occur are monitored, for example, through product selection, update cycles, and product and system configuration.
	Deactivation & disposal	At the end of the software life cycle, deactivation and disposal may influence the sustainability of the product. When introducing a new product, existing data must be backed up and converted. Especially the backup size plays a role from an ecological point of view.
Quality model [11]	Product sustainability	Considers in particular the effect of a software on other products and services and thus includes usage effects as well as systematic effects.
	Process sustainability	Evaluates the impact of product development or software development process to sustainable development.
	Social aspects	Comprise factors that affect society as well as individual users or developers.

(continued)

Table 2.1 (continued)

Categories	Designation	Description
	Portability	Summarizes product features that can be used under changed conditions (e.g., hardware, requirements).
	Efficiency	Refers to the economy of resources, computing time, and storage space used to solve a specified problem.
25010+S [20]	Energy efficiency	Degree to which a software product consumes energy while it is active.
	Resource optimization	Degree to which the resources used by a software product are used optimally.
	Capacity optimization	Degree to which the maximum of a product optimally meets the needs while only using the parameters that are necessary.
	Perdurability	Degree to which a software product can be used and modified over a long period of time.
Labeling categories [21]	Efficiency	Refers to the economy of resources, computing time, and storage space used to solve a specified problem.
	Resources-oriented feasibility	Includes aspects that affect resource consumption caused by software (focus: environmental impacts).
	Well-being-oriented feasibility	Includes aspects that affect society caused by software (focus: social impacts).
	Longevity (permanence)	Describes how software is modified, adapted, and reused to be able to execute specific functions under specific conditions for as long as possible.
Green factors [22]	Feasibility	Describes how projects and processes of software development, maintenance, and use follow the principles of sustainable development.
	Efficiency	Refers to the economy of resources, computing time, and storage space used to solve a specified problem.
	Sustainability	Describes how software supports sustainable development.

them make use of some kind of system under test (SUT), on which the software product is executed. During the execution, measurements are taken to assess the additional energy consumption of the SUT when the software is run. Meanwhile, the hardware usage is usually monitored with a software tool that aggregates the usage of the main components (CPU, RAM, etc.) of the SUT.

There are two main categories regarding measuring the energy consumption: hardware-based measurements and software-based estimations. Table 2.3 lists and categorizes current approaches.

As can be seen, the approaches can be categorized into their corresponding SUT hardware (be it servers, PCs, mobile devices, or embedded systems) and the measurement approach. The hardware-based measurement methods all consist of at least one SUT and a power meter. For example, in the measurement framework described

Table 2.2 Criteria for sustainable software products

Reference	Paper	Criteria (examples)
Albertao et al. [19]	Proposal for criteria	Modifiability, usability, accessibility, supportability, efficiency, portability
Bouwers et al. [30]	Literature review	—
Bozzelli et al. [2]	Literature review	—
Capra et al. [31]	Proposal for criteria	Framework entropy, functional types, number of methods, energy efficiency, energy, age
Calero et al. [7, 32]	Proposal for criteria	Adaptability, maintainability, availability, recoverability, fault tolerance, maturity
Condori-Fernandez et al. [33]	Proposal for criteria	Functional completeness, coexistence, capacity, time behavior, learnability, user error protection, confidentiality, integrity, installability
Hilty [34]	Proposal for criteria	Demand adaptivity, user-oriented configuration, power awareness, flexibility
Kern et al. [11]	Proposal for criteria	Fit for purpose, memory usage, idleness, organization sustainability, carbon footprint, hardware obsolescence
Lago et al.	Proposal for criteria	Employment, pollution, energy savings, performance, education, configurability
Radu [35]	Literature review	—
Taina et al. [22]	Proposal for criteria	Beauty, reduction, feasibility, energy consumption, waste, memory usage, efficiency

in [41], they use a EVM430-F6736[2] electricity meter and a database server. The load is generated using an open-source benchmark suite for databases. They measure and compare the execution time and energy consumption of three DBMSs and perform an analysis of variance. They found differences (in some instances large ones) in energy consumption and execution time among the three systems and even within one system, depending on the data model. Similarly, our approach, described in [3], uses one or more SUTs (server and PCs), a power meter, and a workload generator to collect the data. The approach is described in detail in Sect. 2.4. Mancebo et al. [44] take the measurement of energy consumption of PCs one step further and propose a measurement setup for recording the energy consumption of individual hardware components (CPU, processor, HDD, and graphics card). We have already compared this method with our own in [48].

Jagroep et al. [43] use a Watts Up Pro power meter (discontinued) and measure the energy consumption of two SUTs (application server and database server). They also use a logging server to record the hardware usage. They compared one software product (called "Document Generator") over two consecutive releases and find that the new functionality did have a negative impact on the energy consumption of all SUTs. Furthermore, they compare the hardware-based measurements with

[2]http://www.ti.com/tool/EVM430-F6736 (March 17, 2020).

Table 2.3 Measurement approaches

Reference	Method	Categories
Becker et al. [36] (2017)	Comparison of hardware-based measurement and software-based estimation of the energy consumption of software	Hardware-based, software-based, PC, energy consumption
Bunse [37] (2018)	Measurement and evaluation of code obfuscation techniques on mobile devices	Software-based, mobile devices, energy consumption
Cherupalli et al. [38] (2017)	Application-specific peak power and energy requirements for ultralow power processors	Hardware-based, embedded systems, energy consumption
Georgiou et al. [39] (2018)	Energy consumption estimation approaches for IoT devices	Software-based, embedded systems, energy consumption
Godboley et al. [40] (2017)	Analysis of the branch coverage and energy consumption using concolic testing	Software-based, PC, energy consumption
Gomes et al. [41] (2020)	Measurements and comparison of energy consumption and execution time of NoSQL database management systems (DBMSs)	Hardware-based, server, energy consumption
Henderson et al. [42] (2020)	Framework (impact tracker) for reporting energy consumption, hardware usage, and carbon footprint of machine learning algorithms	Software-based, machine learning, SUT not described, energy consumption, hardware usage
Jagroep et al. [43] (2016)	Measurement and comparison of energy consumption and hardware usage of commercial, distributed document generator across consecutive releases	Hardware-based, software-based, server, energy consumption, hardware usage
Kern et al. [3] (2018)	Measurement and comparison of software products within different groups (word processors, web browsers, content management systems, and database systems) with the goal of assessing a criteria catalog for sustainable software	Hardware-based, PC, server, energy consumption, hardware usage
Mancebo et al. [44] (2018)	Measurement setup for assessing the energy consumption of software, based upon sensors that measure the consumption of individual hardware components	Hardware-based, PC, energy consumption
Palomba et al. [45] (2019)	Measurement and estimation of the impact of code smells on mobile applications	Software-based, mobile devices, energy consumption
Sahin et al. [46] (2016)	Impact of code obfuscation on energy usage	Hardware-based, mobile devices, energy consumption
Strubell et al. [47] (2019)	Estimation of the energy consumed in training artificial neural networks	Software-based, machine learning, SUT not described in detail; they only speak of up to three GPUs that were used to train the networks, energy consumption

simultaneously obtained estimations, using Microsoft Joulemeter (deprecated) and state that "on process level [...] we are still unable to explain a relatively large amount [over 60% in this case] of the energy overhead of software execution." Similar conclusions can be drawn from [36], where the software-based estimation was also below the hardware-based measurements.

However, in addition to the hardware-based methods, software-based approaches exist that use mathematical models to estimate the energy consumption of components from an SUT. Godboley et al. [40] also use Joulemeter to estimate the energy consumption of a wide variety of Java programs, increasing the branch coverage. Similarly, Acar et al. [49] propose a tool to estimate the power consumption of a given software at runtime by taking into account CPU, memory, and hard disk power consumption.

In regard to mobile devices, attempts to assess the energy consumption of the software (or apps) seem to be even more appropriate, because of the limited battery life. Because of the compact hardware architecture of the devices, hardware-based measurements are difficult. Additionally, the much lower energy consumption of the specialized hardware requires accurate measuring methods. Measuring the hardware usage in case of mobile devices is also very complicated, because the overhead of a logging software is often quite high and can, in some cases, outweigh the app being measured.

Nevertheless, some approaches with software-based (e.g., [45]) estimations and hardware-based (e.g., [46]) measurements show the viability of assessing and improving mobile applications. Palomba et al. [45] used PETrA[3] to estimate the energy consumption based upon the execution time of methods from the app under consideration. Sahin et al. [46] used Google Nexus and Samsung Galaxy smartphones, which they modified to power them with an external supply instead of the battery. Using the Android Debug Bridge (ADB), they triggered the execution of the scenario on the devices.

Embedded or IoT-devices face similar issues in regard to powering them using batteries or energy harvesting. Of course, the usage of low-energy hardware plays a key role in these cases; however, there are methods, e.g., busy-waiting vs. deep-sleep, that are triggered by the software, which can have a large influence on the energy consumption. Georgiou et al. [39, 50] present several approaches to estimate energy consumption and promote energy transparency as "a concept that makes a program's energy consumption visible from hardware up to software." Cherupalli et al. [38] measured application-specific loads and power profiles of TI MSP430 microcontrollers, used for IoT devices. They measured the peak and average power consumption for several benchmarks.

In terms of the field of machine learning, Strubell et al. [47] quantify the cost and environmental impact of training off-the-shelf neural network models. They estimate the energy consumption of the CPU and GPU of their SUT using Intel's RAPL tool[4]

[3]Available at https://doi.org/10.6084/m9.figshare.4233767.v1 (March 15, 2020).

[4]https://01.org/rapl-power-meter (March 17, 2020).

and nvidia-smi[5] while training the networks and then approximate the CO_2 emissions. Similarly, Henderson et al. [42] also use RAPL and nvidia-smi and propose an *experiment-impact-tracker* framework which encapsulates the assessment of the energy consumption to make it easier to use. They also include hardware usage measurements in their framework.

Considering these approaches, we find that the components necessary for the assessment of software sustainability are a SUT and measurement devices or estimation models. In order to produce repeatable experiments, a mode of automating the workload (workload generator or automated scenario playback) is also recommended. With this setup, the energy consumption and hardware usage of a software product can be recorded. This, in turn, can be used for comparisons between software products that perform similar tasks, across releases of one software product, or for different features of one software product.

2.2.3 Energy-Efficient Programming

Software development is a broad and diverse branch of computer science. Accordingly, energy measurements in this field are mostly specifically bound to application cases. For example, there are approaches in app development for smartphones as well as in software development for embedded systems and web development. In the following, examples of such work will be given, which treat the topic both in general and in application-specific terms. As shown in [51], the choice of programming language already shows visible differences in the energy consumption of software and its development. Pereira et al. [52] also show significant differences in the translation time of code depending on the language. Furthermore, program code can contain several critical sections that can lead to increased power consumption. To identify these sections, Pereira et al. have presented an approach in [53]. Another model is suggested in Baek et al. [54], which allows programmers to approximate expensive functions and loops. Looking further at different program constructs and models of different programming languages, differences in energy consumption can also be identified, as shown in [55]. Here, we examined different concurrency models of the programming languages C#, GO, and Clojure.

Table 2.4 gives an overview of different approaches in the literature. Li et al. [56] take a more in-depth approach when considering hybrid programming models that use both messaging and shared memory, as large systems with multi-core and multi-socket nodes are increasingly being used. In order to improve power consumption, Li et al. [56] propose new software-driven execution schemes that consider the effects of dynamic concurrency throttling (DCT) and dynamic voltage and frequency scaling (DVFS) in the context of hybrid programming models.

[5]https://developer.nvidia.com/nvidia-system-management-interface (March 17, 2020).

Table 2.4 Energy-efficient programming

Reference	Title	Categories
Cuoto et al. [51] (2017)	Towards a Green Ranking for Programming Languages	Energy consumption, programming languages
Pereira et al. [52] (2017)	Energy efficiency across programming languages: how do energy, time, and memory relate?	Energy consumption, programming languages
Pereira et al. [53] (2017)	Helping Programmers Improve the Energy Efficiency of Source Code	Energy consumption, programming languages
Baek et al. [54] (2010)	Green: a framework for supporting energy-conscious programming using controlled approximation	Energy-conscious programming, controlled approximation
Kreten et al. [55] (2017)	Resource Consumption Behavior in Modern Concurrency Models	Concurrency control, energy consumption
Li et al. [56] (2013)	Strategies for Energy-Efficient Resource Management of Hybrid Programming Models	Energy-efficient resource management, concurrency control
Chauhan et al. [57] (2013)	A Green Software Development Life Cycle for Cloud Computing	Energy consumption, software development, cloud computing
Kreten et al. [58] (2018)	An Analysis of the Energy Consumption Behavior of Scaled, Containerized Web Apps	Energy consumption, cloud computing
Bunse [37] (2018)	Measurement and evaluation of code obfuscation techniques on mobile devices	Software-based, mobile devices, energy consumption
Li et al. [59] (2014)	An investigation into energy-saving programming practices for Android smartphone app development	Programming patterns, energy consumption
Memeti et al. [60] (2017)	Benchmarking OpenCL, OpenACC, OpenMP, and CUDA: Programming Productivity, Performance, and Energy Consumption	Performance, energy consumption
Balladini et al. [61] (2011)	Impact of parallel programming models and CPUs clock frequency on energy consumption of HPC systems	Parallel programming, energy consumption

Just by looking at the programming language, programming models, and code components alone, the energy consumption of software development is not fully investigated. Software development involves an entire software life cycle, which Chauhan et al. [57] are concerned with. They include not only the code but also the integration of new features and the delivery of the software. An important part of the software life cycle are tests and code pipelines, which nowadays are often executed in containers. Therefore, in [58] we considered the question of whether software that is delivered in a container is less efficient overall than without it, whereas only web apps were considered here.

Furthermore, there are different approaches in the different sub-areas of programming. While Bunse [37] and Li et al. [59] are specifically concerned with the energy consumption of code for smartphones and mobile devices, Memeti et al. [60] focus on the programming of machine learning algorithms using different frameworks.

Overall, only a brief overview of the approaches to improving energy efficiency in software development can be given here, as this field is extensive, as the selection of the presented works shows. However, the approaches shown above also depict how useful it is to consider energy consumption in software development, although current developers are generally not forced to pay attention to the resource efficiency of the programs.

2.3 Criteria for Sustainable Software Products

Even if there are different approaches to defining and characterizing sustainable software products, a standardized environmental label for sustainable software product was missing. This has changed—in Germany—with the publication of the Blue Angel for resource and energy efficient software products in early 2020 [62]. As the world's oldest environmental label, the Blue Angel is a trustworthy label that distinguishes particularly environmentally friendly products and services and is published by the German Federal Ministry for the Environment [63].

In contrast to other products awarded the Blue Angel, software products are immaterial products which trigger the consumption of energy and resources and thus, in general, the environmental impacts of the hardware they drive. Due to this special characteristic of software products, the development of the criteria differed from the criteria development of other Blue Angels. The process of establishing criteria for sustainable software products is described in detail in [3]. In general, the criteria for sustainable software products are divided into three categories.[6] For the Blue Angel, several criteria were omitted, because they were either of little importance for the target group or too complex to assess. In the following, we give an overview of the categories. We then discuss sensitivities, which criteria were omitted, the process of putting the criteria into practice, and possible further aspects to be taken into account, in addition to the development and usage phase.

2.3.1 Criteria Categories

The aim of the Blue Angel for software products is to reduce the overall energy consumption of ICT and to increase resource efficiency. The eco-label especially highlights products whose manufacturers disclose information about their products for this transparency. In addition, a product whose manufacturer is actively committed to improving the resource and energy efficiency of its software products is also labeled. Therefore, the criteria were grouped into three categories: resource and

[6]The whole catalog is available at https://www.umwelt-campus.de/en/research/projekte/green-software-engineering/set-of-criteria/introduction

energy efficiency, potential hardware operating life, and user autonomy. Considering resource and energy efficiency, software products have to provide their functionality with a minimum of resource effort and energy requirements. Thus, for the Blue Angel, the following criteria were selected:

- Required minimum system requirements
- Hardware utilization and electrical power consumption when idle
- Hardware usage and energy requirements when running a standard usage scenario
- Support of the energy management

For the hardware usage and power consumption measurements, the setup is detailed in Sect. 2.4 and based upon the research described in Sect. 2.2.2. The assessment of all other criteria is performed by the developers in accordance with the description in the Basic Award Criteria in [62].

Potential hardware operating life describes how software must not contribute to renewing existing hardware due to higher performance requirements. In particular, software updates should not lead to hardware updates. Users should have the option to decide about software and hardware renewal. Thus, the criterion here is "downward compatibility," which states that software products must be able to run on a reference system that is at least 5 years old. Regarding user autonomy, a software product should not limit the autonomy of users in handling the product or create dependencies. This resulted in the following criteria:

- Data formats
- Transparency of the software product
- Continuity of the software product
- Uninstallability
- Offline capability
- Modularity
- Free of ads
- Documentation of the software product, the license, and usage conditions

To reduce the dependance of the users, developers have to publish the used data formats and APIs with an adequate documentation to enable the interoperability of the software product. Publishing the source code is optional, but a long-term use of the product must be ensured. This is also true for continuity, especially in terms of security updates. Making possible a modular installation of the software product can lead to lower consumption, because the users can choose only the functionality they need. The users must also be enabled to completely remove the software product from their system without leaving unnecessary data. Offline capability not only saves resources for the data transfer and remote processing, but also does not encourage users to deactivate standby modes in their system, for fear of losing data when standby is activated. Advertising increases the resource and energy demand, especially for data transfer of the ads [64]. Thus, external ads are not allowed in a labeled software product.

Eventually, to be awarded the Blue Angel, the developer must present a document containing all the results from the resource and energy measurements, as well as the proof of compliance with the other criteria. The compliance verification is checked for plausibility by an unassociated auditor.

2.3.2 Discussion of the Criteria

From the perspective of research on sustainable software products, the label is an example of how scientific concepts can be put into practice. It also shows the relevance of research for the development of practice-relevant methods and procedures, which are needed, for example, to prove the presented criteria. During the development of the methods for assessing the resource and energy efficiency, the following sensitivities were identified:

- *Selection of the software*: considering the amount of available software types (ranging, e.g., from hardware drivers through system and application software, to distributed AI-systems), the available software products within a category (e.g., "word processors") and the way users interact with a software product
- *Configuration of the software*: considering the possible ways in which users can configure the product, e.g., different modes of displaying data, available functionality, etc., and if and how users make use of those possibilities
- *Usage scenarios*: considering the way users interact with a software product, e.g., duration of activities, workflows, menu layouts, etc.
- *Reference system and software stack*: considering underlying hardware and system software, e.g., the SUT, operating system, libraries, databases, etc.

In order to deal with these factors influencing the measurement results, corresponding requirements are placed on the verification process. The Basic Award Criteria currently only refer to application software that can be run on one of the specified reference systems. Furthermore, the exact software product, including the version number, for which the Blue Angel is applied and for which the compliance verification is provided, must be specified. The exact details of the product and the results of the verifications must be provided both when the application is submitted and during the term of the contract, e.g., in the case of further development and updates.

The standard usage scenario used for the verification must include the functionalities typically used for the software product to be evaluated. It is developed by the applicant in compliance with the measurement instructions provided in the Basic Award Criteria. All energy and resource measurements must be carried out on one of the reference systems.

Considering the criteria from the original set, as described in [3], some criteria were omitted. They include:

- "Platform independence and portability": Can the software product be executed on different currently prevalent productive system environments (hardware and software), and can users switch between them without disadvantages?
- "Hardware sufficiency": Does the amount of hardware capacity used remain constant over time as the software product is developed further and additional functions are added?
- "Transparency of task management": Does the software product inform users that it is automatically launching or running tasks in the background that are possibly not being used?
- "Capability to erase data": Does the user receive sufficient support when erasing data generated during operation of the software product as desired?
- "Maintenance functions": Does the software product provide easy-to-use functions permitting users to repair damage to data and programs?

These have not been included in the Blue Angel Award Criteria, either because when the reference systems were defined, they were subsumed in other criteria (e.g., through requesting the software product to be able to run on five-year-old reference systems), or due to the complexity of the assessment methods and feedback during field tests. The aspects "modularity" and "freedom of advertising" were added.

Furthermore, because the Blue Angel is usually awarded for a time period of several years, the energy requirement over this time was also taken into account: Updates, new functionalities, etc. may not increase the energy consumption of the product by more than 10% compared to the values at the time of application. This was done because the energy consumption is an aggregating criterion that effectively subsumes the hardware usage.

In addition to the criteria developed in the criteria catalogue for sustainable software and those transferred to the Blue Angel for software, further environmental and sustainability aspects for the evaluation of products are conceivable—especially if the focus is not solely on ecological aspects during the product use phase. With reference to a study from 2016 [65], questions concerning the manufacturer of the product are often of interest to users:

- To whom does the profit of the manufacturing company go?
- Does the manufacturing company have an environmental management system?
- How are the environmental impacts of (a) the manufacturing company including its infrastructure and (b) the development process to be assessed, e.g., energy consumption, ecological footprint, use of green electricity?

A holistic sustainability assessment of software products can also raise questions about prevailing working conditions. This refers to the entire value chain: extraction of resources, conditions at the manufacturer and involved subcontractors (keyword "corporate social responsibility"), interaction with people and nature during product manufacture, or place where the product was significantly developed. A further aspect is consideration of the ecological or social commitment of the software company (e.g., Is the software product made available to social projects at a lower

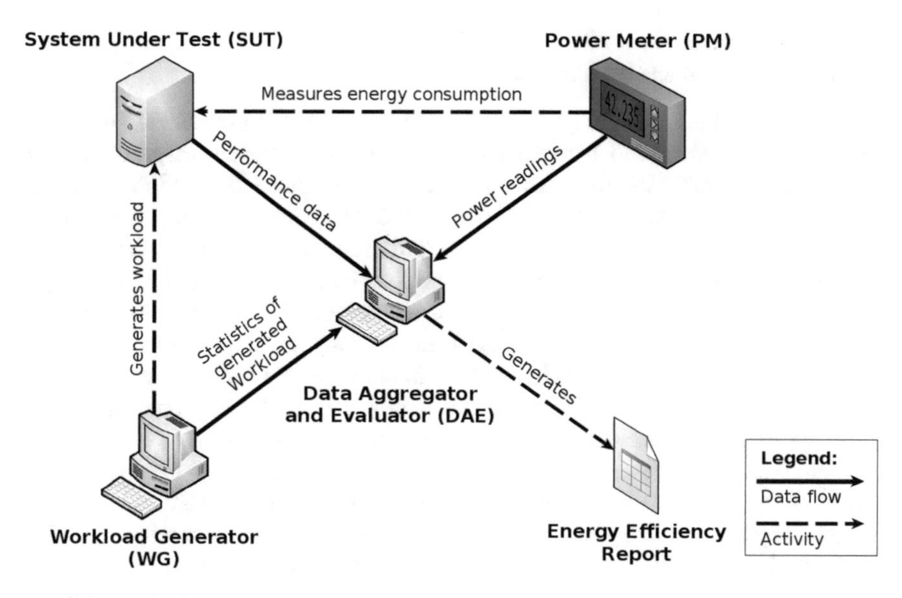

Fig. 2.1 Setup for measuring the energy consumption of software (cf. [68])

price?). Other possible social criteria include usability and user-friendliness. The following questions can be taken up in this respect: Does the software have options to facilitate usability? How much effort is required to learn how to use the software? Is the software intuitive and easy to use?

If the entire life cycle of a software product is considered [5], it is also interesting to look at software delivery: Can the software be obtained via download or via data carriers (physical product)? What is the raw material consumption for the production of the data carrier on which the software is delivered (environmental friendliness of packaging, medium, transport)? Or what are the environmental impacts (e.g., resource consumption, data center infrastructure, energy consumption, type of energy used) of making the software available for download? How is the download made available (design/structure of the website, keyword “Green Web Engineering” [66, 67])?

2.4 Measurement Method

In 2011, we started the development of our measurement method in Dick et al. [68]. The setup is based upon ISO/IEC 14756, as introduced in [69]. It is depicted in Fig. 2.1. As it provides hardware-based measurements, it consists of the following components:

- System under test (SUT)
- Power meter (PM)

- Workload generator (WG)
- Data aggregator and evaluator (DAE)

The SUT is the hardware (PC, server, mobile device, IoT device, etc.) on which the software product will be executed. To assess the hardware usage during the execution, the SUT itself records its own performance data (CPU and RAM usage, network traffic, and hard disk activity). The PM (in our case a Janitza UMG 604[7]) records the energy consumption of the SUT during the execution of the software product. An optional WG generates the workload on the SUT, e.g., by executing a script, repeatedly calling an API, website, database, etc. The WG can also only be a tool, running on the SUT itself that performs user inputs on the software product. This functions much like a benchmark test. In the setup, we call it a "scenario." All generated data (performance from the SUT, power from the PM, and the log-file from the WG) is aggregated with the DAE to produce a report.

Before any measurements can be conducted, several conditions have to be met. To be able to assess the excess energy consumption and hardware usage that is triggered by the software product, a baseline has to be measured. Therefore, the SUT is run without the software product and the measurements are taken. The average consumption of the baseline is then subtracted from the scenario measurement averages. This also allows for the inclusion of the WG in the SUT, as its consumption is included in the baseline and later subtracted. To ensure that the measurements do not influence each other, the SUT is reset after each software product to a state before the software was installed. For this, we advise the usage of disk imaging software such as Clonezilla.[8]

Finally, the scenarios must be recorded or scripted, in order to generate reproducible measurements that are statistically sound. The sample size should not be below 30 measurements [3, p. 206]. The implementation of the automation depends on the scenario itself. Scripts on an external WG can be used to call websites or databases; there are tools such as Monkey runner[9] or appium[10] for automating mobile apps, and tools such as WinAutomation,[11] Pulover's Macro Creator,[12] or Actiona[13] for automating PCs. We are also currently implementing a mouse and keyboard emulator that runs on an arduino, functioning as a human interface device (HID) that can then replay recorded keyboard and mouse inputs, thus externalizing the WG for PCs.

Once the scenario can be recorded, it is also necessary to develop the actions to be executed on the software product. Therefore, it is installed on the SUT with a

[7]https://www.janitza.de/umg-604-pro.html (April 20, 2020).

[8]https://clonezilla.org/ (April 20, 2020).

[9]https://developer.android.com/studio/test/monkeyrunner (April 20, 2020).

[10]http://appium.io/ (April 20, 2020).

[11]https://www.winautomation.com/ (April 20, 2020).

[12]https://www.macrocreator.com/ (April 20, 2020).

[13]https://wiki.actiona.tools/ (April 20, 2020).

standard configuration, i.e., the user does not change any settings during the installation process. We decided to create two separate scenarios: *idle* and *standard*. In the idle scenario, the software is only started and then left to run for 10 min. This measurement reveals how much hardware and energy the software is using when it is not being used and if and when it switches, e.g., to a sleep mode. The usage scenario then should execute all functions of the software like a user would typically do, when working with the software product.

Once the scenarios are created, the measurements can be conducted, and the data gathered and analyzed. Figure 2.2 shows the measurement result from two media players. As can be seen, the two media players require different amounts of power for the execution of the same task. The figure also shows an overview of the actions executed during the usage scenario.

This measurement method provides an accessible way to assess software products in accordance with the Basic Award Criteria for the Blue Angel. The compliance verification requires the calculation of the sum of the additional load on the hardware due to loading the software product and a percentage share of the baseline load (the calculation guidelines can be found in the Basic Award Criteria, Appendix B in [62]).

2.5 Energy-Efficient Software Development and Deployment

In recent years, the range of software requirements of consumers has changed significantly. Software must be highly flexible, which often results in a connection of the software to the internet. Traditional desktop applications that used to rely on licensing models, such as word processors, are moving to the cloud and becoming widely available from different types of applications. As a result, software developers must adapt to the new requirements. Programming takes place in faster cycles and software is developed modularly. Furthermore, elements such as continuous integration (CI) and continuous delivery (CD) have become an integral part of software development [70]. In addition, software is deployed faster, supported by models such as Infrastructure as a Service (IAAS) and Function as a Service (FAAS), which are also based in the cloud. This is evident in the increasing data traffic from cloud data centers. By 2016, cloud traffic already accounted for 88% of all data traffic. The number of hyperscale data centers grew from 338 to 448 between 2016 and 2018 [71].

For these reasons, the requirements for energy-efficient software development are increasing rapidly and include more than just energy-efficient programming, from the choice of programming language, testing, and CI and CD pipelines, to the actual deployment of the application and the choice of the platform, as well as monitoring. All these points are directly related to the energy consumption of the software development. In the following section, we will therefore discuss the difficulties

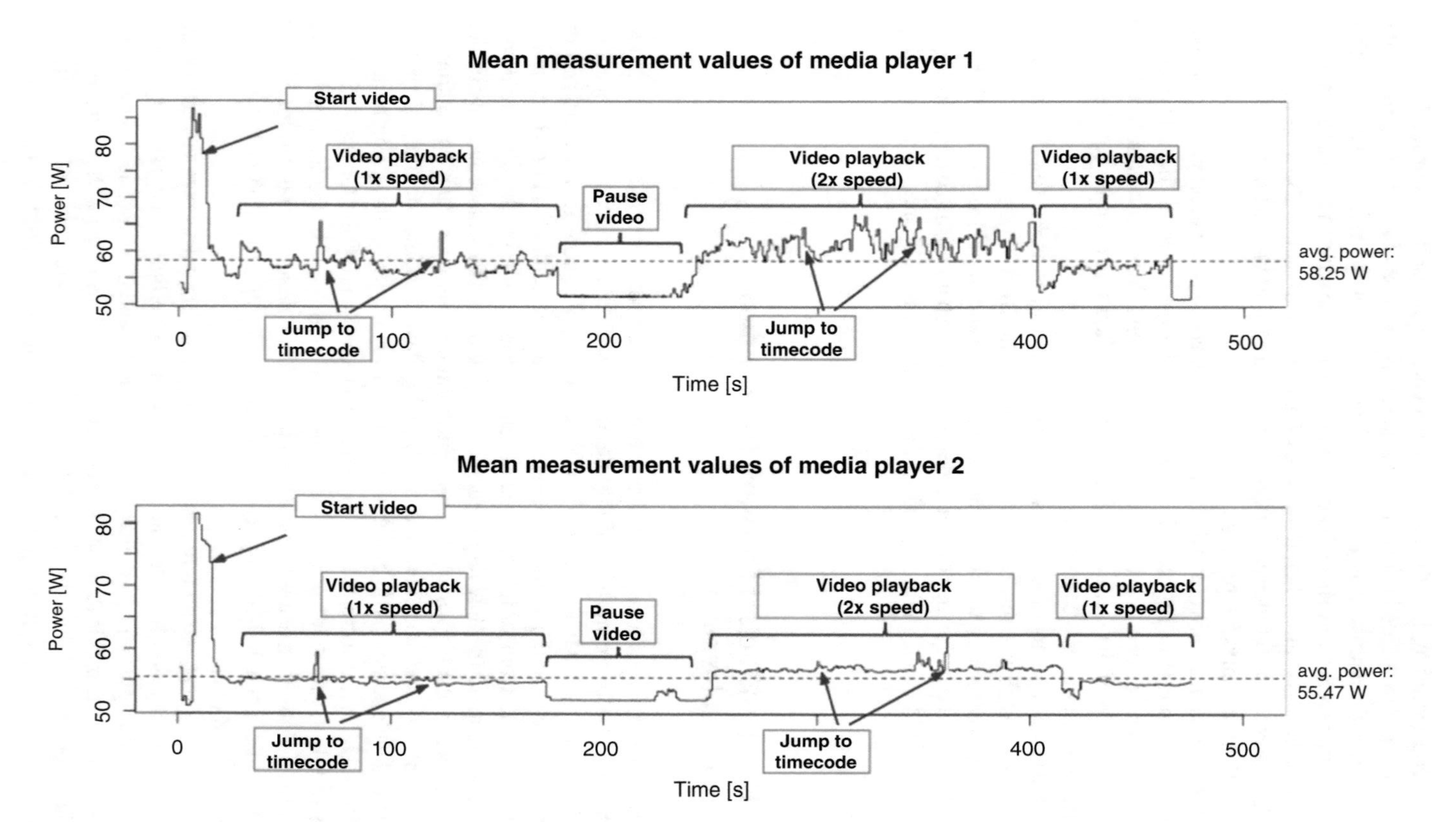

Fig. 2.2 Comparison of two media player software products

that arise in these areas in terms of energy efficiency, starting with the choice of programming language. To support the information presented, practical examples are given.

At the beginning of a software project, the first decision regarding energy efficiency should be made by choosing a suitable programming language depending on the requirements. As [55] shows, for example, programming languages can have different energy efficiency levels depending on the application. Since there are often different programming languages which are particularly suitable for a purpose, this should be considered here as a precaution. Even if programming languages such as C are resource-saving due to their proximity to the hardware, the application purpose should still be in the foreground here. As [55] also shows, the language Go, which is specially designed for development for the cloud, can almost keep up with the energy efficiency of C, if we consider concurrency models. Furthermore, attention must also be paid to resource efficiency during the programming itself. The choice of data and control structures can have an influence on energy consumption, as can the use of different network protocols, as these may alter the software's runtime [72].

The next important area to be considered where energy can be saved during software development is testing. Classical unit tests can be extended by the integration of containers, as in the example of test containers [73]. On the one hand, this leads to greater possibilities in testing (e.g., direct testing of a web application in the server), but on the other hand, additional costs arise here due to runtime, as well as starting, stopping, and deleting the containers used in the tests. Such tests are often executed in automated CI/CD pipelines that are triggered after committing the software to versioning platforms such as Github[14] or Gitlab.[15] Within such pipelines, a variety of tests can be executed, starting from containers to test the build of the software and individual functions. For this reason, not only the energy consumption of the software must be considered during software development, but also the test itself and the container in which the test is performed. Strictly speaking, the energy consumption of the CI/CD platform is also included in the energy consumption of the software development as well as the infrastructure on which the platform is running (see also [74]).

After committing a new software version, the software is delivered. Since the new requirements often mean that this software must be highly available, the finished software is also delivered in a container. This is another reason why it often makes sense to use containers in the CI/CD pipeline test in order to carry out integration tests directly. Containers have the advantage that software can be easily scaled and exchanged in case of errors, and can be moved to other systems. For example, [58] and [75] show that the use of containers alone generates an overhead.

When considering energy efficiency, further factors must be taken into account in software development, for example when software is delivered in containers. The first thing to consider here is the cloud platform. Although it reduces administrative

[14]https://github.com/

[15]https://about.gitlab.com/

tasks, which is why software delivery is increasingly becoming part of software development, the use of the platform also generates energy consumption, which should not be neglected. Furthermore, there are aspects that often cannot be influenced by the developer and still have to be included in the energy considerations. If users use Hosted Container Instances, Virtual Machines (VMs), or even Hosted Container Cluster, only limited configurations can be made. Especially for containers, the choice of different container runtimes or logging drivers can have an influence on the energy consumption. Users of cloud services also have little influence on networking. Although this overhead is small per container, it adds up due to the sheer mass of these containers and therefore offers a possible way to save energy within software development and delivery that should not be neglected. Netflix alone launched over 3 million containers per week in 2018 [71].

However, it should also be mentioned that the cloud in itself increases the energy efficiency of data centers. According to Amazon, a switch to the cloud can reduce electricity consumption by 84%. This has to do with the better utilization of the servers and the extended possibilities in horizontal and vertical scaling [76]. Horizontal scaling describes the consolidation of unused VMs or containers or the addition of new calculation units. Vertical scaling describes the addition or removal of resources such as CPU or RAM.

Overall, it should be noted here that modern software development goes beyond the actual programming and that a large number of factors must be taken into account, particularly with regard to energy consumption in software. In particular, the introduction of automatic testing and CI/CD increases the number of factors to be considered. It must also be taken into account that the problems that developers face when programming AI systems have not yet been addressed. Of course, this also includes big data analysis, machine learning, and high-performance computing in general. Programming for embedded systems or IoT as well as programming apps for mobile devices are also not fully considered here. What they all have in common, however, is that most of them are also developed with the help of a software versioning platform and a code pipeline consisting of CI and CD.

2.6 Conclusion and Outlook

In summary, criteria for sustainable software products and corresponding methods help in informing about environmental issues regarding software. Thus, it is important to bring this information to all actors who are part of the software life cycle, mainly developers, purchasers, and users. While an environmental label primarily addresses the latter, software developers need guidelines for sustainable programming, as presented in Sect. 2.5. Apart from these actors, stakeholder groups who can push the proposed strategies to reach software developers and users need to be addressed in order to close the gap between an intention to do something and the actual behavior [77]. Stakeholders within the context of green and sustainable IT are summarized by Penzenstadler et al. [78] and Herzog et al. [79]. Penzenstadtler et al.

structure their list of stakeholders, relevant in implementing sustainability issues in the software context, along the five dimensions of sustainability: individual, technical, social, environmental, and economic [78]. Focusing on environmental issues related to software, the stakeholders are the legislation (state authority), CSR managers, and nature conservation activists and lobbyists. All of these stakeholders are called to promote the reduction of resource consumption caused by ICT, especially software. Herzog et al. set "Actors Developing Innovation in Green IT" apart from "Actors Supporting Innovation for Green IT." Here, the first group seems to be more important, including standardization bodies, influential groups, universities and academic institutes, and company members [79].

Thus, education is of high relevance in the context of informing about the major influence of software on the environmental impact of information and communication technologies. It is therefore necessary for future software developers to know the requirements and methods of environmentally sound software. The knowledge should be integrated in curricula, especially for students of computer science. Students should be enabled to design environmentally friendly and resource-efficient software and to understand the environmental and sustainability effects of software and its interactions. Finishing their studies, they can bring their knowledge to the software companies. First proposals for this were developed by Issa et al. [80], Penzenstadtler et al. [81], and Gil [82].

From the point of view of communication, environmental associations play a major role. To support the transfer from science to practice, appropriate information and learning material should be created. Environmental associations are experienced in bringing information into society, and could take up the topic of "software and its environmental effects" here.

Procurement in companies as well as in agencies is a best practice that can positively influence the consideration of environmental aspects in the software sector. Here it can be shown how it is possible to integrate these aspects into procurement guidelines. In addition, this will generate a corresponding demand for resource- and energy-efficient software.

Additionally, certification bodies play a significant role in this context. The Blue Angel for resource- and energy-efficient software is a first step to create awareness of the topics in software companies. However, the profile of the environmental label needs to be heightened. In addition, the scope of the label needs to be extended.

Apart from information and education, clear statements are needed from politicians: procurement guidelines for software products that include ecological criteria and oblige software companies to enter into license agreements that promote long-term product use; requirements for software development that, similar to the hardware sector, enable long-term use, so that changes to the software product do not mean that the hardware has to be replaced in order to continue using the software and functional extensions become possible, taking into account resource efficiency; and pricing models that promote energy-saving products. Another important target group for implications in this context are researchers.

In our chapter, we described an overview of how ICT, and especially software products, can be attached to criteria regarding energy consumption and environmental impact. Since the energy consumption of ICT is still increasing, and software has a big influence on this consumption, we developed criteria to measure and analyze software products. These include aspects of energy-efficient programming as well as questions on how container solutions in data centers such as Docker can be optimized. These solutions and criteria help stakeholders such as developers, users, admins, and also decision makers from politics and society regarding "greener" ICT solution. Looking at increasing network traffic, devices, and usage, it is necessary to find and test concepts regarding how the energy and greenhouse impact of ICT can be reduced or even flattened in its growth. The next step will be to test several other products and especially the sensitivity of the influence factors. Furthermore, the criteria and programming concepts have to be extended to other architectures (client server, Software as a Service, etc.), programming languages, and concepts. Possible next steps would be to enlarge the view of observed products and also to refine and test the criteria with several software products. A vision for the future is that labeling software products and publishing their energy consumption is as common as stating system requirements or used libraries.

References

1. Jones N (2018) The information factories. Springer Nature 561(7722):163–166. https://media.nature.com/original/magazine-assets/d41586-018-06610-y/d41586-018-06610-y.pdf. Accessed 19 Apr 2020
2. Bozzelli P, Gu Q, Lago P (2013) A systematic literature review on green software metrics. VU University, Amsterdam
3. Kern E, Hilty LM, Guldner A, Maksimov YV, Filler A, Gröger J, Naumann S (2018) Sustainable software products—towards assessment criteria for resource and energy efficiency. Future Gen Comput Syst 86:199–210. https://www.sciencedirect.com/science/article/pii/S0167739X17314188. Accessed 19 Apr 2020
4. Dick M, Naumann S (2010) Enhancing software engineering processes towards sustainable software product design. In: EnviroInfo. pp 706–715
5. Naumann S, Dick M, Kern E, Johann T (2011) The GREENSOFT model: a reference model for green and sustainable software and its engineering. Sustain Comput Inf Syst 1(4):294–304
6. Penzenstadler B (2013) Towards a definition of sustainability in and for software engineering. In: Proceedings of the 28th Annual ACM Symposium on Applied Computing. pp 1183–1185
7. Calero C, Moraga M, Bertoa MF (2013) Towards a software product sustainability model. arXiv preprint arXiv:1309.1640
8. Ahmad R, Baharom F, Hussain A (2014) A systematic literature review on sustainability studies in software engineering. In: Knowledge Management International Conference (KMICe), Langkawi, Malaysia
9. Calero C, Piattini M (2015) Introduction to green in software engineering. In: Green in software engineering. Springer, pp 3–27
10. Hilty LM, Aebischer B (2015) ICT for sustainability: an emerging research field. In: ICT innovations for sustainability. Springer, pp 3–36
11. Kern E, Dick M, Naumann S, Guldner A, Johann T (2013) Green software and green software engineering–definitions, measurements, and quality aspects. Hilty et al 2013:87–94

12. Kern E, Guldner A, Naumann S (2019) Including software aspects in green it: how to create awareness for green software issues. In: Green IT engineering: social, business and industrial applications. Springer, pp 3–20
13. Kharchenko V, Illiashenko O (2016) Concepts of green it engineering: taxonomy. Principles and implementation
14. Mahaux M, Heymans P, Saval G (2011) Discovering sustainability requirements: an experience report. In: International Working Conference on Requirements Engineering: Foundation for Software Quality. Springer, pp 19–33
15. Venters CC, Capilla R, Betz S, Penzenstadler B, Crick T, Crouch S, Nakagawa EY, Becker C, Carrillo C (2018) Software sustainability: research and practice from a software architecture viewpoint. J Syst Softw 138:174–188
16. Kern E, Naumann S, Dick M (2015) Processes for green and sustainable software engineering. In: Green in software engineering. Springer, pp 61–81
17. Berkhout F, Hertin J (2001) Impacts of information and communication technologies on environmental sustainability: speculations and evidence. Report to the OECD, Brighton, 21
18. Brundtland G, Khalid M et al (1987) Our common future: Report of the World Commission on Environment and Development (United Nations General Assembly, the Brundtland Commission).
19. Albertao F, Xiao J, Tian C, Lu Y, Zhang KQ, Liu C (2010) Measuring the sustainability performance of software projects. In: 2010 IEEE 7th International Conference on E-Business Engineering. IEEE, pp 369–373
20. Calero C, Moraga MÁ, Bertoa MF, Duboc L (2015) Green software and software quality. In: Green in software engineering. Springer, pp 231–260
21. Kern E, Dick M, Naumann S, Filler A (2015) Labelling sustainable software products and websites: ideas, approaches, and challenges. In: EnviroInfo and ICT for sustainability 2015. Atlantis Press
22. Taina J (2011) Good, bad, and beautiful software-in search of green software quality factors. Cepis Upgrade 12(4):22–27
23. Penzenstadler B (2013) What does sustainability mean in and for software engineering? In: Proceedings of the 1st International Conference on ICT for Sustainability (ICT4S), vol 94
24. Lago P, Koçak SA, Crnkovic I, Penzenstadler B (2015) Framing sustainability as a property of software quality. Commun ACM 58(10):70–78
25. Penzenstadler B, Femmer H (2013) A generic model for sustainability with process- and product-specific instances. In: Proceedings of the 2013 workshop on Green in/by software engineering. pp 3–8
26. Razavian M, Procaccianti G, Tamburri DA et al (2014) Four-dimensional sustainable e-services. In: EnviroInfopages. pp 221–228
27. Goodland R et al (2002) Sustainability: human, social, economic and environmental. Encyclopedia Glob Environ Change 5:481–491
28. Becker C, Chitchyan R, Duboc L, Easterbrook S, Penzenstadler B, Seyff N, Venters CC (2015) Sustainability design and software: the karlskrona manifesto. In: 2015 IEEE/ACM 37th IEEE International Conference on Software Engineering, vol 2. IEEE, pp 467–476
29. Betz S, Caporale T (2014) Sustainable software system engineering. In: 2014 IEEE Fourth International Conference on Big Data and Cloud Computing. IEEE, pp 612–619
30. Bouwers E, van Deursen A, Visser J. Evaluating usefulness of software metrics: an industrial experience report. In: 2013 35th International Conference on Software Engineering (ICSE). IEEE, pp 921–930
31. Capra E, Francalanci C, Slaughter SA (2012) Is software "green"? Application development environments and energy efficiency in open source applications. Inf Softw Technol 54(1):60–71
32. Calero C, Bertoa MF, Moraga MÁ (2013) Sustainability and quality: icing on the cake. In: RE4SuSy@ RE
33. Condori-Fernandez N, Lago P (2018) Characterizing the contribution of quality requirements to software sustainability. J Syst Softw 137:289–305

34. Hilty LM, Lohmann W, Behrendt S, Evers-Wölk M, Fichter K, Hintemann R (2015) Final report of the project: Establishing and exploiting potentials for environmental protection in information and communication technology (green it). Technical report, Federal Environment Agency, Berlin. Förderkennzeichen 3710 95 302/3
35. Radu L-D (2018) An ecological view on software reuse. Informatica Economica 22(3):75–85
36. Becker Y, Naumann S (2017) Software based estimation of software induced energy dissipation with powerstat. In: From science to society: the bridge provided by environmental informatics. Shaker Verlag, pp 69–73
37. Bunse C (2018) On the impact of code obfuscation to software energy consumption. In: From science to society. Progress in IS. Springer International Publishing
38. Cherupalli H, Duwe H, Ye W, Kumar R, Sartori J (2017) Determining application-specific peak power and energy requirements for ultra-low-power processors. ACM Trans Comput Syst 35(3)
39. Georgiou K, Xavier-de Souza S, Eder K (2018) The IOT energy challenge: a software perspective. IEEE Embed Syst Lett 10:53–56
40. Godboley S, Panda S, Dutta A, Mohapatra DP (2017) An automated analysis of the branch coverage and energy consumption using concolic testing. Arab J Sci Eng 42(2):619–637
41. Gomes C, Tavares E, Junior MO (2020) Energy consumption evaluation of NOSQL DBMSs. In: Anais do XV Workshop em Desempenho de Sistemas Computacionais e de Comunicaçäo, Porto Alegre, RS, Brasil. SBC, pp 71–81
42. Henderson P, Hu J, Romoff J, Brunskill E, Jurafsky D, Pineau J (2020) Towards the systematic reporting of the energy and carbon footprints of machine learning
43. Jagroep EA, van der Werf JM, Brinkkemper S, Procaccianti G, Lago P, Blom L, Van Vliet R (2016) Software energy profiling: comparing releases of a software product. In: Proceedings of the 38th International Conference on Software Engineering Companion – ICSE 16. pp 523–532
44. Mancebo J, Arriaga HO, García F, Moraga M, de Guzmán IG-R, Calero C (2018) EET: a device to support the measurement of software consumption. In: Proceedings of the 6th International Workshop on Green and Sustainable Software. ACM, pp 16–22
45. Palomba F, Di Nucci D, Panichella A, Zaidman A, De Lucia A (2019) On the impact of code smells on the energy consumption of mobile applications. Inf Softw Technol 105:43–55
46. Sahin C, Wan M, Tornquist P, McKenna R, Pearson Z, Halfond WGJ, Clause J (2016) How does code obfuscation impact energy usage? J Softw Evol Process 28(7):565–588
47. Strubell E, Ganesh A, McCallum A (2019) Energy and policy considerations for deep learning in NLP
48. Mancebo J, Guldner A, Kern E, Kesseler P, Kreten S, Garcia F, Calero C, Naumann S (2020) Assessing the sustainability of software products — a method comparison. In: Schaldach R, Simon K-H, Weismüller J, Wohlgemuth V (eds) Advances and new trends in environmental informatics ICT for sustainable solutions. Springer International Publishing, 1–16
49. Acar H, Alptekin G, Gelas J-P, Ghodous P (2016) Teec: Improving power consumption estimation of software. In: EnviroInfo 2016
50. Georgiou K, Kerrison S, Chamski Z, Eder K (2017) Energy transparency for deeply embedded programs. ACM Trans Architect Code Optimization 14:03
51. M Marco Couto, Pereira R, Riberio F, Rua R, Saraiva J (2017) Towards a green ranking for programming languages. In: Proceedings of the 21st Brazilian Symposium on Programming Languages. ACM Proceedings
52. Pereira R, Couto M, Ribeiro F, Rua R, Cunha J, Fernandes JaP, Saraiva Ja (2017) Energy efficiency across programming languages: How do energy, time, and memory relate? In: Proceedings of the 10th ACM SIGPLAN International Conference on Software Language Engineering, SLE 2017, New York, NY. Association for Computing Machinery, pp 256–267
53. Pereira R (2017) Locating energy hotspots in source code. In: Proceedings of the 39th International Conference on Software Engineering Companion. IEEE Press, pp 88–90
54. Baek W, Chilimbi TM (2010) Green: a framework for supporting energy-conscious programming using controlled approximation. In: Proceedings of the 31st ACM SIGPLAN Conference

on Programming Language Design and Implementation, PLDI '10, New York, NY. Association for Computing Machinery, pp 198–209
55. Kreten S, Guldner A (2017) Resource consumption behavior in modern concurrency models. In: EnviroInfo 2017 – From science to society: the bridge provided by environmental informatics. Shaker
56. Li D, de Supinski BR, Schulz M, Nikolopoulos DS, Cameron KW (2013) Strategies for energy-efficient resource management of hybrid programming models. IEEE Trans Parallel Distrib Syst 24(1):144–157
57. Chauhan NS, Saxena A (2013) A green software development life cycle for cloud computing. IT Prof 15(1):28–34
58. Kreten S, Guldner A, Naumann S (2018) An analysis of the energy consumption behavior of scaled, containerized web apps. Sustainability 10(8)
59. Li D, Halfond WGJ (2014) An investigation into energy-saving programming practices for android smartphone app development. In: Proceedings of the 3rd International Workshop on Green and Sustainable Software, GREENS 2014, New York, NY. Association for Computing Machinery, pp 46–53
60. Memeti S, Li L, Pllana S, Ko-lodziej J, Kessler C (2017) Benchmarking OpenCL, OpenACC, OpenMP, and CUDA: programming productivity, performance, and energy consumption. In: Proceedings of the 2017 Workshop on Adaptive Resource Management and Scheduling for Cloud Computing, ARMS-CC '17, New York, NY. Association for Computing Machinery, pp 1–6
61. Balladini J, Suppi R, Rexachs D, Luque E (2011) Impact of parallel programming models and cpus clock frequency on energy consumption of hpc systems. In: 2011 9th IEEE/ACS International Conference on Computer Systems and Applications (AICCSA). pp 16–21
62. RAL gGmbH (2020) Blue angel – resource and energy-efficient software products. Website. https://www.blauer-engel.de/en/get/productcategory/171. Accessed 16 Mar 2020
63. Horne RE (2009) Limits to labels: the role of eco-labels in the assessment of product sustainability and routes to sustainable consumption. Int J Consum Stud 33(2):175–182
64. Pärssinen M, Kotila M, Cuevas R, Phansalkar A, Manner J (2018) Environmental impact assessment of online advertising. Environ Impact Assess Rev 73:177–200
65. Kern E (2018) Green computing, green software, and its characteristics: awareness, rating, challenges. In: Otjacques B, Hitzelberger P, Naumann S, Wohlgemuth V (eds) From science to society. Springer International Publishing, Cham, pp 263–273
66. Dick M, Kern E, Johann T, Naumann S, Gülden C (2012) Green web engineering-measurements and findings. In: EnviroInfo. pp 599–606
67. Dick M, Naumann S, Held A (2010) Green web engineering. A set of principles to support the development and operation of "Green" websites and their utilization during a website's life cycle. Filipe, Joaquim, pp 7–10
68. Dick M, Kern E, Drangmeister J, Naumann S, Johann T (2011) Measurement and rating of software induced energy consumption of desktop pcs and servers. In: Pillmann W, Schade S, Smits P (eds) Innovations in sharing environmental observations and information: Proceedings of the 25th International Conference on Environmental Informatics October 5–7, 2011, Ispra, Italy. Shaker Verlag, pp 290–299
69. Dirlewanger W (2006) Measurement and rating of computer systems performance and of software efficiency. Kassel University Press, Kassel
70. Krishna R, Jayakrishnan R (2013) Impact of cloud services on software development life cycle. In: Mahmood Z, Saeed S (eds) Software engineering frameworks for the cloud computing paradigm. Springer London, London, pp 79–99
71. Cisco (2018) Cisco Global Cloud Index: Forecast and Methodology, 2016-2021. https://www.cisco.com/c/en/us/solutions/collateral/service-provider/global-cloud-index-gci/white-paper-c11-738085.html. Accessed 15 July 2019
72. Cormen TH, Leiserson CE, Rivest RL, Stein C (2009) Introduction to algorithms, 3rd edn. The MIT Press

73. Wittek K (2019) Auf dem Prüfstand - Testen mit Docker und Testcontainers. In: iX - Magazin für professionelle Informationstechnik, 7
74. Drangmeister J, Kern E, Dick M, Naumann S, Sparmann G, Guldner A (2013) Greening software with continuous energy efficiency measurement. In: Horbach M (ed) INFORMATIK 2013 – Informatik angepasst an Mensch, Organisation und Umwelt. Gesellschaft für Informatik e.V., Bonn, pp 940–951
75. Tadesse SS, Malandrino F, Chiasserini C (2017) Energy consumption measurements in docker. In: 2017 IEEE 41st Annual Computer Software and Applications Conference (COMPSAC), vol 2. pp 272–273
76. AWS and Sustainability. https://aws.amazon.com/about-aws/sustainability/. Accessed 15 July 2019
77. Carrington MJ, Neville BA, Whitwell GJ (2010) Why ethical consumers don't walk their talk: towards a framework for understanding the gap between the ethical purchase intentions and actual buying behaviour of ethically minded consumers. J Bus Ethics 97(1):139–158
78. Penzenstadler B, Femmer H, Richardson D (2013) Who is the advocate? Stakeholders for sustainability. In: 2013 2nd International workshop on green and sustainable software (GREENS). IEEE, pp 70–77
79. Herzog C, Lefêvre L, Pierson J-M (2015) Actors for innovation in green it. In: ICT innovations for sustainability. Springer, pp 49–67
80. Issa T, Issa T, Chang V (2014) Sustainability and green it education: practice for incorporating in the Australian higher education curriculum. Int J Sustain Educ 9(2):19–30
81. Penzenstadler B, Fleischmann A (2011) Teach sustainability in software engineering? In: 2011 24th IEEE-CS Conference on Software Engineering Education and Training (CSEE&T). IEEE, pp 454–458
82. Gil D, Fernández-Alemán JL, Trujillo J, García-Mateos G, Luján-Mora S, Toval A (2018) The effect of green software: a study of impact factors on the correctness of software. Sustainability 10(10):3471

Chapter 3
GSMP: Green Software Measurement Process

Javier Mancebo, Coral Calero, and Félix García

Abstract To improve the sustainability of software it is necessary to be able to measure the energy efficiency of the software. For this purpose, there are several measuring instruments, but for these measurements to be as correct and reliable as possible there must be a process to guide researchers in this effort.

The objective of this chapter is to define the activities to be carried out during the software energy efficiency analysis process, so as to obtain greater control over the measurements performed, ensuring the reliability and consistency of the information obtained regarding energy efficiency. To this end, we have collected a set of good practices in the measurement of energy consumption found in the literature and, together with our own experience, we have defined the Green Software Measurement Process (GSMP) that details all the activities and roles necessary to carry out the measurement and analysis of the energy consumption of the software executed. The GSMP ensures the reliability and consistency of the measurements, and also allows the repetition and comparison of the studies carried out. Furthermore, to validate the process, it was applied to a case study in which energy consumption was analyzed using two measuring instruments.

3.1 Introduction

Improving software sustainability is not a trivial project. To do so, it is essential to be aware of how efficient software is from an energy point of view when it is running.

To measure the energy efficiency of the software, several measuring instruments allow us to know, with greater or lesser accuracy, the energy that is consumed by the software. However, having these measuring instruments that allow us to fully analyze consumption may not be enough. In order to conduct a successful evaluation of the energy efficiency of software, it is necessary to define a process to guide

J. Mancebo · C. Calero (✉) · F. García
Alarcos Research Group, Institute of Technologies and Information Systems, University of Castilla-La Mancha (UCLM), Ciudad Real, Spain
e-mail: Javier.Mancebo@uclm.es; Coral.Calero@uclm.es; Felix.Garcia@uclm.es

C. Calero et al. (eds.), *Software Sustainability*,
https://doi.org/10.1007/978-3-030-69970-3_3

researchers and practitioners as they seek to carry out measurements of the software's energy consumption. A well-defined and established process allows greater control over the measurements performed, ensuring their reliability and consistency. It also allows the studies performed to be easily replicated and the results obtained to be comparable with those of other studies [1, 2].

Despite the importance of the existence of a process to evaluate the energy efficiency of software, if we analyze the available empirical studies that perform software energy consumption analysis, it is possible to identify a lack of a generally-agreed-on methodology that would guide software energy consumption measurements.

Different approaches that could be useful for this purpose can be found in the existing literature. On the one hand, there are software measurement frameworks and standards, which, although they are not related to energy measurement, have the aim to provide guidelines for carrying out the software measurement process effectively and systematically, based on the defined objectives. Some of the best-known standards and methods are the Goal/Question/Metric (GQM) method [3], the Practical Software Measurement (PSM) method [4], and the ISO/IEC/IEEE 15939 standard [5]. On the other hand, to the best of our knowledge, there is also a unique proposal that specifically measures the energy consumed by the software. This is the research carried out by Hindle, which proposes an abstract methodology to measure and correlate the energy consumption of a software application. This methodology is known as the "Green Mining Methodology" [6]. The Green Mining methodology describes the process of measuring, extracting, and analyzing energy consumption information from the software that is running. The main defect in this methodology is that it does not provide any protocol or good practices regarding how to carry out the measurement in a way that is valid and reliable. For this reason, Jagroep et al. [7] present a measurement protocol, in which an extension of "Green Mining" is performed, detailing the specific measurement tasks to be carried out.

Bearing in mind the lack of a specific method to help researchers to analyze the energy efficiency of software, and in the endeavor to ensure that the study can be replicated and the results obtained compared with other studies, we have defined a process, known as the Green Software Measurement Process (GSMP), that integrates all of the activities that must be carried out to measure and analyze the energy consumption of the software being evaluated. The GSMP is composed of seven different phases, covering the main steps to be performed, from the measurement of energy consumption right through to the analysis of the results, including actions such as defining the scope of the study, details on how to conduct valid and reliable measurements, and how the results obtained should be reported.

The following section of this chapter details each of the phases and activities of the GSMP. In addition, Sect. 3.3 presents the application of the process described in the case study in which energy consumption is analyzed using two measuring instruments. Finally, Sect. 3.4 provides some conclusions.

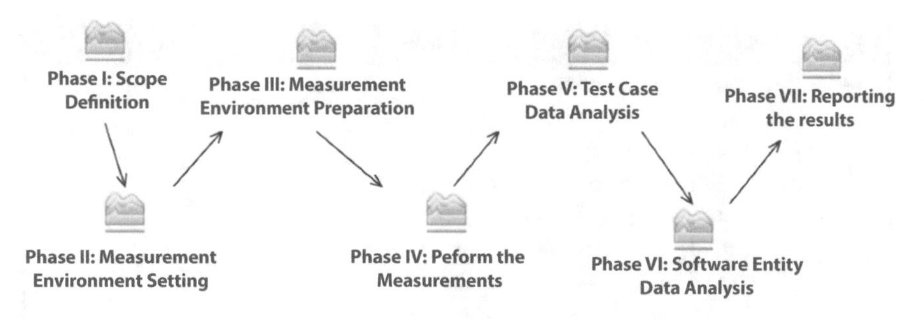

Fig. 3.1 GSMP phases

3.2 Green Software Measurement Process

The aim of the Green Software Measurement Process (GSMP) is to guide researchers and practitioners as they seek to carry out measurements of software energy consumption.

To define the GSMP, we have followed the method engineering approach [8], using the SPEM 2.0 specification [9], and the EPF Composer tool to model the defined process. In addition, to define some aspects or artifacts of the process, we have taken as our basis well-known approaches to software measurement and good practices related to green software that have been proposed by other authors.

The GSMP consists of seven phases (Fig. 3.1), which are divided into different activities (Fig. 3.2). The GSMP is described in detail below, including roles, phases, and activities (with inputs, outputs, and guidelines). A more detailed and comprehensive version of the process, along with its elements, can be consulted at https://alarcos.esi.uclm.es/FEETINGS.

3.2.1 Roles

In this subsection, the different roles involved in the different phases and activities of the process are described below:

- *Client*. The person interested in the results obtained from the measurement of the energy consumption of the selected software. The client is also responsible for providing information about the software to be evaluated and the requirements needed to carry out the energy consumption measurement (Fig. 3.3).
- *Measurement Analyst*. The person responsible for defining in detail the scope of measurements and the configuration of the measurement environment. This role is also responsible for reporting and documenting the results obtained (Fig. 3.4).

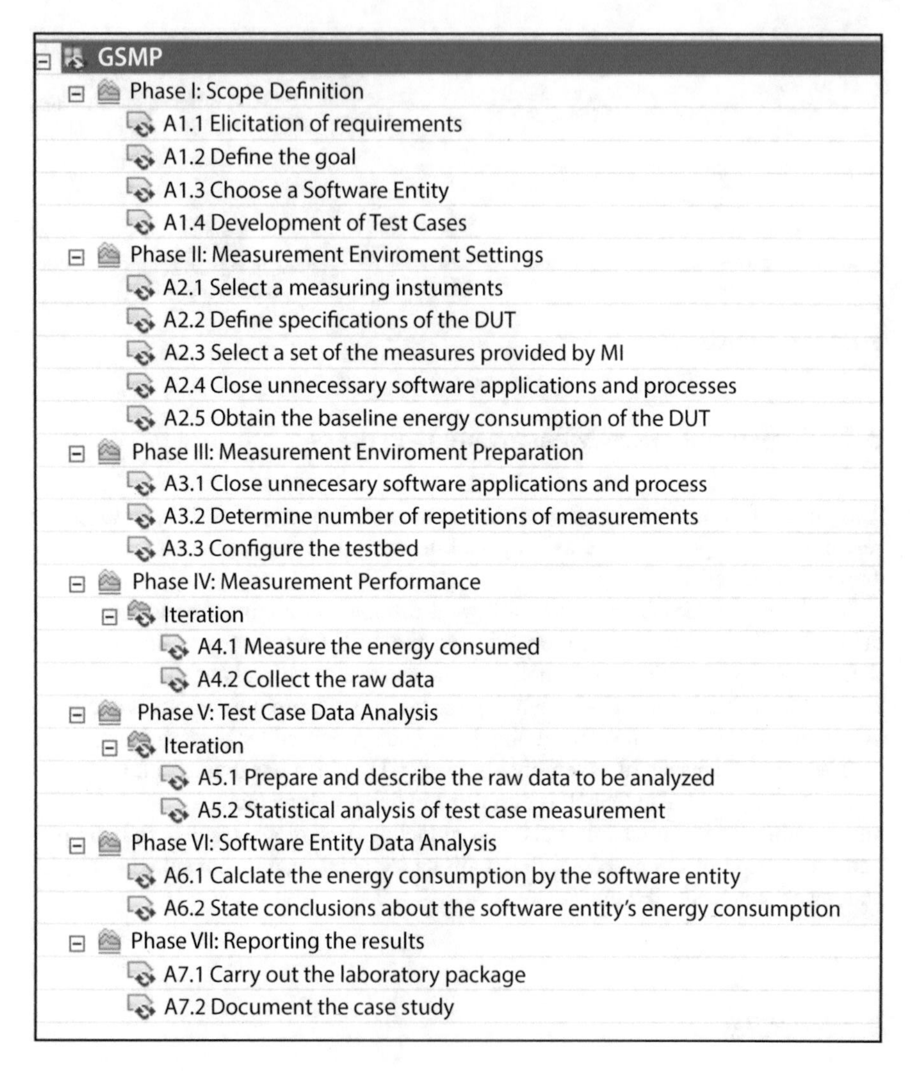

Fig. 3.2 GSMP work breakdown structure

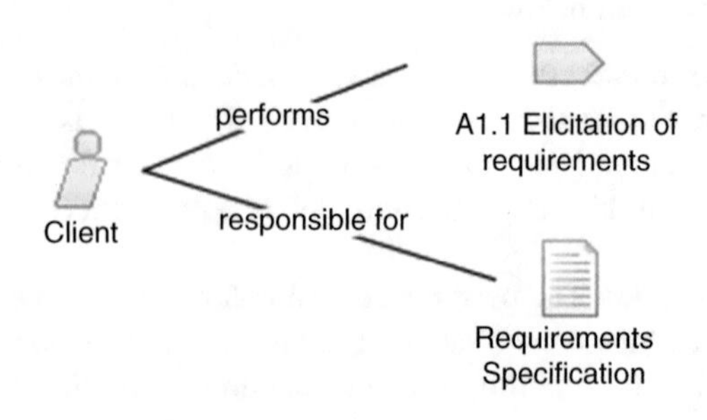

Fig. 3.3 GSMP: Client

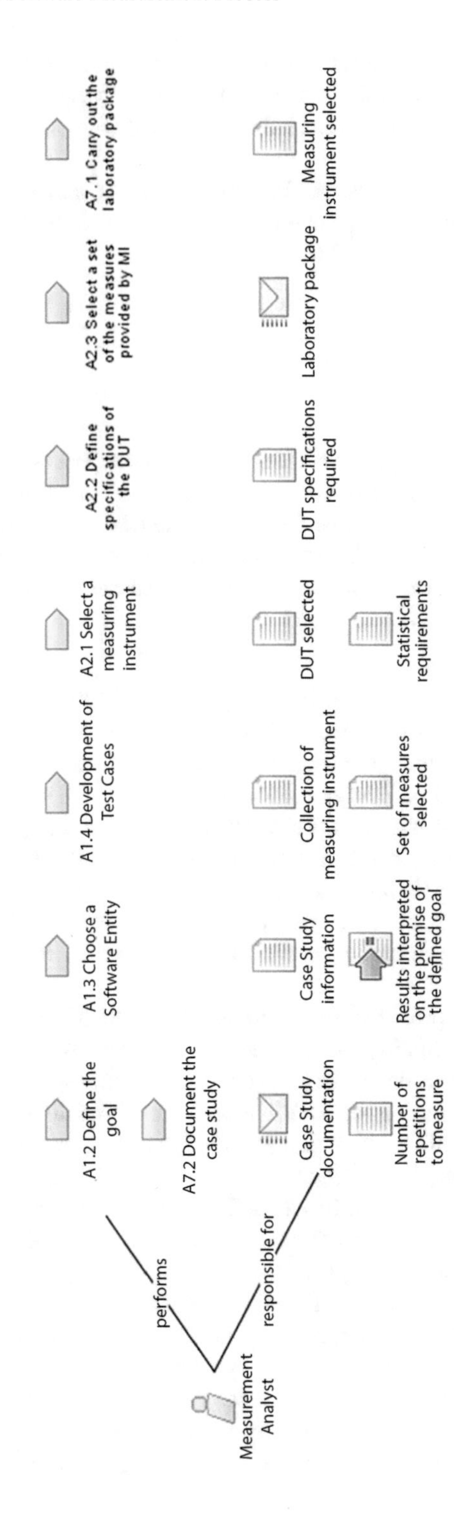

Fig. 3.4 GSMP: Measurement analyst

- *Measurement Performer.* Prepares the measurement environment and sets up the testbed. Furthermore, this role is responsible for carrying out the energy consumption measurements in the selected environment (Fig. 3.5).

- *Data Analyst.* The person responsible for processing and analyzing the data extracted from the measuring device and converting it into software energy consumption information (Fig. 3.6).

3.2.2 Phases

The GSMP is intended to be performed iteratively, so the phases are closely related to each other. The initial phase focuses primarily on the definition of the requirements and the software system to be evaluated. The next two phases focus on the configuration and preparation of the measurement environment. In the fourth phase, energy consumption measurement activities are carried out. Finally, the last phases are the analysis of the data obtained and reporting.

Phase I. Scope Definition

The main goal of this phase is to obtain a complete specification of the requirements for the evaluation of energy efficiency. Moreover, the software to be analyzed must be defined. To achieve this purpose, this phase is composed of four different activities, with inputs and outputs as shown in Fig. 3.7.

The first activity of this phase is the elicitation of requirements (Activity A1.1) for the analysis of software energy consumption. To do this, the Client provides the Measurement Analyst with information about the software to be evaluated. In addition, all the requirements for carrying out the energy consumption measurement must be detailed. This information needs to be documented in the Requirements Specification.

Once the Client has provided all the necessary information, the Measurement Analyst performs the definition of the objective (Activity A1.2) and chooses the collection of all the entities that satisfy the determined purpose, known as Software Entity Class. We suggest using the recommendations of Wohlin et al. [2], based on the application of the Goal/Question/Metric (GQM) method, so as to correctly define the Goal and the Software Entity Class.

After choosing the Software Entity Class, the software entity must be chosen; this is the software that is to be characterized by measuring its attributes. This corresponds to the third activity in this phase (Activity A1.3). It is essential to check that the entire selected software entity is available and can be installed, and/or to run the Device Under Test (DUT). In the effort to facilitate the selection of the Entity's software, a template is included and can be consulted on the process website.

Finally, the fourth activity is dedicated to the development of test cases to execute and measure energy consumption (Activity A1.4). Based on the software entities defined in the previous activity, a representative test case must be built that will

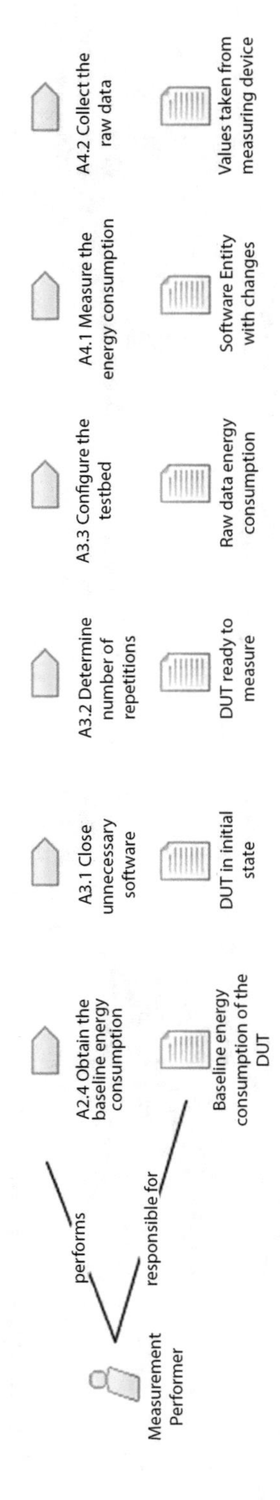

Fig. 3.5 GSMP: Measurement performer

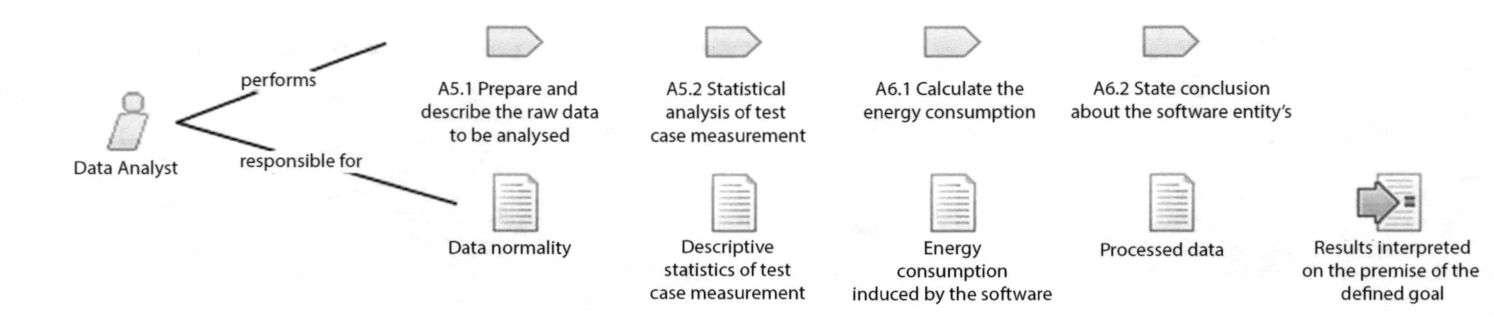

Fig. 3.6 GSMP: Data analyst

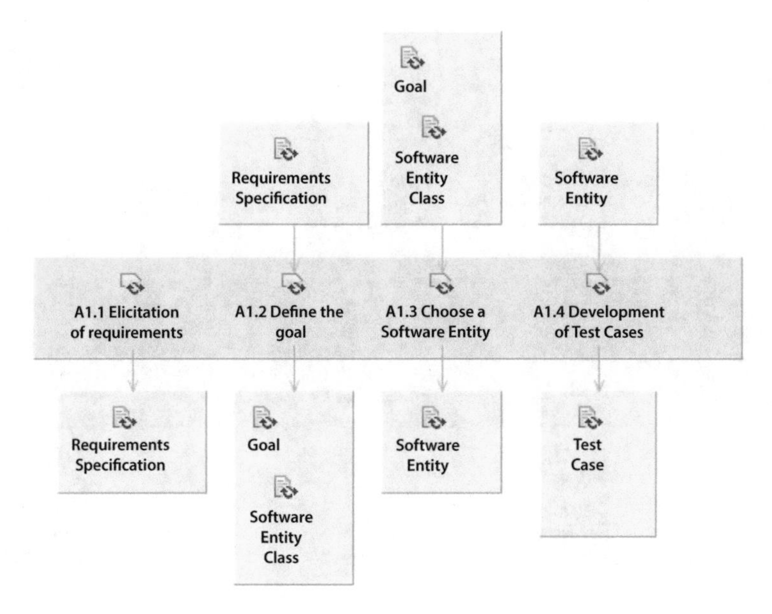

Fig. 3.7 Phase I. Scope definition

exercise the necessary functionality of the software product whose energy consumption is to be measured. The test case is expected to be independent and should not affect the following test case [6]. A test case could simulate user input, or focus on specific software tasks or on the execution of an algorithm. Moreover, if several software entities have been chosen, the defined test cases should be able to be tested in all software entities. This activity is very important, because if the test cases are not well defined, this can cause problems in the analysis of the energy consumption of the software product.

Phase II. Measurement Environment Setting

The purpose of the second phase is the definition of the measurement environment that will be used to satisfy the goal established in the first phase.

As can be observed in Fig. 3.8, the first activity carried out in this phase is the selection of the measuring instrument (Activity 2.1). The measuring instrument is used to perform the power consumption measurements of the software being analyzed. This measuring instrument may be a hardware device or a software tool; depending on whether or not we want to obtain very precise measurements, and depending on the availability of the measuring instrument, we will follow one of two approaches.

The second activity consists of defining the specifications that the Device Under Test (DUT) must have (Activity A2.2). The test cases defined in Activity A1.4 will be executed in the selected DUT in order to carry out the energy consumption measurements. To choose the right DUT, we have to consider the features of the software entity, as it has to be able to be installed and run on the DUT. Moreover,

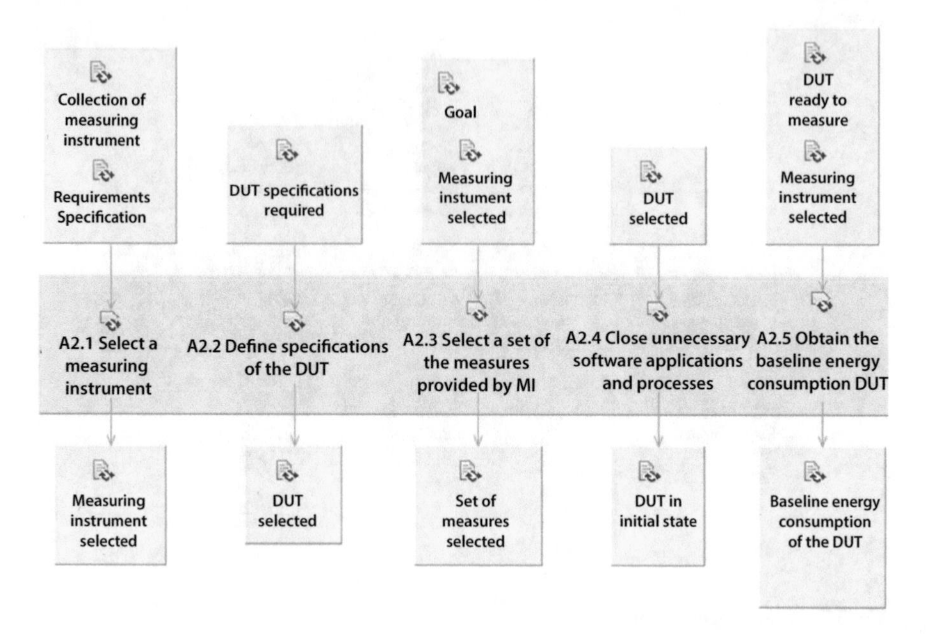

Fig. 3.8 Phase II. Measurement environment setting

depending on the results we want to obtain, the DUT will have different specifications. For example, if we want the results obtained to be more generalizable, we must choose a DUT without special processing or storage capabilities and with a conventional configuration. However, if we want to know the energy consumption in a specific environment where the software will usually be executed, we must simulate this environment by configuring the DUT to be as similar as possible to it.

The next step is to decide on the set of measures to be used for the analysis (Activity A2.3). The main measure of interest is obviously the energy consumption (EC), which is obtained by the measuring instrument. Sometimes it is necessary to recover other measures, however, such as the performance of some hardware components or different kinds of measures that are required for further analysis, e.g., information about the executed source code (Total Lines of Code or Complexity).

Activity 2.4 consists of checking that no other software is running in the background, while also interrupting all services and processes that may affect the baseline measurement of consumption.

Finally, the fifth activity is to obtain baseline energy consumption (Activity 2.5). The baseline measurement determines the idle energy consumption for the DUT that is used. As the idle energy consumption depends mainly on the hardware used, this value must be determined separately for each DUT used, by carrying out measurements while the DUT is running without any active software [10]. The baseline energy consumption allows us to calculate the energy consumption induced by the execution of the selected test cases, under the assumption that the increase in the energy consumed by the DUT depends exclusively on running the software entity under test.

Phase III. Measurement Environment Preparation

This phase focuses on the preparation of the energy consumption measurements to be performed, and on the configuration of the measurement environment that was defined in the second phase (Fig. 3.9).

The first activity to carry out before starting the energy consumption measurements is to check that no other software is running in the background (Activity A3.1). Then, we must interrupt any services or processes that are not required by the software under test, seeking to minimize the effect they may have on the power consumption of the DUT (e.g., the automatic update service or virus scans).

The next activity (Activity A3.2) to perform is that of determining the number of times each measurement should be repeated. We consider a measurement to be a set of energy consumption samples from a single test case run. There is no exact and correct number of repetitions to be measured. The choice of this value depends on the objective we have defined, as well as on the resources available. Some authors [11] recommend that, for measurements of software energy consumption in a controlled environment, 30 measurements are usually a sufficient sample size for an analysis of each of the test cases devised, as the sampling distribution will tend to be normal.

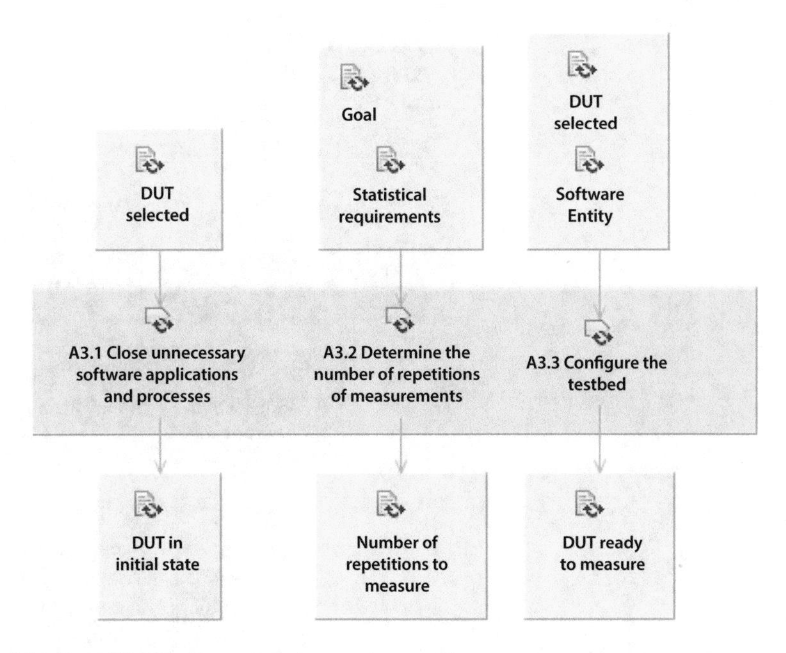

Fig. 3.9 Phase III. Measurement environment preparation

The last activity (Activity A3.3) in the preparation of the measurement environment is the configuration of the testbed. The software entity and the services required in the DUT need to be installed. Once the measurements for one of the software entities are completed, the DUT is restored, such that it returns to its initial state. This procedure is repeated for the different software entities that are going to be assessed. In this activity, the chosen software entity must also be prepared, so that it can execute the test cases defined.

Phase IV. Measurement Performance
During this phase, energy consumption measurements will be carried out, as shown in Fig. 3.10. The set of both activities of this phase are iteratively performed, as many times as test cases were defined in the first phase, and the measurements will also be repeated this number of times, as defined in Phase III. Measuring the energy consumed (Activity 4.1) for the selected software entities is the first activity of this phase. Once the measurement is completed, the testbed should be cleaned, so as to avoid affecting the power consumption when another test case is run.

It is then time to collect the raw energy consumption data taken from the measuring instrument (Activity A4.2). Later, the data obtained will be processed to make its analysis easier. When storing the results of each test, the relevant information, such as the details of the DUT, should be recorded, as should the

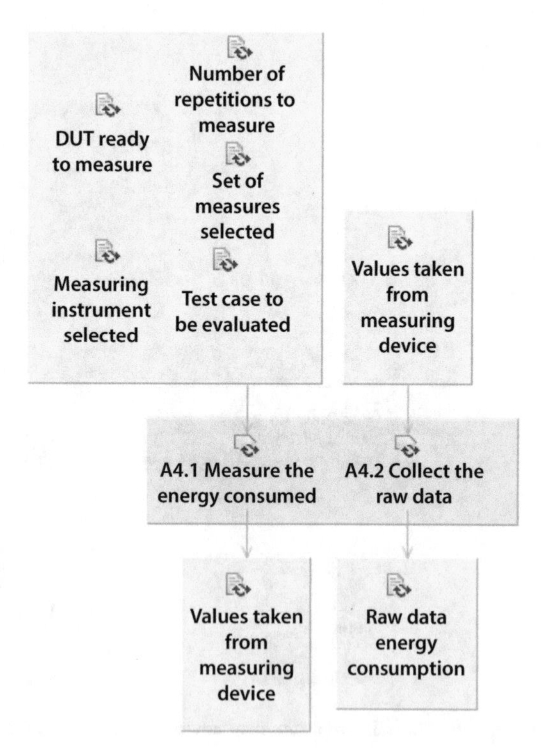

Fig. 3.10 Phase IV. Measurement performance

definition of the test cases, the current configuration, the start and end times, and the power monitor trace itself.

Phase V. Test Case Data Analysis

From this phase onwards, the analysis of the energy consumption data obtained by the measuring instrument begins. The main goal of Phase V is the processing and analysis of the energy consumption data of each of the test cases that were defined in the first phase. This phase is composed of two different activities, which are summarized in Fig. 3.11.

The first activity focuses on the preparation of the raw data obtained by the measuring instrument (Activity A5.1). The steps to be performed in this activity depend on the source of the data, but it is crucial to achieve a transformation of the raw data into useful information for performing an analysis. This process of data transformation is known as Data Wrangling. The most outstanding tasks to be performed in Data Wrangling, according to Kandel, S. et al. [12] are:

- *Data formatting*: reformatting and integrating data from different sources so that they can be analyzed correctly.
- *Correcting erroneous values:* Once the data has been formatted, data preparation begins. Data preparation includes the detection of outliers, the imputation of

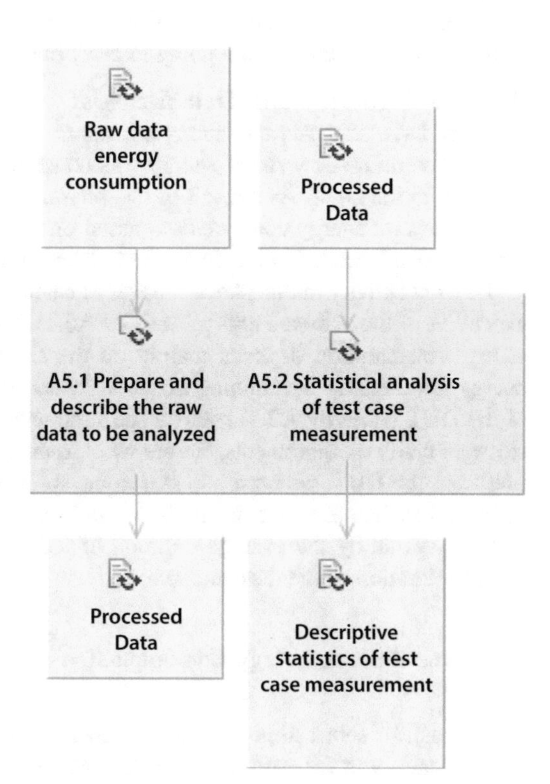

Fig. 3.11 Phase V. Test case data analysis

missing values, and the resolution of duplicate records. For the identification of possible outliers that may be present in the samples of the measurements, we recommend the use of robust parametric methods, such as the median of the absolute deviations from the median [13, 14].

- *Validating the measurements*: Check that each of the measurements performed is correct. To find unusual measurements, you can use the interquartile range method (IQR) [14]. With this method, all values that fall below Q1 − 1.5 * IQR or above Q3 + 1.5 * IQR, where Qi is the quartile, are considered extraneous or incorrect. Another method of identifying incorrect measurements is to use a confidence interval. However, the problem is that to define a confidence interval, it is necessary to have made a large number of measurements beforehand.

The next step to be performed, once the data has been processed, is the statistical analysis of the values obtained from the measurements of the defined test cases (Activity A5.2). To carry out the analysis, the descriptive statistics for each test case analyzed need to be calculated. To obtain the most complete information available on energy consumption, we suggest the calculation of the following descriptive statistics: on the one hand, standard descriptive statistics (maximum and minimum value, range, mean, standard deviation, variance, or interquartile range), and on the other hand, the robust descriptive statistics such as median, trimmed mean, winsorized mean, or median absolute deviation. It is not compulsory to calculate all the descriptive statistics mentioned. We must choose those that adapt to the statistical analysis that we are going to carry out.

Phase VI. Software Entity Data Analysis

Once we have analyzed the energy consumption data of the test cases, we will be able to determine how much energy was consumed when the software entity was executed in the DUT. As a result of this phase, we will carry out an analysis of the information on energy consumption, based on the goal defined at the beginning of the process of measuring the energy efficiency of a software (Fig. 3.12).

The first activity in this phase consists of calculating the energy consumed by the execution of the software entity (Activity A6.1). As mentioned above, the software energy consumption depends mainly on the DUT used. Hence, to calculate the energy required for the running of the software, the baseline energy consumption of the DUT (Activity A2.4) needs to be subtracted from the average energy of the software entity measurements. Before we can subtract the baseline energy consumption from the DUT, we must adjust it to the software measurement performed. The adjusted baseline energy consumption is calculated by dividing the average energy of the baseline by the average duration of the baseline and multiplying it by the average duration of the measurement:

$$\text{Adjusted Baseline Energy Consumption} = \frac{\bar{\text{EC}}\ \text{Baseline}}{\bar{\text{T}}\ \text{Baseline}} * \bar{\text{T}}\ \text{Measurement} \quad (3.1)$$

The task of subtracting the baseline energy consumption of the DUT from the average energy of the software entity's measurements may not be performed if we

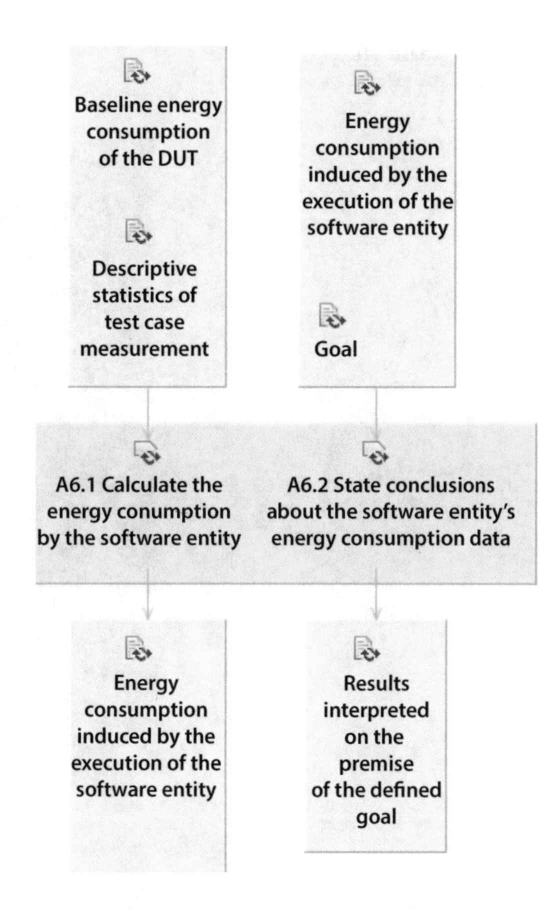

Fig. 3.12 Phase VI. Software entity data analysis

provide relative information on energy consumption. That is, if we classify or sort according to the energy consumptions of each scenario that has been measured in the same DUT, all the results will have been equally affected by the baseline energy, and the classification will not vary.

The last activity of this phase deals with interpreting the data of the energy consumed by the software entity analyzed and establishes some conclusions (Activity A6.7). As a result of this activity, information is obtained on energy efficiency in response to the objective defined. It is essential to have fulfilled all the requirements proposed by the Client at the beginning of the process if the objective is to be completely satisfied.

Phase VII. Reporting the Results

Finally, the last phase involves documenting the study performed, describing the entire process followed, along with the results on the energy consumption of the software that has been extracted. Figure 3.13 contains all the activities, inputs, and outputs of this phase.

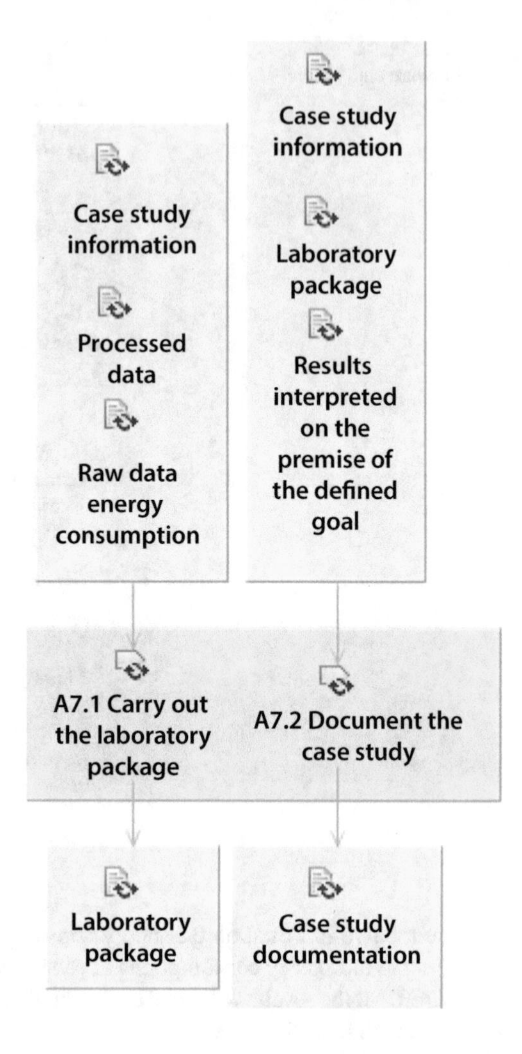

Fig. 3.13 Phase VII. Reporting the results

The first activity in this phase focuses on the development of a laboratory package (LP) intending to achieve repeatability of the experiment performed (Activity A7.1). The main objective of an LP is to be an instrument for supporting knowledge transfer, as well as for conducting replications; it should support all activities in the experimental process and not only the implementation. Laboratory packages should contain all the information and materials required to replicate an experiment or case study [15, 16]. The content of an LP should not be static; it needs to be adapted to the needs of the researcher and the limitations of the experiment. Indeed, proposals for the development of correct LPs, such as the one put forward by [17], can be followed, in which the content and structure of the laboratory packages for software engineering experiments are indicated. Considering the indications of these authors, the LP should include the following information:

- *Planning*: a description of each of the activities to be carried out and the order in which they are to be performed. It is also recommended that the estimated workload for the replicant experimenter be indicated.
- *Study conception*: a description of the high-level attributes that are studied by the experiment, together with its goals. In addition, the variables used in the experiment should be shown.
- *Experimental design*: information about the design of the experiment. It should include details on what the subject of the evaluation will be, and in what cases.
- *Operation*: information for the creation of the laboratory environment to be used. This includes specific software engineering objects (such as programs, specifications, or test cases) and instruments used for measurement and analysis of the data.
- *Analysis*: a specification of the data wrangling process followed, as well as the analysis methods applied. A report of the experiment should be included, and the analysis should conclude with a high-level interpretation of the results. In addition, the raw data should be included in the standard format, so as to allow other researchers to repeat all the analysis activities of the results.

The last activity of the process for the measurement of the energy efficiency of the software is the production of detailed documentation, in which the whole process is explained, along with the results obtained in the study (Activity A7.2). The main difference with the laboratory package is that while it is oriented to other researchers who want to replicate the experiment, the documentation is directed at the Client and other stakeholders, with the information that has been obtained. The LP can be considered to be a piece of this documentation. To report a study where we evaluate the energy consumption of the software, we can use the guidelines proposed by Jedlitschka and Pfahl [18].

3.2.3 *Summary of Roles Involvement in GSMP*

In line with SPEM2, roles can operate in the process in two different ways: Primarily Performs (PP), which refers to the roles that participate in the realization of the activity (see Sect. 3.2.1); and Additionally Performs (AP), which are the roles that must be informed or which are in some way interested in the realization of the activity. Table 3.1 presents a summary of the type of involvement of the four defined roles in the GSMP.

3.2.4 *Considerations for the Validity of Energy Consumption Measurements of Software*

Although the process described above provides a solid basis for carrying out energy consumption measurements, the assumptions that may occur, and which jeopardize

Table 3.1 Roles and their responsibilities in GSMP

		Roles			
Phase	Activity	Client	Measurement analyst	Measurement performer	Data analyst
Phase I	A1.1	PP	AP		
	A1.2		PP		
	A1.3		PP		
	A1.4		PP		
Phase II	A2.1		PP		
	A2.2		PP		
	A2.3		PP		
	A2.4		PA	PP	
	A2.5		PA	PP	
Phase III	A3.1			PP	
	A3.2		PA	PP	
	A3.3			PP	
Phase IV	A4.1			PP	
	A4.2			PP	
Phase V	A5.1				PP
	A5.2				PP
Phase VI	A6.1				PP
	A6.2		AP		PP
Phase VII	A7.1		PP	AP	AP
	A7.2		PP		

the validity of the measurements, must be identified. The assumptions that may threaten the validity of software energy consumption measurements are shown below:

- *Sampling interval:* The frequency with which samples of the power consumed are provided must be taken into consideration. If the frequency is too low, this might lead to an underestimation of the energy consumed, due to the high frequency of the hardware components [10].
- *OS effects and interaction with other software:* The energy used by the operating system is usually included in the energy consumption measurements. In addition, other applications or services of the operating system may be activated during the measurement. We mitigate this threat by performing a large number of measurements, and by obtaining the baseline of DUT consumption.
- *Laboratory temperature:* Not having direct control over the temperature in the laboratory where measurements are performed can be harmful when measuring accurate energy consumption. This risk can be mitigated by repeating the measurements several times.
- *Experiment settings:* The choice of the software entity to be analyzed, together with the creation of the test cases to be run to measure energy consumption, can be considered a limitation of the experiment. Hence, we cannot generalize the

results obtained for other software entities, although they may be useful for future experiments.

- *Measuring instrument:* There is an inevitable dependence on the measuring instrument in terms of accuracy and detail of measurements, as these may vary when a different measuring instrument is used. However, when possible it is always useful to provide comparisons about different instruments by clearly stating their settings.
- *DUT specificity:* One of the main factors that can influence energy consumption measurements is the configuration of the DUT in which the software being evaluated is running, since the energy consumption obtained is specific to the DUT used. It is therefore possible to use the results as absolute values, if the DUT used is similar to the one on which the software will normally be run. Otherwise, the values obtained must be considered relative, and serve to determine in which situations there is a greater or lesser consumption of energy.

3.3 Application of the GSMP

This section presents the application of the process for evaluating the energy efficiency of the software, which was defined in the previous section. To demonstrate that the GSMP can be adapted to any study in which energy consumption is evaluated, a case study is presented, following the protocol template defined by Brereton et al. [19] and the guidelines proposed by Runeson and Höst [20].

3.3.1 Design

The aim of this case study is to find out if the GSMP is useful for measuring and analyzing software energy consumption. To this end, the following research question is addressed: Is the GSMP comprehensive and detailed enough to guide researchers and practitioners in performing software energy measurements?

3.3.2 Subject and Analysis Units

In this case study, the application of the GSMP for the measurement of energy consumed when running different sorting algorithms is evaluated.

To demonstrate that this process can be adapted to any study of energy consumption analysis, the measurements have been carried out with two different measuring instruments. On the one hand, we have used the Energy Efficiency Tester (EET) measuring instrument [21], which will be described in detail in the next chapter of this book. On the other hand, the same energy measurements were replicated by the measuring instrument proposed by the Institute for Software Systems (ISS) [11].

Therefore, the units of analysis in this case study are the GSMP and the energy consumption of the sorting algorithms.

3.3.3 Field Procedure and Data Collection

The field procedure and data collection for this case study are closely related to the activities, roles, and templates of the GSMP described in the previous section.

Furthermore, other data collected are the energy consumption that have been obtained by the EET and the ISS proposal.

3.3.4 Intervention in Case Study

This section presents the application of the GSMP for the measurement and analysis of the energy efficiency of different sorting algorithms. Each of the phases and activities performed is detailed below.

Phase I. Scope Definition

The main goal of this case study is to determine which sorting algorithm consumes the least amount of energy (Activity 1.1 and A1.2). In addition, it aims to demonstrate that the GSMP is validated to carry out energy consumption measurements with any measuring instrument.

For this purpose, we used five sorting algorithms (bubble sort, cocktail sort, insertion sort, quicksort, and mergesort), which were chosen as the software entities to be measured (Activity 1.3).

The last activity that has to be carried out in this phase is the creation of test cases to be executed (Activity A1.4). In this case study, the sorting algorithms were executed multiple times, and in each execution an item array with 50,000 random numbers from 1 to 1000 was sorted, so the execution of each classification algorithm takes approximately 2 min. Therefore:

- Bubble sort was executed 18 times (900,000 items sorted)
- Cocktail sort was executed 30 times (1.5 million items sorted)
- Insertion sort was executed 280 times (14 million items sorted)
- Quicksort was executed 20,000 times (1 billion items sorted)
- Mergesort was executed 10,000 times (500 million items sorted)

Phase II. Measurement Environment Setting

As stated above, the test cases defined will be measured using two measuring instruments (Activity 2.1). The first measuring instrument used was EET [21]. Afterwards, the energy measurements were replicated using the measuring instrument proposed by the ISS [11].

Table 3.2 Specifications of the measurement environments used

Component	DUT specifications used with EET	SUT specifications used with the proposal by ISS
Processor	AMD Athlon 64 X2 Dual Core 5600+ 2.81 GHz	Intel Core 2 Duo E6750 2.66 GHz
Memory	4×1GBDDR2	4×1GBDDR2
Hard disk	Seagate barracuda 7200 500 Gb	320 GB WD 3200YS-01PBG0
Mainboard	Asus M2N-SLI Deluxe	Intel Desktop Board DG33BU
Graphics card	Nvidia XfX 8600 GTS	Nvidia GeForce 8600 GT
Power Supply	350 W AopenZ350-08Fc	430 W Antec EarthWatts EA-430D
Operating System	Windows 10 Enterprise	Windows 10 Pro
Java version	Oracle Java 8u201	Oracle Java 8u201

Table 3.3 Measurement results of baseline energy consumption

Measuring instrument		Mean	SD	Median	Min	Max	Range	IQR
Baseline power [W]	ISS	78.78	0.58	78.82	78.63	78.91	0.28	0.16
	EET	73.40	3.71	72.86	66.04	78.80	12.76	2.53
Baseline energy [Wh]	ISS	13.13	0.02	13.13	13.10	13.17	0.07	0.03
	EET	6.18	0.31	6.10	5.57	6.66	1.09	0.19

The execution of the sorting algorithms was carried out on two different computers, but with similar specifications (Activity 2.2). Table 3.2 shows the specifications of the DUT and system under test (SUT).

Concerning the set of measures to be used in this case study, we can identify the energy consumption obtained by each of the measuring instruments (Activity 2.3).

Finally, it is determined whether any other software is running in the background, while also interrupting all services and processes that may affect energy consumption (Activity 2.4), and the baseline energy consumption of each of the computers where the test cases measurements are to be carried out is measured (Activity 2.5). Table 3.3 shows the results of the baseline measurement with EET and with the ISS proposal.

Phase III. Measurement Environment Preparation

Before starting the measurement, it is determined whether any other software is running in the background. If any process or software not related to the software entity to be analyzed is running, it must be closed (Activity A3.1).

Another aspect to be defined in this phase is the number of repetitions to be performed for each measurement of a test case (Activity A3.2). We consider that each test case should be measured 30 times, since being in a controlled environment is enough to mitigate the effect of other processes that may be executed at the same time.

Table 3.4 Measurement results of the test run

Measuring instrument		Mean	SD	Median	Min	Max	Range	IQR
Test run power [W]	ISS	109.61	2.41	109.44	109.24	113.54	4.30	0.120
	EET	104.57	6.26	103.41	91.58	130.68	39.10	8.46
Test run energy [Wh]	ISS	18.38	0.21	18.32	18.23	19.33	1.10	0.05
	EET	22.63	1.99	22.37	20.04	27.97	7.93	1.95

Once it has been established, the DUT (or SUT) is configured, and the preparation of the chosen software entity is carried out (Activity A3.3).

Phase IV. Measurement Performance

In this phase, power consumption measurements will be made for each of the chosen sorting algorithms (Activity A4.1). Between every sorting algorithm there was a break of 10 s and after every loop run there was a break of 60 s. The pauses between the execution of the algorithms was added to allow the computer to return to its idle state, before starting the next task, in order to capture irregular patterns in the consumption. While the loop was running, two log files were generated for further analysis with the power consumption data (Activity A4.2). In these log files, the starting and ending timestamps of every test run and every sorting algorithm loop were recorded.

Phase V. Test Case Data Analysis

During this phase, the analysis of the energy consumption data for each of the test cases is carried out. The first activity is the preparation of the raw data, which has been obtained from the measuring instruments in the previous phase. In this activity, the average values of each of the measurements of the test cases are calculated. The outliers are also identified and eliminated, and the values obtained are checked to ensure they are valid (Activity A5.1).

Once the data have been processed and prepared, the descriptive statistics of the values obtained are calculated, as shown in Table 3.4 (Activity A5.2).

Phase VI. Software Entity Data Analysis

In this phase, the results obtained for each of the software entities (sorting algorithms) are analyzed (Activity A6.1). To do so, the energy consumed induced by the software run is calculated by subtracting the average adjusted baseline energy from the average energy of the test run measurements, as shown in Table 3.5.

In addition, the energy consumption and the induced energy for each classification algorithm have been obtained independently. Table 3.6 shows the consumed energy values obtained by each of the measurement instruments, and the energy consumption induced by each of the algorithms taking into account the reference consumption.

Based on this information, we can draw some conclusions (Activity 6.2). As shown in Fig. 3.14, the most energy-consuming sorting algorithm is Insertion Sort. The quicksort algorithm is the most energy efficient.

Table 3.5 Energy consumption induced

	EET	ISS proposal
Energy consumption induced	6.74 Wh	5.17 Wh

Table 3.6 Energy consumption by each software entity (sorting algorithm)

	Energy consumed		Energy consumption induced	
Sorting algorithm	ISS	EET	ISS	EET
Bubble Sort	3.35 Wh	3.45 Wh	0.90 Wh	0.83 Wh
Cocktail Sort	3.47 Wh	3.42 Wh	0.92 Wh	0.85 Wh
Insertion Sort	5.55 Wh	5.90 Wh	1.48 Wh	1.51 Wh
Quicksort	2.42 Wh	2.37 Wh	0.66 Wh	0.61 Wh
Mergesort	3.40 Wh	3.30 Wh	1.05 Wh	1.18 Wh

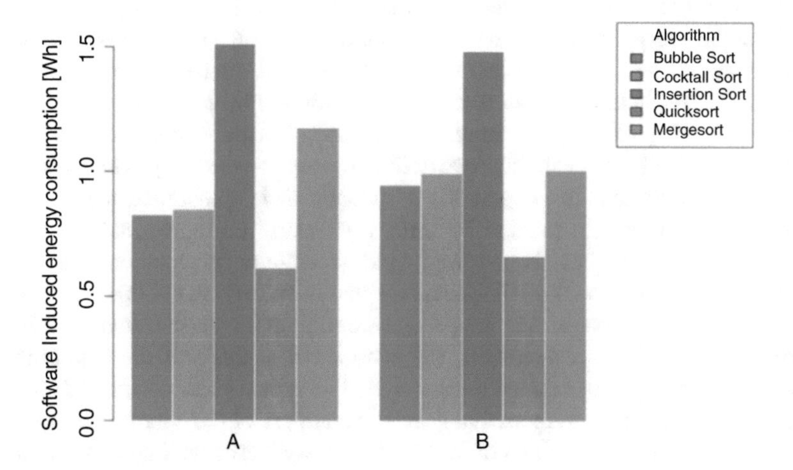

Fig. 3.14 Energy consumption induced by each software entity (sorting algorithm)

Phase VII. Reporting the Results
Finally, all the results obtained, along with the process followed to achieve them, are documented (Activity A7.1). In addition, the laboratory package[1] of the study is created, so that it can be analyzed and replicated by other researchers (Activity A7.2). The LP includes all the raw energy consumption data obtained from EET and by the ISS proposal.

3.3.5 *Case Study Analysis and Lessons Learned*

After applying the GSMP to carry out software energy consumption measurements with two different measuring instruments, we can conclude that the GSMP serves as

[1]https://doi.org/10.5281/zenodo.3257517

a guide in the activities that must be followed to carry out the measurement and analysis of the software's energy consumption, regardless of the measuring instrument used.

Moreover, the use of this process helps to have greater control over the measurements performed, since it enables us to identify when a measurement is incorrect, or whether it is necessary to repeat the measurements, ensuring their reliability and consistency.

3.4 Conclusions

The development of environmentally friendly software is not a trivial project. In fact, to determine the energy efficiency of the software, it is necessary to be able to evaluate the energy consumed when it is running. For this, it is necessary to follow some suitable steps to ensure that the results obtained are correct and appropriate. That has led us, in this chapter, to present our proposal for a process to evaluate the energy efficiency of software, known as the Green Software Measurement Process. Using the GSMP, we can improve reliability and consistency when measuring energy consumption. To support the systematic development, management, and growth of the proposed process by using a standardized representation, we have used SPEM 2.0; this has also allowed us to generate documentation in a standard format. The GSMP covers all the necessary phases in carrying out this type of study, such as the definition of the scope and configuration of the environment, the performance of the measurements, the subsequent analysis of the data obtained, and the reporting of the results. Furthermore, this process was designed to be valid with any measuring instrument that is used, regardless of the approach adopted. To define some aspects or artifacts of the GSMP, we have used well-known approaches to software measurement and good practices related to green software proposed by other authors.

The GSMP defined allows us to analyze the energy efficiency of software. This process enables researchers to obtain greater control over the measurements made, guaranteeing their reliability and consistency. It also means that the studies carried out can be easily replicated, and the results obtained are comparable to those of other studies.

From the point of view of software professionals, this contribution helps them to be aware that there are processes and tools to evaluate the energy efficiency of the software applications they develop. They can thus develop software that is environmentally friendly.

References

1. Fenton N, Bieman J (2014) Software metrics: a rigorous and practical approach. CRC press
2. Wohlin C, Runeson P, Höst M, Ohlsson MC, Regnell B, Wesslén A (2012) Experimentation in software engineering. Springer Science & Business Media
3. Basili VR, Weiss DM (1984) A methodology for collecting valid software engineering data. IEEE Trans Softw Eng 6:728–738
4. Defens USDo (2000) PSM: Practical software and systems measurement – a foundation for objective project management vol version 4.0c
5. Standard IIII (2017) ISO/IEC/IEEE 15939:2017 – Systems and software engineering-Measurement process
6. Hindle A (2015) Green mining: a methodology of relating software change and configuration to power consumption. Empir Softw Eng 20(2):374–409
7. Jagroep EA, van der Werf JM, Brinkkemper S, Procaccianti G, Lago P, Blom L, van Vliet R (2016) Software energy profiling: comparing releases of a software product. In: Proceedings of the 38th International Conference on Software Engineering Companion, pp 523–532
8. Henderson-Sellers B (2003) Method engineering for OO systems development. Commun ACM 46(10):73–78
9. OMG Software & Systems Process Engineering Metamodel specification (SPEM) Version 2.0
10. Jagroep E, Procaccianti G, van der Werf JM, Brinkkemper S, Blom L, van Vliet R (2017) Energy efficiency on the product roadmap: an empirical study across releases of a software product. J Softw Evol Process 29(2):e1852
11. Kern E, Hilty LM, Guldner A, Maksimov YV, Filler A, Gröger J, Naumann S (2018) Sustainable software products—towards assessment criteria for resource and energy efficiency. Futur Gener Comput Syst 86:199–210
12. Kandel S, Heer J, Plaisant C, Kennedy J, Van Ham F, Riche NH, Weaver C, Lee B, Brodbeck D, Buono P (2011) Research directions in data wrangling: visualizations and transformations for usable and credible data. Inf Vis 10(4):271–288
13. Kitchenham B, Madeyski L, Budgen D, Keung J, Brereton P, Charters S, Gibbs S, Pohthong A (2017) Robust statistical methods for empirical software engineering. Empir Softw Eng 22 (2):579–630
14. Wilcox RR (2011) Introduction to robust estimation and hypothesis testing. Academic Press
15. Basili VR, Selby RW, Hutchens DH (1986) Experimentation in software engineering. IEEE Trans Softw Eng 7:733–743
16. Brooks A, Daly J, Miller J, Roper M, Wood M (1996) Replication of experimental results in software engineering. International Software Engineering Research Network (ISERN) Technical Report ISERN-96-10, University of Strathclyde 2
17. Solari M, Vegas S, Juristo N (2018) Content and structure of laboratory packages for software engineering experiments. Inf Softw Technol 97:64–79
18. Jedlitschka A, Pfahl D (2005) Reporting guidelines for controlled experiments in software engineering. In: 2005 International Symposium on Empirical Software Engineering. IEEE, p 10
19. Brereton P, Kitchenham B, Budgen D, Li Z (2008) Using a protocol template for case study planning. In: 12th International Conference on Evaluation and Assessment in Software Engineering (EASE) 12, pp 1–8
20. Runeson P, Höst M (2009) Guidelines for conducting and reporting case study research in software engineering. Empir Softw Eng 14(2):131
21. Mancebo J, Arriaga HO, García F, Moraga MÁ, de Guzmán IG-R, Calero C (2018) EET: a device to support the measurement of software consumption. In: Proceedings of the 6th International Workshop on Green and Sustainable Software, pp 16–22

Chapter 4
FEETINGS: Framework for Energy Efficiency Testing to Improve eNvironmental Goals of the Software

Javier Mancebo, Coral Calero, Félix García, Mª Ángeles Moraga, and Ignacio García-Rodríguez de Guzmán

Abstract Energy consumption and carbon emissions caused by the use of software have been increasing in recent years, and it is necessary to increase the energy awareness of both software developers and end users.

The objective of this chapter is to establish a framework that provides a solution to the lack of a single and agreed terminology, a process that helps researchers evaluate the energy efficiency of the software, and a technology environment that allows for accurate measurements of energy consumed. The result is FEETINGS (Framework for Energy Efficiency Testing to Improve eNvironmental Goals of the Software), which promotes the reliability of capture, analysis, and interpretation of software energy consumption data.

FEETINGS is composed of three main components: an ontology to provide precise definitions and harmonize the terminology related to software energy measurement; a process to guide researchers in carrying out the energy consumption measurements of the software, and a technological environment which allows the capture, analysis, and interpretation of software energy consumption data.

In addition, an example of the application of FEETINGS is presented, as well as a guide to good practice for energy efficiency of software, based on different experiments carried out with this framework.

The results obtained demonstrate that FEETINGS is a consistent, valid, and useful framework to analyze the energy efficiency of software, promoting the accuracy of its energy consumption measurements. Therefore, FEETINGS serves as a tool to make developers and users aware of the impact that software has on the environment.

J. Mancebo · C. Calero (✉) · F. García · M. Á. Moraga · I. G.-R. de Guzmán
Alarcos Research Group, Institute of Technologies and Information Systems, University of Castilla-La Mancha (UCLM), Ciudad Real, Spain
e-mail: Javier.Mancebo@uclm.es; Coral.Calero@uclm.es; Felix.Garcia@uclm.es; MariaAngeles.Moraga@uclm.es; Ignacio.GRodriguez@uclm.es

C. Calero et al. (eds.), *Software Sustainability*,
https://doi.org/10.1007/978-3-030-69970-3_4

4.1 Introduction

Tablets, computers, smartphones, smartwatches, and a multitude of technological devices have invaded our daily lives. All of these devices require energy to operate, which has led to a huge annual growth in energy consumption. According to recent studies, energy used for global information and communications technology (ICT) could exceed 20% of total energy, and emit up to 5.5% of the world's carbon emissions by 2025 [1, 2]. These data on the growth of energy consumption and global emissions have raised issues of great concern for both software professionals and users.

Although hardware is generally seen as the main culprit for ICT energy usage, software also has a tremendous impact on the energy consumed [3]. Unfortunately, to date, little attention has been given to this topic by the information and communications technology (ICT) community [4]. However, in recent years, trends such as "Green Software" have gained importance [5]. The purpose of green software is to promote improvements in the energy efficiency of software, minimizing the impact it may have on the environment [6, 7].

To improve the energy efficiency of software, it is first necessary to raise energy awareness among all stakeholders [4]. On the one hand, developers must be aware of the energy that the software consumes when used, so that they can develop more energy-efficient and environmentally friendly software. And software professionals in general must treat energy efficiency as a quality attribute of the software, in the same way that usability or security is treated [8–10].

On the other hand, awareness also needs to be raised among end users as to how much energy is required by the software they use on a daily basis, so that they are aware of the impact that software can have on the environment [4]. Ideally, end users could compare the software applications that meet their needs, and choose the option that consumes the least energy, and should also know how a given software application can be used in a more efficient manner from the point of view of energy consumption.

In order to raise awareness among stakeholders or to develop a sustainable and environmentally friendly software product, it is first necessary to know the energy consumption induced by the software when it is running, since if the energy consumption is not measured, it cannot be managed [11, 12]. As the European Union report indicates [13], "the existence of a methodology for measuring the energy or CO2 of the ICT infrastructure is extremely important for this sector, as it will allow the development of much more robust estimates of the impact of ICT."

However, there is currently a lack of both knowledge and tools to reliably and accurately analyze software energy consumption [5]. We consider that in this regard one can identify three main problems:

- Several inconsistencies and terminological conflicts appear [14] due to the fact that researchers have defined their methods of work using their own terms or concepts, provoking numerous examples of both synonymy (same concepts with different term associated) and homonymy (different concepts with the same term). This lack of formal consensus makes it difficult to understand the main concepts involved when performing a software energy consumption assessment.

- There is a lack of a generally-agreed-on methodology that would guide software energy consumption assessments. This implies that the rigor of the studies carried out cannot be guaranteed, meaning that it is more complicated to replicate or compare the results obtained [15].
- Several measuring instruments are available for the analysis of software energy consumption. It is important to note that each measurement instrument has its own particular characteristics, and that it is necessary to choose the one that best adapts to the particular evaluation requirements concerned [16].

To contribute to the mitigation of these problems, and to be able to raise energy awareness among all stakeholders, we have developed a framework to promote the reliability of capture, analysis, and interpretation of software energy consumption data, known as "FEETINGS" (Framework for Energy Efficiency Testing to Improve eNvironmental Goals of the Software). FEETINGS aims to provide: (1) a solution to the lack of a unique and agreed terminology; (2) a process that helps researchers to evaluate the energy efficiency of the software, allowing greater control over the measurements made, thereby ensuring their reliability and consistency; and (3) a technological environment that supports the process and allows for realistic measurements of the energy consumed by the software and its subsequent analysis.

In this chapter, we present FEETINGS and an example of how to use it, so as to guide end users. The contents of the chapter are structured as follows. First, we present the different studies and proposals that have served as a basis for our framework. Then, in Sect. 4.2, we present the FEETINGS framework, detailing each of its components. In Sect. 4.3, an example of the application of FEETINGS is presented. In Sect. 4.4, a best practice guideline for software energy efficiency is proposed, based on different experiments carried out using FEETINGS. Finally, Sect. 4.5 sets out some conclusions of this work.

4.2 FEETINGS

In this section, we will describe FEETINGS, a framework to promote more reliable capture and analysis of software energy consumption data. This framework is made up of three main components, classified according to their nature as conceptual, methodological, and technological components (see Fig. 4.1), as described in the following subsections.

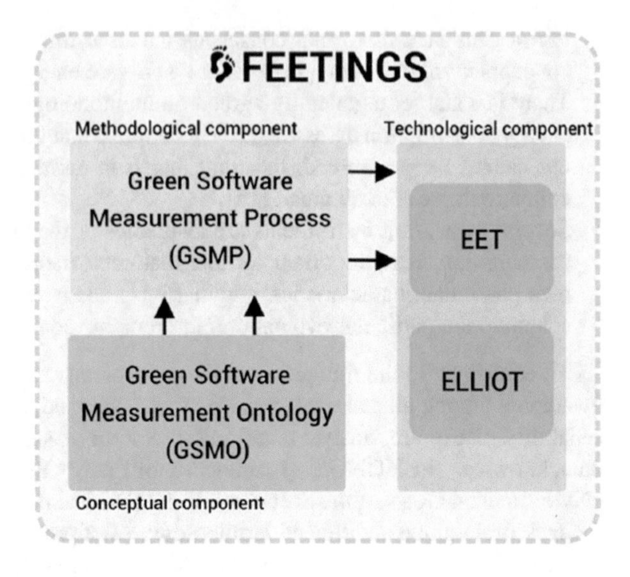

Fig. 4.1 Overview of FEETINGS

4.2.1 Conceptual Component

As commented upon in the introduction, one of the common problems is confusion and inconsistencies in the main concepts used in software energy assessment. This lack of formal consensus makes it difficult to understand the main concepts involved when performing a software energy consumption measurement.

The conceptual part of FEETINGS seeks to solve the lack of a unique and agreed terminology. For this purpose, an ontology has been elaborated which contains the concepts related to the software energy measurement. According to Chandrasekaran et al. [17], the unification of terms and concepts in an ontology allows knowledge to be shared, while ontological analysis clarifies the structure of knowledge.

The ontology proposed is known as "Green Software Measurement Ontology" (GSMO), and its purpose is to provide precise definitions of all terms related to software energy measurement and to clarify the relationships between them, removing terminological conflicts and fostering the consistent application of the framework by other researchers and practitioners with reference to a common vocabulary. The Green Software Measurement Ontology (GSMO) is an extension of the SMO ontology proposed by Garcia et al. [18] for green software measurement.

Figure 4.2 shows the graphical representation of the terms and relationships of the GSMO, using UML (Unified Modeling Language). The highlighted concepts are the new concepts which extend/adapt the SMO [18].

The conceptual component (GSMO ontology) aims to solve the problem of terminology consistency in software energy measurement, since it proposes a common vocabulary extracted from several international standards and research proposals. More details about the GSMO ontology can be found at [19].

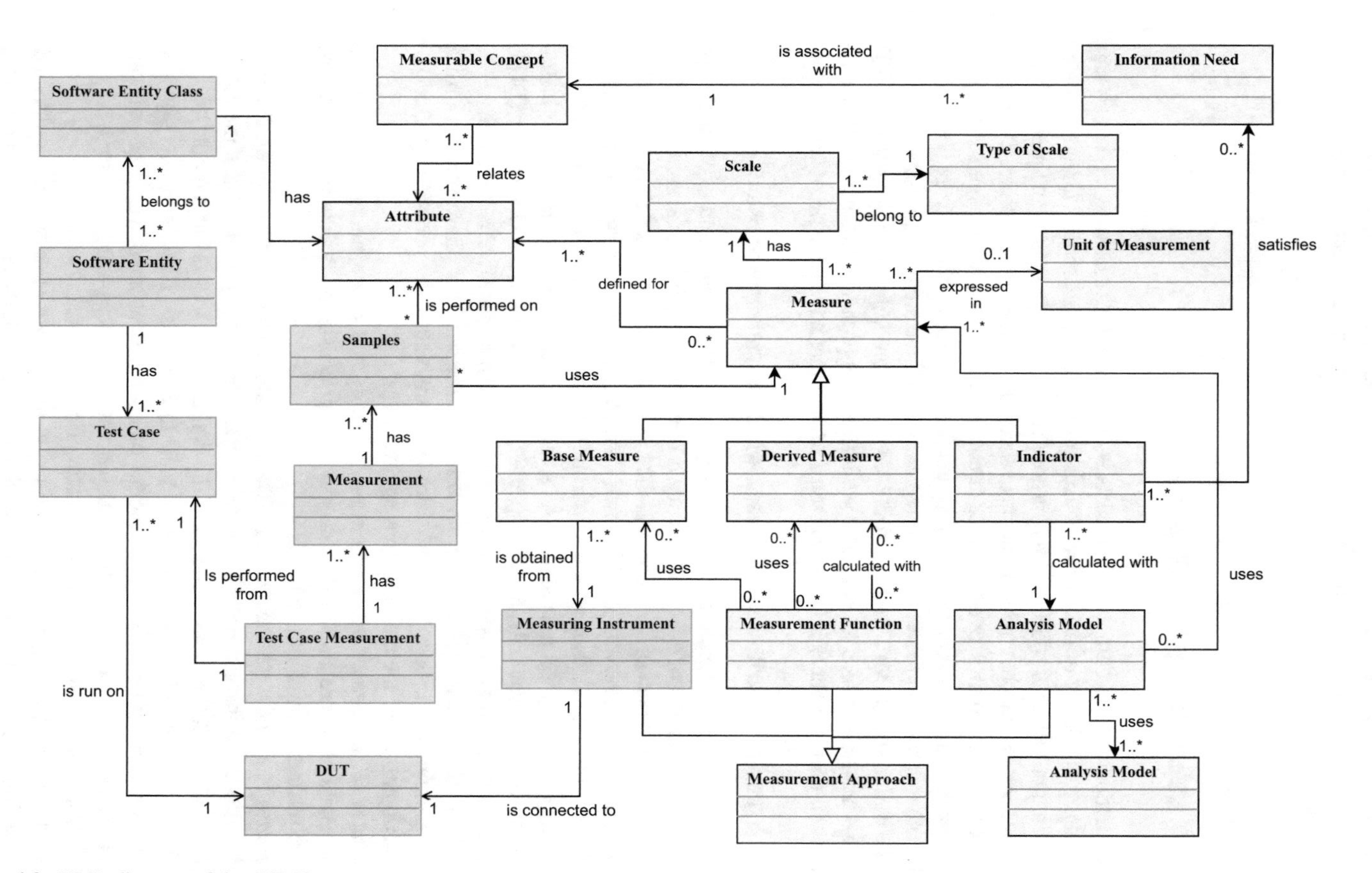

Fig. 4.2 UML diagram of the GSMO

This ontology has, moreover, served as a basis for the development of the methodological component of FEETINGS, which is presented in the following subsection.

4.2.2 Methodological Component

The methodological component consists in a process for measuring and analyzing the energy efficiency of the software. This process is known as the "Green Software Measurement Process" (GSMP). Its purpose is to guide researchers and practitioners as they seek to carry out measurements of software energy consumption. The GSMP ensures greater control over the measurements made, improving their reliability, consistency, and coherence. It also ensures that the results obtained are comparable with other studies and facilitates the replicability of the analyses performed.

To define the GSMP, we have followed the method engineering approach [20], and we have also taken as our basis well-known approaches to software measurement and good practices related to green software that have been proposed by other authors.

The process consists of seven phases, which are summarized below:

- *Phase I. Scope Definition:* In this phase, a complete specification of requirements for the evaluation of energy efficiency is obtained. In addition, the software which is to be the subject of the study and the test cases to be analyzed must be defined.
- *Phase II. Measurement Environment Settings:* The purpose of this phase is the definition of the measurement environment that is to be used in the software energy consumption assessment. As a result of this phase, the measuring instrument, its measurements, and the specifications of the Device Under Test (DUT) are defined and the baseline energy consumption of the DUT is obtained.
- *Phase III. Measurement Environment Preparation:* This phase focuses on the preparation of the energy consumption measurements to be performed and on the configuration of the measurement environment.
- *Phase IV. Measurements Performance:* During this phase, energy consumption measurements are carried out and raw energy consumption data taken from the measuring instrument is collected.
- *Phase V. Test Case Data Analysis*: The raw data of energy consumption obtained by the measuring instrument is processed, and the statistical analysis of the values obtained from the measurements of the defined test cases is carried out.
- *Phase VI. Software Entity Data Analysis:* In this phase, with the results obtained from the previous phases, the amount of energy consumed when the software entity was executed in the DUT is determined and interpreted, and some conclusions about the software energy consumption are stated.
- *Phase VII. Reporting the Results:* Finally, the study carried out is documented, describing the entire process followed, and setting out the results obtained on the energy consumption of the software.

In the previous chapter (Chap. 3) of this book, a more detailed and complete version of the GSMP is presented, including a description of the roles, phases, and activities (input, output, and guidelines).

4.2.3 *Technological Component*

In this section, the technological environment is presented. The main objective of the technological component of the FEETINGS framework is to perform more realistic measurements of the energy consumed by the software, and to use these results to

- Analyze the consumption of the software
- Learn about the behavior of the software and its different versions, to find out if these versions worsen the software energy consumption or not
- Identify the consumption patterns that can guide the improvement of the energy efficiency of software applications
- Recommend changes to software to improve energy efficiency

The technological component, as can be seen in Fig. 4.3, is composed of two artifacts: EET (Energy Efficiency Tester) and ELLIOT.

4.2.3.1 EET (Energy Efficiency Tester)

EET [21, 22] is a measuring instrument that enables the accurate capture of the energy consumption of the computer (DUT) on which the software is running. In addition to the total energy consumption of the DUT, this measuring instrument supports the measurement of four different hardware components: processor, hard disk, graphic card, and monitor. Figure 4.4 shows the EET measuring instrument in working use.

As can be seen in Fig. 4.4, the EET is connected to the DUT where the software is executed, and is composed of three main components:

- A system microcontroller, whose task is to gather the information extracted from the different sensors and store it in a MicroSD memory. It also allows the frequency with which the device performs the measurements to be adjusted.
- A set of sensors, which are responsible for taking energy consumption measurements of the hardware components (processor, hard disk, graphics card, and monitor) of the DUT connected to the EET.
- A power supply, which must be connected to the device under test where the software is executed, replacing the power supply of the DUT; the sensors are connected to the energy distribution lines from the power supply to the different hardware components.

In a nutshell, EET is a measuring instrument, which is considered a core component of FEETINGS. It allows us to capture and record the energy efficiency of

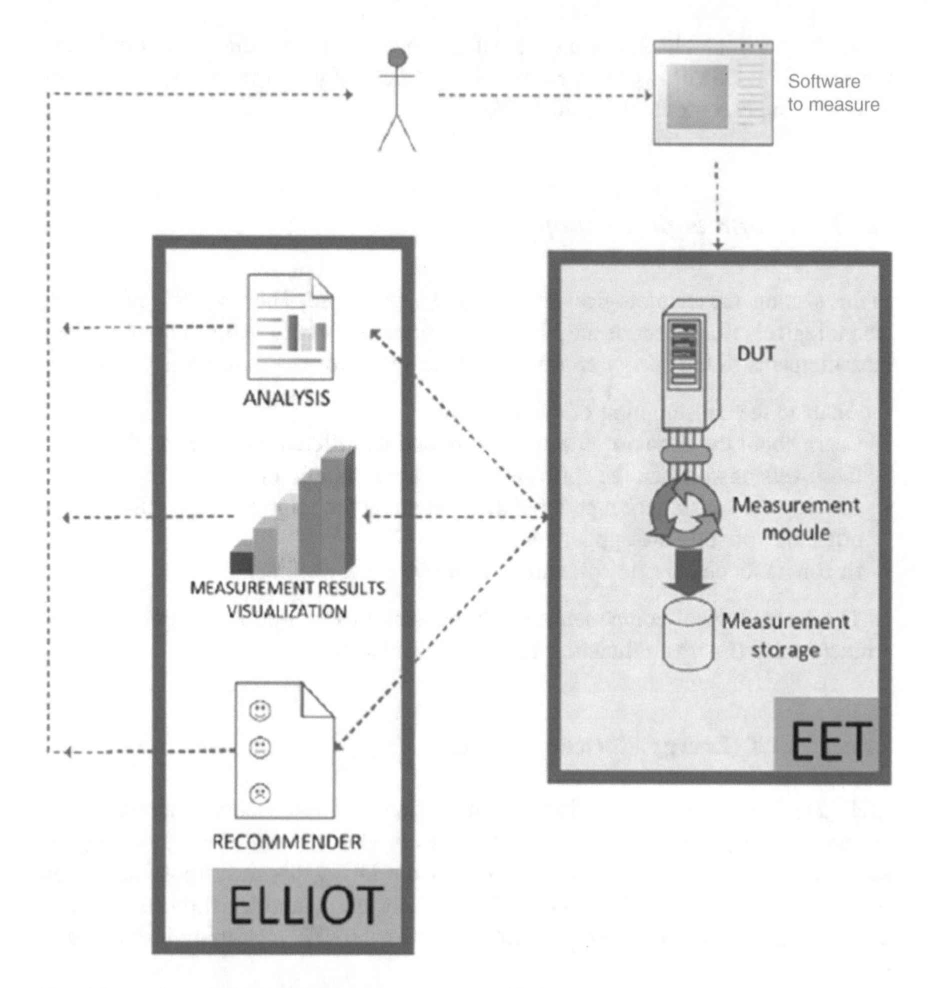

Fig. 4.3 Artifacts of the technological component of FEETINGS

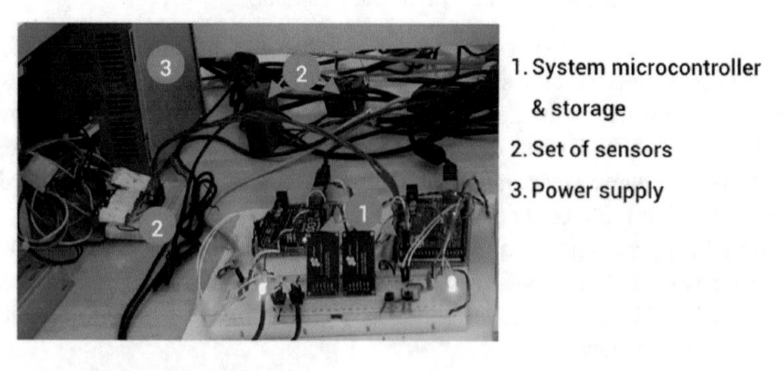

Fig. 4.4 EET measuring instrument

software when it is running. EET provides a realistic measurement of energy consumption and, moreover, is capable of obtaining detailed energy measurements from different components of the DUT (processor, graphic card, hard disk, and monitor). Another advantage of this measuring instrument is its sampling frequency, around 100 Hz, which provides very reliable consumption information.

As the EET produces a huge amount of data on energy consumption, it is necessary to support the processing and analysis of these data with a suitable software tool. For this reason, the ELLIOT tool was developed, which is described in the following section.

4.2.3.2 ELLIOT

ELLIOT [19] is a software tool tasked with processing the data collected by the EET, analyzing these data, and providing a visual environment that allows researchers to process the software energy consumption data. Furthermore, the ELLIOT tool is aligned with the GSMP described in the previous section.

The main functionalities supported by ELLIOT are outlined below:

- Processes all measurements carried out with the EET measuring instrument.
- Calculates different statistical variables of the energy consumption measurements according to the user's needs.
- Identifies possible outliers that may be present in the measurement samples, using robust parametric methods such as median absolute deviations from the median (MADN).
- Visualizes the results through graphs and data tables which contain information on the measurements of the energy consumption of the software.
- Compares the results obtained from the different energy consumption measurements.
- Generates reports that include all the information on the energy efficiency of the software analyzed.

The ELLIOT tool is composed of four modules (see Fig. 4.5) that support these functionalities. The modules are: (1) user management, which allows one to manage the permissions and roles of ELLIOT users; (2) system management, to add and modify information about the instruments and the DUT in which the measurements are carried out; (3) measurement management, which is the central module of ELLIOT since it supports all the tasks of processing, data wrangling, measurements analysis, and visualization of the energy consumption information; and (4) report management, which generates reports and allows comparisons between the measurements.

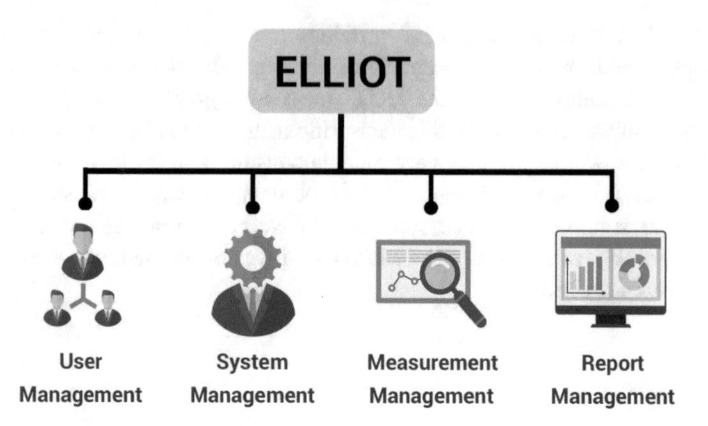

Fig. 4.5 ELLIOT tool modules

4.3 Application of FEETINGS: A Case Study of the Energy Consumed by Translators

This section presents an application of the FEETINGS framework to measure and analyze software energy consumption, as defined in the previous section. One of the most widely used software applications today is online translation, which includes, additionally, several options for automatic text translation, such as Google translate and DeepL.

In this study, our aim is to analyze the energy consumption of the main online translation tools in order to raise users' awareness of the environmental impact of their use, and also to try to provide them with a set of guidelines so that, when they use these tools, such use is as efficient as possible.

Following the GSMP, Phase I defines the scope of the study. As mentioned above, this study aims to determine the energy consumption of the main online translators (Software Entity Class). The chosen translators (Software Entity) were Google Translate, DeepL, Bing Translator, Tradukka, Systran Translate, and Yandex. To evaluate the selected software entities, five test cases were defined:

- Translate a text with 10 characters.
- Translate a text with 100 characters.
- Translate a text with 1000 characters.
- Translate a text with 3000 characters.
- Translate a text with 5000 characters.

All defined test cases were executed in two different browsers (Google Chrome and Firefox). Thus, we can also study the efficiency of the browser in which each of the translators is used.

In the second phase of the process, we selected the FEETINGS technological environment to analyze the energy consumption, choosing the EET as the measuring instrument and ELLIOT to analyze and process the energy consumption data. The

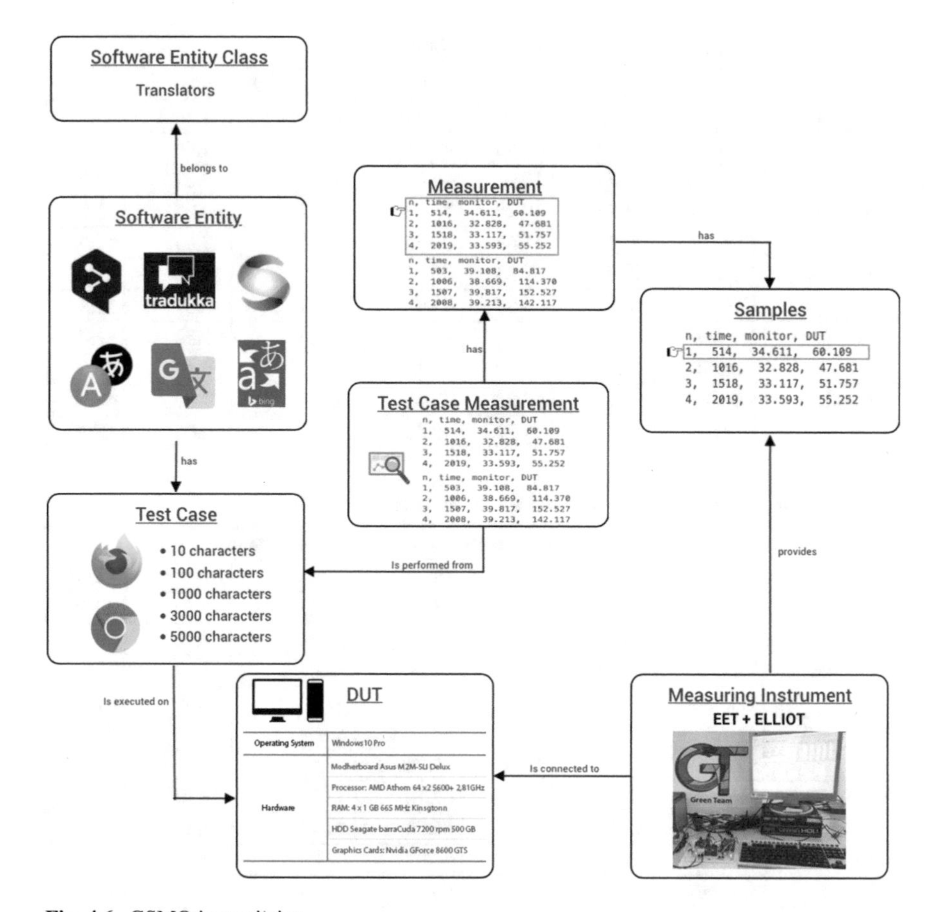

Fig. 4.6 GSMO instantiation

specification of the DUT in which the test cases were executed was also defined. For this study, we decided that from the measurements provided by the EET, we would take into account only the energy measurements of the monitor and the total consumption of the DUT. We have not recovered and analyzed data from the hard disk, graphics card, or processor because, as all the translators were executed in a web browser, the use of these components was minimal.

In accordance with the third phase of the process, we determined that each of the test cases was to be run and measured (with EET) 35 times. Being a controlled test environment, 35 measurements is usually a sufficient sample size to mitigate the impact of outliers (such as energy consumption devoted to operating system tasks).

Figure 4.6 shows the instantiation of the concepts of this study, defined in the GSMO ontology for this study, which also serves as a summary of the outputs obtained in the first phases of the GSMP.

Tables 4.1 and 4.2 show the energy consumed once the measurement and analysis tasks have been performed, by the DUT and the monitor respectively, in the execution of each of the defined test cases.

Table 4.1 DUT energy consumption for each test case

Test cases		DUT energy consumption (Watts per second)					
		DeepL	Bing	Google	Tradukka	Systran	Yandex
10 char.	Firefox	46.12	64.65	46.21	41.56	61.09	45.11
	Chrome	50.77	53.07	45.09	61.07	40.27	43.26
100 char.	Firefox	53.43	40.20	76.23	44.47	64.03	53.61
	Chrome	46.23	25.69	38.66	50.22	47.61	43.96
1000 char.	Firefox	57.34	39.83	52.25	54.12	59.72	66.52
	Chrome	48.91	28.99	45.42	32.57	40.67	51.37
3000 char.	Firefox	56.38	45.46	47.36	70.64	95.20	62.72
	Chrome	41.91	37.75	44.19	39.47	50.18	58.04
5000 char.	Firefox	67.31	57.62	49.45	62.98	77.29	66.56
	Chrome	51.77	52.53	45.30	41.53	56.05	69.93

Table 4.2 Monitor energy consumption for each test case

Test cases		Monitor energy consumption (Watts per second)					
		DeepL	Bing	Google	Tradukka	Systran	Yandex
10 char.	Firefox	63.06	70.40	65.54	53.27	65.50	60.14
	Chrome	66.88	56.41	63.92	68.34	57.62	59.96
100 char.	Firefox	75.83	58.51	62.28	61.07	64.77	70.15
	Chrome	61.23	50.64	59.34	57.48	57.09	61.17
1000 char.	Firefox	56.93	59.02	63.59	66.67	58.30	87.79
	Chrome	57.82	49.66	69.96	58.42	56.31	63.01
3000 char.	Firefox	56.61	54.90	67.10	52.87	73.70	72.07
	Chrome	53.35	52.97	66.95	56.43	60.38	64.14
5000 char.	Firefox	65.83	60.19	64.97	68.88	69.92	70.62
	Chrome	61.98	63.34	66.58	57.24	63.31	77.27

To analyze the data in the above tables, we will study the test cases executed in each of the browsers independently, and then compare the results obtained in both browsers.

First, focusing on the test cases executed in Firefox, in Fig. 4.7, we can observe that for translating texts of intermediate size (between 100 and 3000 characters) the option that requires the least energy consumption in the DUT is the Bing translator. However, Bing is the worst option for very small texts (around 10 characters), with Tradukka being the most efficient option in this case. For large texts (5000 characters) Google Translate is the best option. The consumption data results obtained from the monitor are very similar to those obtained from the DUT. The Yandex and Systran translators are the worst choice in almost every test case.

Analyzing the consumption data of the tests executed in Google Chrome, we can see in Fig. 4.8 that the energy consumption of the translators behaves in a similar way as in Firefox. In this case, Bing is also the best choice for intermediate-length text. But unlike the results obtained in Firefox, the Tradukka translator is the least efficient option for small texts (fewer than 100 characters).

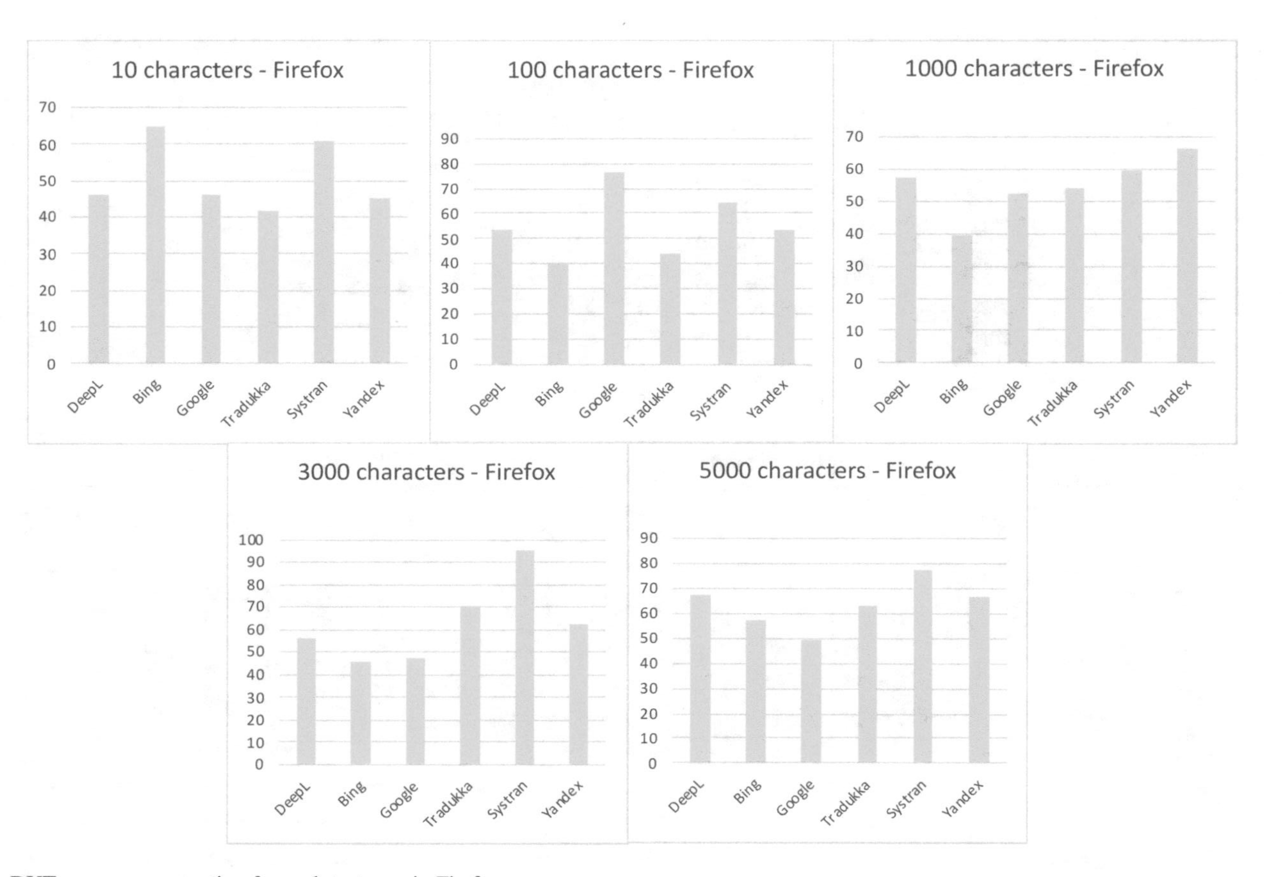

Fig. 4.7 DUT energy consumption for each test case in Firefox

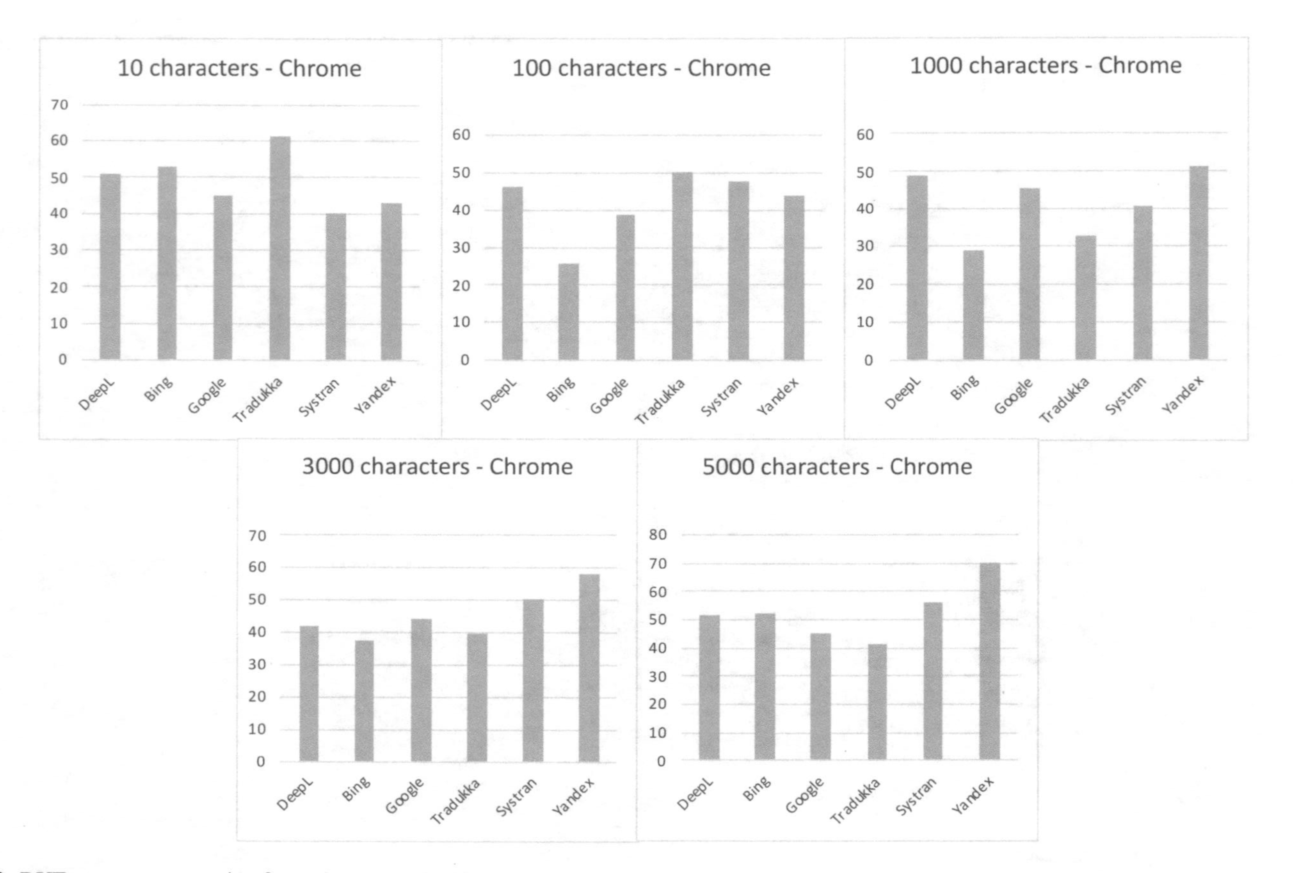

Fig. 4.8 DUT energy consumption for each test case in Chrome

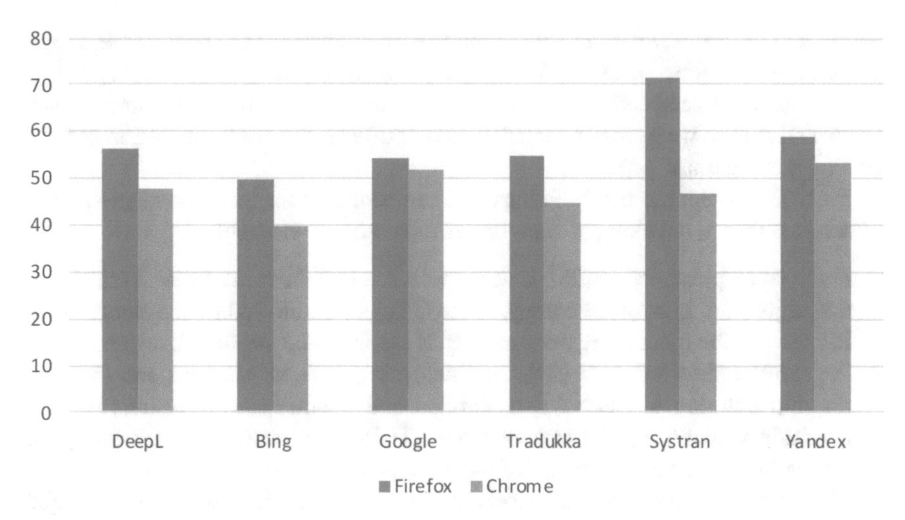

Fig. 4.9 Comparison of the mean energy consumption by the DUT

To determine in which browser (Firefox or Chrome) it is better to use the translators, we have calculated the average power consumption of the executed test cases. As shown in Fig. 4.9, all the translators analyzed have a lower DUT power consumption in the Chrome browser than in Firefox. Regarding the monitor's energy consumption, the results are similar to those of the DUT, although here the variation between browsers is less.

Considering the results obtained, the main conclusions that we can draw from this study are as follows:

- The most energy-efficient scenario is to use the Tradukka translator with the Google Chrome browser for texts of an intermediate length (1000 characters), resulting in a consumption of 32.56 Ws.
- The most inefficient option (95.19 Ws) is to use the Systran translator in the Firefox browser for large texts (3000 characters).
- All translators show a direct relationship when translating texts of more than 1000 characters, in which case an increase in the number of characters also increases energy consumption.
- It is more efficient to use the Google Chrome browser to translate than Firefox.

4.4 Best Practices Guideline on Software Sustainability

As explained in the introduction, FEETINGS can be used for several purposes. One of the main objectives of FEETINGS is to measure software energy efficiency so that researchers and software professionals can develop software that is environmentally friendly. Another purpose is focused on the other key perspective in software: the end user. This framework can help to make society aware of the responsible use of software applications, so as to take care of the environment.

Keeping both perspectives in mind, this section presents some guidelines, based on several studies we have conducted, to make software more sustainable, in both its development and use.

First, we present the main findings that can be useful for researchers and practitioners wishing to develop software that is more energy-efficient:

- In the study presented in [23], it was concluded that the most efficient classification algorithm, in terms of energy, is the Quicksort, followed by the Bubble sort. In contrast, the most energy-demanding is the Insertion sorting algorithm.
- For the Redmine software [24], after analyzing different versions and the relationship between its energy consumption and maintenance measures, we can conclude that the Total Lines of Code (TLOC) maintainability measurement affects the energy consumption of the processor and the DUT.
- The study presented in [25] shows that text compression, using End-Tagged Dense Code (ETDC) or Tagged Huffman algorithms, not only reduces search space and time, but also leads to lower energy consumption. In addition, the use of search algorithms in compressed text, such as the Horspool algorithm, when run on compressed data, also requires less CPU power than when run over uncompressed data.
- In [26], recommendations were made to improve the energy efficiency of applications. The best practices in energy-efficient computing drawn from this study are summarized below:

  - A balance should be struck between the energy efficiency of the graphical user interface and a good experience from the user's perspective.
  - Efficient UIs should be designed to allow a task to be completed quickly and easily.
  - Data redundancy should be reduced.
  - Devices when not in use should be powered down by batch I/O.

- We have worked on the comparison of the time and energy consumption required by three releases of the same application, developed with and without Spring. In conclusion, it seems that products developed without using Spring are better in all the conditions and for all the measures. This could indicate that, although Spring has some advantages for programmers, once the product starts to run, this advantage disappears to the benefit of the non-Spring development.

Second, we present the following set of guidelines to help end users make use of the software in a more environmentally friendly way:

- To publish a tweet or a post on Facebook, you should consider that the most energy-efficient option is to publish a single emoji or a picture. Furthermore, if you want to be respectful of the environment, you should avoid publishing a GIF, since it is the option that consumes the most energy [19].

- The most energy-efficient personal health record (PHR) is the NoMoreClipboard [26].
- As regards navigating the internet: if you are looking for browsers which are environmentally friendly, choose Edge or Firefox; if you are looking for maximum privacy and energy efficiency use the DuckDuckGo search engine, especially when used with Edge and Firefox; if you are looking for lower emissions in your searches use the Ecosia search engine; and if in any event you do wish to use Chrome as your browser, then do so with DuckDuckGo.

4.5 Conclusions

Software plays an important role in the global energy consumption of a PC. For this reason, it is very important that both professionals and users be aware that the use of software has a great impact on the energy consumed by the devices on which it is executed.

In order to raise awareness of energy consumption among stakeholders, it is necessary to quantify its impact. Bearing this in mind, this chapter has presented the FEETINGS framework, which aims to promote reliable measurement, analysis and interpretation of software energy consumption data. FEETINGS is composed of three main components: (1) a GSMO ontology, to provide precise definitions of all concepts and their relationships related to software energy measurement; (2) a GSMP, to guide researchers in carrying out the energy consumption measurements of the software; and (3) a technological component, which is composed of two artifacts: EET, a measuring instrument, and the software tool ELLIOT to process and analyze data collected by the EET.

Thus, the use of FEETINGS serves two purposes. The first is to enable researchers and professionals to measure and make them aware of the energy that the software they develop consumes when in use, and thus be able to develop more energy-efficient software. The second is to show end users just how much energy is required by the software we use every day, and to make them aware of the impact that software can have on the environment.

In this work, we have also demonstrated an application of the FEETINGS framework to analyze the energy consumption involved in the use of different online translators. In addition, we have presented a set of best practice guides on sustainable software design based on our experience using the FEETINGS framework.

References

1. Andrae A (2019) Prediction studies of electricity use of global computing in 2030. Int J Sci Eng Invest 8:27–33

2. Vidal J (2017) Tsunami of data'could consume one fifth of global electricity by 2025. Climate Home News 11
3. Pereira R, Carção T, Couto M, Cunha J, Fernandes JP, Saraiva J (2020) Spelling out energy leaks: aiding developers locate energy inefficient code. J Syst Softw 161:110463
4. Fonseca A, Kazman R, Lago P (2019) A manifesto for energy-aware software. IEEE Softw 36 (6):79–82
5. Pinto G, Castor F (2017) Energy efficiency: a new concern for application software developers. Commun ACM 60(12):68–75
6. Calero C, Piattini M (2015) Introduction to green in software engineering. In: Green in software engineering. Springer, pp 3–27
7. Calero C, Piattini M (2017) Puzzling out software sustainability. Sustain Comput Informatics Syst 16:117–124
8. Calero C, Moraga MÁ, Bertoa MF, Duboc L (2014) Quality in use and software greenability. In: RE4SuSy@ RE. pp 28–36
9. Condori-Fernandez N, Lago P (2018) Characterizing the contribution of quality requirements to software sustainability. J Syst Softw 137:289–305
10. Penzenstadler B, Raturi A, Richardson D, Tomlinson B (2014) Safety, security, now sustainability: the nonfunctional requirement for the 21st century. IEEE Softw 31(3):40–47
11. Briand LC, Morasca S, Basili VR (1996) Property-based software engineering measurement. IEEE Trans Softw Eng 22(1):68–86
12. Moraga MÁ, Bertoa MF (2015) Green software measurement. In: Green in software engineering. Springer, pp 261–282
13. European Union (2011) The contribution of ICT to Energy Efficiency: local and regional initiatives
14. Pinto G, Castor F, Liu YD (2014) Understanding energy behaviors of thread management constructs. In: Proceedings of the 2014 ACM International Conference on Object Oriented Programming Systems Languages & Applications, pp 345–360
15. Fenton N, Bieman J (2014) Software metrics: a rigorous and practical approach. CRC Press
16. Moura I, Pinto G, Ebert F, Castor F (2015) Mining energy-aware commits. In: IEEE/ACM 12th Working Conference on Mining Software Repositories. IEEE, pp 56–67
17. Chandrasekaran B, Josephson JR, Benjamins VR (1999) What are ontologies, and why do we need them? IEEE Intell Syst Their Applications 14(1):20–26
18. García F, Bertoa MF, Calero C, Vallecillo A, Ruiz F, Piattini M, Genero M (2006) Towards a consistent terminology for software measurement. Inf Softw Technol 48(8):631–644
19. Mancebo J, Calero C, García F, Moraga MÁ, García-Rodríguez De Guzmán I (2020) FEETINGS: Framework for Energy Efficiency Testing to Improve eNvironmental Goal of the Software. Paper presented at the The Eleventh International GREEN and Sustainable Computing (under review)
20. Henderson-Sellers B (2003) Method engineering for OO systems development. Commun ACM 46(10):73–78
21. Mancebo J, Arriaga HO, García F, Moraga MÁ, de Guzmán IG-R, Calero C (2018) EET: a device to support the measurement of software consumption. In: Proceedings of the 6th International Workshop on Green and Sustainable Software, pp 16–22
22. Piattini M, Calero C, García F, Moraga MÁ, de Guzmán IGR, Mancebo J, Arriaga HO, Tabaco R (2018) Aparato para medición del consumo eléctrico de equipos informáticos (PC). ES 1199234 Y

23. Mancebo J, Guldner A, Kern E, Kesseler P, Kreten S, Garcia F, Calero C, Naumann S (2020) Assessing the sustainability of software products—a method comparison. In: Advances and new trends in environmental informatics. Springer, pp 1–15
24. Mancebo J, Calero C, García F (2021) Does maintainability relate to the energy consumption of software? A case study. Softw Qual J 29(1):101–127
25. Mancebo J, Calero C, Garcia F, Brisaboa N, Fariña A, Pedreira O (2019) Saving energy in text search using compression. Paper presented at the GREEN 2019: The Fourth International Conference on Green Communications, Computing and Technologies, Nice, France
26. García-Berná JA, Fernández-Alemán JL, Carrillo-de-Gea JM, Toval A, Mancebo J, Calero C, García F (2020) Energy efficiency in software: a case study on sustainability in Personal Health Records. J Clean Prod

Chapter 5
Patterns and Energy Consumption: Design, Implementation, Studies, and Stories

Daniel Feitosa, Luís Cruz, Rui Abreu, João Paulo Fernandes, Marco Couto, and João Saraiva

Abstract Software patterns are well known to both researchers and practitioners. They emerge from the need to tackle problems that become ever more common in development activities. Thus, it is not surprising that patterns have also been explored as a means to address issues related to energy consumption. In this chapter, we discuss patterns at code and design level and address energy efficiency not only as the main concern of patterns but also as a side effect of patterns that were not originally intended to deal with this problem. We first elaborate on state-of-the-art energy-oriented and general-purpose patterns. Next, we present cases of how patterns appear naturally as part of decisions made in industrial projects. By looking at the two levels of abstraction, we identify recurrent issues and solutions. In addition, we illustrate how patterns take part in a network of interconnected components and address energetic concerns. The reporting and cases discussed in this chapter emphasize the importance of being aware of energy-efficient strategies to make informed decisions, especially when developing sustainable software systems.

D. Feitosa (✉)
University of Groningen, Groningen, The Netherlands
e-mail: d.feitosa@rug.nl

L. Cruz
Delft University of Technology, Delft, The Netherlands
e-mail: l.cruz@tudelft.nl

R. Abreu
Faculty of Engineering, University of Porto & INESC-ID, Porto, Portugal
e-mail: rui@computer.org

J. P. Fernandes
CISUC and University of Coimbra, Coimbra, Portugal
e-mail: jpf@dei.uc.pt

M. Couto · J. Saraiva
HASLab/INESC TEC and University of Minho, Braga, Portugal
e-mail: marco.l.couto@inesctec.pt; jas@di.uminho.pt

C. Calero et al. (eds.), *Software Sustainability*,
https://doi.org/10.1007/978-3-030-69970-3_5

5.1 Introduction

The existence of *patterns* cannot be dissociated from our daily life. We may reason about patterns as concrete observations that are grouped into coherent categories. Patterns help us understand and describe our world. As an example, the evolutionary theory proposed by Charles Darwin was synthesized based on his understanding of patterns emerging from the observations he conducted during his voyage. Patterns can also be found in music, and in this context it has been shown that only a few musical notes sustain the essential melody of landmark music pieces.

Patterns are also well known to both researchers and practitioners in the software development world. In between the various definitions and types of patterns, there is a common understanding that they encapsulate solutions to recurrent problems [1]. A collection of recurrent problems that have become ever more apparent involves energy efficiency, as the growing energy demand associated with ICT usage is already a concern [2]. Notably, energy consumption is an issue with data/computation centers and their massive energy footprint [3], and, nowadays, the ubiquitous use of battery-powered devices such as smartphones [4].

Within ICT, energy consumption is an issue that needs to be addressed not only at hardware and firmware level, but also at the software, or application, level. Indeed, energy efficiency is a multifaceted problem, which encompasses networks, hardware, drivers, operating systems, and applications. In this chapter, we focus on applications and address the problems as systems of forces that can be fully or partially addressed by patterns [1]. In this context, software optimizations have been discussed at source code, design, and architecture level, from which we focus on the first two.

At the code level, we find solutions that are platform-specific and also commonly language-specific, which benefit from being more straightforward to apply. As the scopes open up, design patterns can be language-agnostic and generalizable to a broader range of software domains.

In this chapter, we aim to demonstrate how patterns with various scopes can help build energy-efficient software. Moreover, we discuss patterns that address energy consumption as the main concern (i.e., energy patterns), and patterns that were not initially intended to serve that purpose but have an energy-related side effect. To that end, the subject matter is organized as depicted in Fig. 5.1. In particular, we present energy-oriented code patterns in Sect. 5.2, move on to energy-oriented design patterns in Sect. 5.3, and elaborate the impact of general purpose design patterns on energy efficiency in Sect. 5.4.

Finally, we also illustrate how patterns appear naturally as part of decisions made in industrial projects. Thus, in Sect. 5.5 we present cases from open source projects where energy efficiency issues were factored in and a pattern was applied as part of the solution.

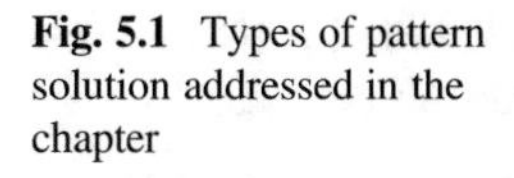

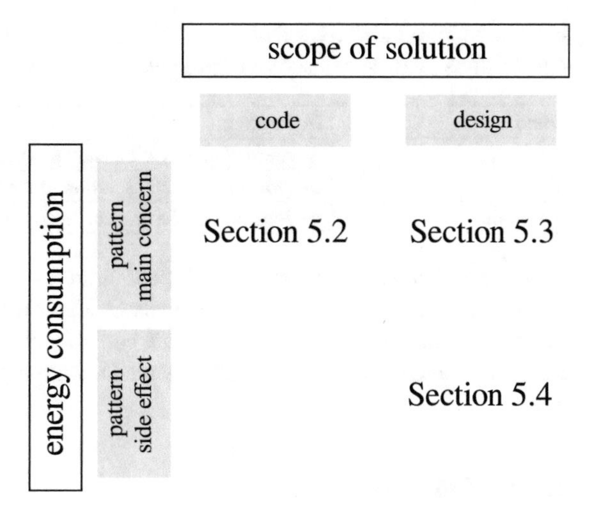

Fig. 5.1 Types of pattern solution addressed in the chapter

5.2 Code-Level Patterns

In this section, we focus on code-level patterns that have been shown to exhibit *greedy* energy consumption behaviors. Identifying patterns at the code level facilitates their transformation into more efficient alternatives, an approach that is widely known as *refactoring*. The potential of code refactorings is maximized when it is possible to automatically realize them, namely using tools that locate code fragments that can be improved and replacing them with the documented alternatives.

The patterns we consider in this section are specific to mobile application development. Mobile devices are these days an essential component of our daily lives, to support both our personal and professional activities. In this context, battery life is one of the principal factors that influence the satisfaction of mobile device users [5], and a recent survey in the US ranked battery life as the most important factor influencing purchasing decisions [6]. Battery life is such a concern that it has been suggested that nine out of ten users suffer from anxiety when their devices are low on battery [7], and this anxiety is under discussion within the Diagnostic and Statistical Manual of Mental Disorders as a potential clinical condition named *nomophobia*, which reflects the fear of not being able to use one's mobile phone [8].

The perception that an application causes excessive battery consumption is actually one of the most common causes for bad app reviews in app stores [9, 10]. This has raised the awareness of mobile application developers regarding the impact their applications have on battery life. In fact, while it has been shown that developers often seek information on how to improve the energy profile of their applications, they rarely receive proper advice [11–13].

Our focus is on code patterns reported as energy greedy within Android. We will refer to these patterns as EGAPs—Energy-Greedy Android Patterns. We synthesize contributions from several works that have documented and validated energy-oriented code refactorings. Specifically, we focus on energy-greedy patterns that have been automatically refactored in a large-scale empirical study involving 600+ applications [14]. We describe each pattern using the template shown next.[1]

Field	Description
Problem	A recurrent energy efficiency problem where the pattern can be used.
Solution	Generic and reusable solution to the problem.
Example	An illustration of a practical usage of the energy pattern.

The problem and solution for each pattern are tentatively provided by high-level descriptions that we believe can be understood by a broad audience of users and developers. Complementarily, we provide concrete instances of each pattern as snippets whose interpretation is specially oriented toward Android application developers.

5.2.1 Patterns

Below we describe each type of pattern that we considered in [14].

Pattern: Draw Allocation

This pattern is detected by Android *lint*,[2] and is the first of five EGAPs whose energy impact analysis was included in [15, 16].

Problem **Draw Allocation** occurs when new objects are allocated along with draw operations, which are very sensitive to performance. In other words, it is a bad design practice to create objects inside the **onDraw** method of a class which extends a View Android component.

Solution The recommended alternative for this EGAP is to move the allocation of independent objects outside the method, turning it into a static variable.

Example The code snippet to the left should be transformed to the one on the right.

```
public class CMView extends View {
  @Override
  protected void onDraw(Canvas c){
    RectF rectFl = new RectF(); ✗
    ...
    If(!clockwise) {
      rectFl.set(X2-r, Y2-r, X2+r,
        Y2+r);
      ...
} } }
```

```
public class CMView extends View {
  RectF rectFl = new RectF(); ✓
  @Override
  protected void onDraw(Canvas c) {
    ...
    if(!clockwise) {
      rectFl.set(X2-r, Y2-r, X2+r,
        Y2+r);
      ...
} } }
```

[1]When the practical usage is obvious, we will exclude the illustrative example.

[2]*Lint* is a code analysis tool, provided by the Android SDK, which reports upon finding issues related to the code structural quality.

Pattern: Wakelock

Wakelock is the second Android *lint* performance issue [15–18].

Problem **Wakelock** occurs whenever a wakelock, a mechanism to control the power state of the device and prevent the screen from turning off, is not properly released, or is used when it is not necessary.

Solution The alternative here would be to simply add a **release** instruction.

Example The code snippet to the left should be transformed to the one on the right.

```
public class DMFSetTempo extends
... {
  PowerManager.WakeLock l;
  public void onClickBtStart(View
v) {
    wakelock.acquire(); ✓
  }
  @Override()
  public void onPause() {
      super.onPause(); ✗ }
  }
```

```
public class DMFSetTempo extends
... {
  PowerManager.WakeLock l;
  public void onClickBtStart(View
v) {
    wakelock.acquire(); ✓
  }
  @Override()
  public void onPause() {
    super.onPause();
    if (l.isHeld()) l.release(); ✓
  }
}
```

There exist other types of wakelocks for resources such as Sensor, Camera, and Media. They differ from the Screen only in the mechanism used to release the lock.

Pattern: Recycle

Recycle is another Android *lint* performance issue [15, 16].

Problem **Recycle** is detected when some collections or database-related objects, such as **TypedArrays** or **Cursors**, are not recycled or closed after being used. When this happens, other objects of the same type cannot efficiently use the same resources.

Solution The alternative in this case would be to include a **close** method call before the method's return.

Example The code snippet to the left should be refactored to the one on the right.

```
public Summoner getSummoner(int id) {
  SQLiteDatabase db =
      this.getReadableDatabase();

Cursor c = db.query( ... );
...
return summoner; ✗
}
```

```
public Summoner getSummoner(int id) {
  SQLiteDatabase db =
      this.getReadableDatabase();

Cursor c = db.query( ... );
...
c.close(); ✓
return summoner;
}
```

Pattern: Obsolete Layout Parameter

The fourth Android *lint* performance issue, **Obsolete Layout Parameter**, is the only one that is not Java-related [15, 16].

Problem The view layouts in Android are specified using **XML**, and they tend to suffer several updates. As a consequence, some parameters that have no effect in the view may still remain in the code, which causes excessive processing at runtime.

Solution The alternative is to parse the **XML** syntax tree and remove these useless parameters.

Example The next snippet shows an example of a view component with parameters that can be removed.

```
<TextView android:id="@+id/centertext"
  android:layout_width="wrap_content" android:
layout_height="wrap_content"
  android:text="remote files"
 Xlayout_centerVertical="true" Xlayout_alignParentRight="true" >
</TextView >
```

Pattern: View Holder

View Holder is the last Android ***lint*** performance issue [15, 16], whose alternative intends to make a smoother scroll in ***List Views***.

Problem The process of drawing all items in a ***List View*** is costly, since they need to be drawn separately.

Solution To reuse data from already drawn items, therefore reducing the number of calls to **findViewById()**, known to be energy greedy [19].

Example Every time **getView()** is called, the system searches on all the view components for both the **TextView** with the id "label" (❶) and the **ImageView** with the id "logo" (❷), using the energy-greedy method **findViewById()**. The alternative version is to cache the desired view components, with the following approach:

```
public View getView(int p, View v, ViewGroup par) {
  LayoutInflater inflater = ...

  v = inflater.inflate(R.layout.apps , par, false);
 TextView txt=(TextView) v.findViewById(R.id.label);   ❶
 ImageView img=(ImageView) v.findViewById(R.id.logo);   ❷
 return row;
}
```

```
static class HolderItem {
  TextView txtView; ImageView imgView;
}
public View getView(int p, View v, ViewGroup par) {
  HolderItem hld; LayoutInflater inflater = ...

 if (v == null) {                ❸
   v = inflater.inflate(...); hld = new HolderItem();
  hld.txtView = (TextView) v.findViewById (...);        ❹
  hld.imgView = (ImageView) v.findViewById (...);        ❺
  v.setTag(hld);
 } else { hld = (HolderItem) v.getTag(); } »        ❻
 TextView txt = hld.txtView; ImageView img = hld.imgView;
 ...
}
```

Condition ❸ evaluates to true only once, which means instructions ❹ and ❺ execute once, i.e., **findViewById()** executes twice, and its results are stored in the **ViewHolderItem** instance. The following calls to **getView()** will use cached values for the view components **txt** and **img** (❻).

Pattern: HashMap Usage

This EGAP is related to the usage of the **HashMap** collection [17, 20–22].

Problem The usage of **HashMap** is discouraged, since the alternative **ArrayMap** is allegedly more energy efficient, without decreasing performance.[3]

Solution To simply replace the type **HashMap**, whenever used, by **ArrayMap**.

Pattern: Excessive Method Calls

Unnecessarily calling a method can penalize performance, since a call usually involves pushing arguments to the call stack, storing the return value in the appropriate processor's register, and cleaning the stack afterwards.

Problem **Excessive Method Calls** was explored by [20, 23], showing that the energy consumption in Android applications can be decreased by removing method calls inside loops that can be extracted from them.

Solution The alternative is to replace the method call by a variable that is declared outside the loop, and is initialized with the return value of the method call extracted.

[3]As stated in the Android *ArrayMap* documentation: http://bit.ly/32hK0y9.

Example An example of an extractable method call would be one which receives no arguments, and is accessed by an object that is not transformed in any way inside the loop.

Pattern: Member Ignoring Method

This EGAP addresses the issue of having a non-static method inside a class, and which could be static instead [17, 20].

Problem Having a method not declared as static, but which does not access any class fields, does not directly invoke non-static methods, and is not an overriding method. This causes multiple instances of the method to be created and used at runtime, which can be avoided.

Solution Use static methods as these are stored in a memory block separated from where objects are stored, and no matter how many class instances are created throughout the program's execution, only an instance of such a method will be created and used. This mechanism helps in reducing energy consumption.

5.3 Energy Design Patterns

In the previous section, we learned about code patterns that are specific to a given platform (i.e., Android) or paradigm (i.e., object oriented) and how they affect energy consumption. In this section, we bring these patterns to a higher level of abstraction: we delve into design patterns that provide reusable solutions that generalize to any software of a given domain and that are not coupled with any particular development framework or paradigm.

The energy patterns in this section do not give any particular advice on coding practices. Rather, they help software engineers create energy-efficient software by design. Nevertheless, these patterns may have a direct impact on the feature set of the application and ultimately on the user experience.

In this particular case, we focus on energy patterns in the mobile domain. We present a catalog of 22 energy patterns that are commonly used for mobile applications. This catalog is the result of an empirical study with more than 1700 mobile applications [24] to document energy patterns that are commonly adopted by iOS and Android software engineers and are expected to generalize to any mobile platform.

We describe each energy pattern with the template used in the previous section, explaining the problem and the solution while providing an illustrative example.

5.3.1 Patterns

Below we pinpoint different design patterns to develop energy-efficient mobile applications.

Pattern: Dark UI Colors

Provide a dark UI color theme to save battery on devices with AMOLED[4] screens [25–28].

Problem One of the major sources of energy consumption in mobile devices comes from the screen. Thus, mobile applications that rely on the screen for all use cases, such as video apps or reading apps, can significantly drain the battery.

Solution Opt for dark colors when designing the UI. Smartphones typically feature screens that are more energy efficient with dark colors. Depending on the application, users can be given the option to choose between a light and a dark UI theme. Alternatively, a special trigger (e.g., when battery is running low) can activate the dark UI theme.

Example In a reading app, provide a theme with a dark background using a light foreground color to display text. When compared to themes using light background colors, a dark background will have a higher number of dark pixels.

Pattern: Dynamic Retry Delay

When trying to access a resource that is failing or not responding, increase the waiting time before attempting a new access.

Problem Mobile apps often need to exchange data with different resources (e.g., connect to a server in the cloud). It may happen that the communication with these resources fails and a new attempt needs to be made. However, if the resource is temporary, the app will repeatedly try to connect to the resource with no success, leading to unwanted energy consumption.

Solution After each failed connection, increase the waiting time before the next attempt. A linear or exponential growth can be used for the waiting interval. Upon a successful connection or a given change in the context (e.g., network status), the waiting time can be set back to the original value.

[4]AMOLED is a display technology used in mobile devices and stands for Active Matrix Organic Light Emitting Diodes.

Example Consider the scenario in which an app with a news feed is not able to communicate with the server to retrieve updates. The naive approach is to continuously poll the server until the connection is successful—i.e., the server is available. Instead, a dynamic retry delay can be used by, for example, adopting the Fibonacci series[5] to increase the time between sequential attempts.

Pattern: Avoid Extraneous Work

Avoid tasks in the mobile application that do not add enough value to the user experience or whose results quickly become obsolete.

Problem Typically, mobile applications execute multiple tasks at the same time. However, there are cases in which the results of these tasks are not immediately presented to the user. For example, when the application is synchronizing real-time data that does not immediately meet the information needs of the user, it may become obsolete before the user actually accesses it.

This is even more evident when apps are running in the background. The phone will be using resources unnecessarily to update data that will never be used.

Solution Define the minimal set of data that is presented to the users. In addition, disable all the tasks that are not affecting the data being displayed to the user.

Example Consider a plot with a time series of real-time data that is being continuously updated. When the user scrolls up/down, the plot might move out of the visible area of the UI. In this case, updating the plot is a waste of energy. Drawing operations related to updating the plot should be ceased and restarted when the plot is visible again.

Pattern: Race-to-idle

Resources or services ought to be closed as soon as possible (e.g., location sensors, wakelocks, screen) [15, 29–31].

Problem Mobile apps resort to different resources and components that need to be stopped after being used. After activating a given resource, it starts operating and is ready to respond to the app's requests. Even if the app is not making any request, the resource will waste energy until it is properly closed.

Solution Make sure resources are inactive when they are not necessary by manually closing them. Static analysis tools may help identify cases of resources that are not being properly closed—e.g., *Facebook Infer*, *Leafactor* [16].

[5]Fibonacci series is a sequence of numbers in which each number is the sum of the two preceding numbers (i.e., 1, 1, 2, 3, 5, 8, etc.).

Example Wakelocks are commonly used by mobile applications to prevent phones from entering sleep mode. Different types of wakelocks can be used; for example, there are wakelocks specific for the screen, CPU, and so on. Always implement event handlers that listen to the application events of the entering or leaving background. Implement handlers for the events that are fired when the app goes to background, and release wakelocks accordingly.

Pattern: Open Only When Necessary

Open/start resources/services immediately before they are required. This is similar to the pattern ***Race-to-idle***.

Problem Resources, such as location sensors or database connections, must be activated before they are ready to use. Once a given resources is opened, it actively consumes more energy. Thus, it should only be opened immediately before its usage. In particular, resources should not be activated upon the creation of the view or activity where it operates.

Solution Activate resources and services immediately before they are needed. This will also prevent the activation of resources that are never used [29].

Example In a video call app, the camera is used to share the faces or images of the different participants in the call. The camera should only start capturing video when it is actually being displayed in the view to the user.[6]

Pattern: Push over Poll

Using push notifications is more energy efficient than actively polling for notifications.

Problem Mobile apps typically resort to notifications to get updates from resources (e.g., from a server). The naive approach to getting updates is by reaching the resource and asking it for updates. The downside is that, by continuously asking a server for updates, it might be making several requests without any update. This leads to unnecessary energy consumption.

Solution Use push notifications to get updates. Note—for Free and Open Source applications this is a big challenge because it requires having a cloud messaging

[6]A real example where the camera was being initiated too early can be found here: https://github.com/signalapp/Signal-Android/commit/cb9f225f5962d399f48b65d5f855e11f146c bbcb (visited on June 15, 2020).

server set-up. For example, in the case of Android there is no good open source alternative to Google's Firebase Cloud Messaging.

Example In a social network app, instead of actively reaching the server to provide relevant notifications to the user, the app should prescribe push notifications.

Pattern: Power Save Mode

Implement an alternative execution mode in which some features are dropped to ensure energy efficiency. In some cases, user experience is hindered.

Problem When the battery level is low, users may want to make sure they will not lose connectivity before reaching a power station and charging their phone.

Solution Implement a power save mode that only provides the minimum functionality that is essential to the user. This mode can be manually activated by the user or through power events (e.g., when battery reaches a given level) raised by the operative system. In some cases, the mobile platform already features this out of the box—e.g., this is enforced in iOS for use cases using the **BackgroundSync** APIs.

Example Reduce update intervals, disable less important features, or disable UI animations.

Pattern: Power Awareness

Features operate in a different way regarding the battery level or depending on whether the device is connected to a power station.

Problem There are some features that, despite improving user experience, are not strictly necessary for users—e.g., UI animations. Moreover, there are low-priority operations that do not need to be executed immediately (e.g., backup data in the cloud).

Solution Adjust the feature set according to its power status. Even when the device is being charged, the battery level may be low and it is better to wait for a higher battery level before executing any intensive task.

Example Postpone intensive tasks, such as cloud syncing or image processing, until the device reaches a satisfactory power level, typically above 20%.

Pattern: Reduce Size

Minimize the size of data being transferred to the server.

Problem Mobile apps typically transfer data with servers over an internet connection. Such operations are battery intensive and should be reduced to a minimum. There are cases in which the size of the data can be reduced without affecting user experience.

Solution Exclusively transmit data that is strictly necessary and compression techniques whenever possible.

Example Enable gzip content encoding when sending data over HTTP requests.

Pattern: WiFi over Cellular

Postpone features that require a heavy data connection until a WiFi network is available.

Problem Mobile apps typically need to synchronize data with a server. However, cellular data connections (e.g., 4G) tend to be energy greedy.

Solution WiFi connections are usually a more energy-efficient alternative to cellular connections [32]. These are use cases that do not require real-time sync and should be postponed until a WiFi connection is available.

Example Consider a music stream application that allows users to play their favorite songs and to organize them in playlists. In addition, the app allows users to play the playlists offline—i.e., when there is no internet connection. When a new song is added to a given offline playlist, the app waits for a WiFi connection before downloading the song.

Pattern: Suppress Logs

Avoid intensive logging as much as possible. Overusing logging leads to significant energy consumption, as found in previous work [33].

Problem Logging is commonly used to simplify debugging. However, there is a trade-off between having the necessary information and energy efficiency that needs to be considered.

Solution Manage logging rates to a maximum of one message per second.

Example In a mobile app that is processing real-time data, avoid logging this behavior. If necessary, enable logging only for debugging executions.

Pattern: Batch Operations

Bundle multiple operations instead of running them separately. This will avoid putting the device into an active state many times in the same time window.

Problem Executing operations separately leads to extraneous energy consumption related to turning a particular resource on and off—this is typically called *tail energy consumption* [23, 34, 35]. Executing a task often induces tail energy consumption related to starting and stopping resources (e.g., starting a cellular connection).

Solution Combine multiple operations in a single one to optimize tail energy consumption. Although background tasks can be expensive, very often they have flexible time constraints. For example, a given background task that needs to be executed eventually does not need to be executed at a specific time. Thus, it can wait for other operations to be scheduled before it is executed.

Example Use Operative System-wide APIs tailored for job scheduling (e.g., 'android.app.job.JobScheduler,' 'Firebase JobDispatcher'). These APIs manage multiple background tasks occurring in a device to guarantee that the device will exit sleep mode (or doze mode) only when the tasks in the waiting list really need to be executed.

Pattern: Cache

This pattern proposes the use of cache mechanisms to avoid unnecessary operations.

Problem A common functionality in mobile apps is to display data fetched from a remote server. A potential issue with that need is that an app may fetch the same data from the server multiple times during the lifetime of the mobile app.

Solution Mobile apps should put in place caching mechanisms to avoid fetching data from the server [36]. Moreover, lightweight strategies to decide whether to refresh the data in the cache need to be implemented to guarantee that the mobile app is displaying the up-to-date data.

Example Consider a social network app that displays profiles of other users. Instead of downloading basic information and profile pictures every time a given profile is opened, the app can use data that was locally stored from earlier visits.

Pattern: Decrease Rate

This pattern proposes to increase the time between syncs/sensor reads as needed.

Problem It is common for mobile apps to perform certain operations periodically. A potential issue is that, if the time between two executions is small, the app will be executing operations more often.

Solution Increase the time-between-operations to find the minimal time interval that would compromise user experience, while having a positive impact on the energy consumption. This time-between-operations can be manually tuned by developers, defined by users, or even found in an empirical way. One could also envisage more sophisticated and dynamic solutions that can also use context (e.g., time of day, history data) to infer the optimal update rate.

Example Consider a news app that gathers news from different sources, doing so by fetching the news of a given source in its own thread. Instead of triggering updates for all threads at the same rate, use data from previous updates to infer the optimal update rate of these threads. Connect to the news source only if updates are expected.

Pattern: User Knows Best

This pattern proposes to offer capabilities to allow users to enable/disable certain features to save energy.

Problem The number of features offered by a mobile app and power consumption is a trade-off generally considered when devising energy-efficient solutions. However, there is no **one–size-fits-all** user as far as this trade-off is concerned. There are users who might be satisfied with fewer features but better energy efficiency, and vice versa.

Solution The possibility for users to customize their preferences regarding energy-critical features is therefore important. This customization should be intuitive and an optimal default set of preferences.

Example Consider a mail client for POP3 accounts as an example. One can imagine that a user may want their mail client to check/poll for new messages every other minute, and others—depending on the time of day—much less often. As there is no automatic mechanism to infer the optimal update interval, the best option is to allow users to define it.

Pattern: Inform Users

This pattern proposes to inform the user when the app is performing any battery-intensive operation.

Problem It is known that there are use cases in mobile apps that require a substantial amount of energy. In turn, one can activate features to be energy efficient at the cost of user experience. We argue that if users do not know the expected behavior from the mobile app, they may flag its operation as failing.

Solution Inform users about battery-intensive operations or energy management features. This could be done by flagging (e.g., via alerts) this information in the user interface.

Example Alert users when (1) a power-saving mode is active or (2) a battery-intensive operation is being executed.

Pattern: Enough Resolution

This pattern proposes that data should be pulled or provided with high accuracy only when strictly necessary.

Problem Users tend to use precise data points when fetching and/or displaying data. An issue with such a strategy is that the collection and manipulation requires more resources, entailing, naturally, high-energy consumption. There are, however, use cases where dealing with low-resolution data suffices.

Solution Developers should find the trade-off between data resolution and app/user needs as well as user experience.

Example Take as an example a running app that is able to record running sessions. The app shows the user the current overall distance to a given location. Instead of using precise real-time processing of GPS or accelerometer sensors, which can be energy greedy, a lightweight method could be used to estimate this information with lower but reasonable accuracy. Evidently, at the end of the session, the accurate results would still be processed, but without real-time constraints.

Pattern: Sensor Fusion

This pattern proposes using data from low-power sensors to decide whether to fetch data from high-power sensors.

Problem Operations to interact with distinct sensors or components may be energy greedy, causing the app to consume a substantial amount of energy. Therefore, such operations should be executed only in case of absolute necessity.

Solution Making use of data sources that entail low power consumption (such as alternative low-power sensors) may prevent the need to execute an energy-greedy operation.

Example As an example, one can imagine using the accelerometer to infer whether the user has changed location, and only interacting with the energyintensive GPS to obtain a more precise location in case of a location change.

Pattern: Kill Abnormal Tasks

This pattern proposes to offer capabilities to interrupt energy-greedy operations (e.g., using timeouts, or users input).

Problem Mobile apps may trigger an operation that unexpectedly consumes more energy than anticipated (e.g., taking a long time to execute).

Solution Offering an intuitive way for end users to interrupt an energy-greedy operation would help to fix this issue. Alternatively, a fair timeout could be included for energy-greedy tasks or wakelocks.

Example As an example, consider a mobile app that features an alarm clock. Implementing a fair timeout for the duration of the alarm, in case the user is not able to turn it off, will prevent the battery from being drained.

Pattern: No Screen Interaction

This pattern proposes to allow interaction without using the display whenever possible.

Problem There are mobile apps that involve constant use of the screen. However, there may be cases in which the screen can be replaced by less power-intensive alternatives.

Solution Enable users to use alternate interfaces (e.g. audio) to communicate with the app.

Example As an example, consider a navigation app. There are use cases in which users may be using audio instructions only, having no need the see updates on the screen. This strategy is commonly adopted by audio players that use the earphone buttons to play/pause or skip songs.

Pattern: Avoid Extraneous Graphics and Animations

Graphics and animations are at the forefront as far as improving the user experience is concerned, but can also be battery intensive. Therefore, this pattern proposes to use them with care [37]. This is well aligned with what is recommended in the official documentation for iOS developers.[7]

Problem Mobile apps often feature impressive visual effects. However, they need to be properly tuned to prevent the battery from being drained quickly. This has been shown to be particularly critical in e-paper devices.

Solution Study the importance and impact of visual effects (such as graphics and animations) to the user experience. The improvement in user experience may not be sufficient to overcome the overhead imposed on the energy consumption. Therefore, developers should consider avoiding using visual effects or high-quality graphics, and should instead resort to low frame rates for animations when viable and/or feasible.

Example For instance, high frame rates may make sense while playing a game, but a lower frame rate may suffice while in the menu screens. In other words, use a high frame rate only when the user experience requires it.

Pattern: Manual Sync, On Demand

This pattern proposes to execute tasks if, and only if, requested by the user.

Problem Some tasks may be energy intensive, but not really needed to give the best user experience of the app. Hence, they could be avoided.

Solution Providing a mechanism in the UI (e.g., button) which allows users to trigger energy-intensive tasks would be helpful in letting the user decide which tasks he wants to trade off for energy consumption.

[7]*Energy Efficiency Guide for iOS* Apps—Avoid Extraneous Graphics and Animations available here: https://developer.apple.com/library/archive/documentation/Performance/Conceptual/EnergyGuide-iOS/AvoidExtraneousGraphicsAndAnimations.html (visited on June 15, 2020).

Example Take as an example a beacon monitoring app. There may be situations in which the user does not need to keep track of her/his beacons. This app could implement a mechanism to let the user (manually) start and stop monitoring.

5.4 Object-Oriented Patterns

In this section, we focus on patterns that are tailored to a certain programming paradigm. In particular, we discuss the Gang of Four (GoF) patterns, a popular catalog of object-oriented (OO) design patterns proposed by Gamma, Helm, Johnson, and Vlissides [38] that describe recurring solutions to common OO problems. Although these patterns do not primarily target energy efficiency, they do have an impact that ought to be considered when designing sustainable systems. To refresh the reader's mind, we present two such patterns.

Pattern: Template Method

An algorithm must accommodate custom steps while maintaining the same overall structure [38].

Problem Software systems oftentimes implement behaviors that are similar, containing only a couple of steps that differ. Maintaining the code for each behavior independently incurs greater effort. Moreover, there is a risk that patches will not be applied uniformly among similar instances, which may unnecessarily (and potentially erroneously) diverge the designs.

Solution The overarching steps among all behaviors should be implemented in a single component. The steps that are implemented differently between the behaviors are accessed via an interface. The individual behaviors must now inherit the general component and only implement the interfaced steps.

Example A library implements several supervised learning classification algorithms. The steps to create and use such an algorithm are similar, e.g., configure model, define features and response variables, train model, and predict new values. In this scenario, template methods can be used on steps such as train and predict, while centralizing the implementation of the overall classification task.

Pattern: State

A single component may alter its states with different behaviors as if the component had been replaced [38].

Problem One or more behaviors of a component depend on a state that is only identifiable at runtime. Although the state is mutable, the set of possible states and the different ways behavior is implemented are well defined.

Solution The component consists of an interface accessible to other components (i.e., clients). Each state implements the interface. The state of the component is reassessed internally upon the execution of an implemented behavior.

Example A sensor component offers the behaviors *read_data*, *turn_on*, *turn_off*, and *get_state*, which are implemented for the states **enabled**, **disabled**, and **defective**. Upon an unsuccessful read in the **enabled** state, the component changes its state to **defective**. Otherwise, the state is defined via *turn_on* and *turn_off*.

The GoF patterns can be grouped according to the purpose they serve, i.e., to create objects, to organize structure, or to orchestrate behavior. A pattern instance comprises the association of one or more classes and interfaces fulfilling the various roles described by the pattern. For example, the instance of a State pattern comprises an interface that is implemented on a set of state classes that can provide a different behavior for the predefined actions, which are in turn accessed by a context (client) class.

As the reader may already know or have noticed by now, the GoF patterns do not address energy problems by intent. However, design pattern instances (like any design) have effects on quality attributes. Moreover, the instantiations of a design pattern are not uniform, nor are their effects on quality attributes [39]. In particular, several studies suggest that the effect of a pattern on a quality attribute depends on factors such as the number of classes, invoked methods, and polymorphic methods [39–41].

Considering the systematic use of OO features (e.g., polymorphism) in pattern instances, one may expect a potential impact (positive or negative) on energy consumption. Furthermore, researchers consistently find that, at least on Java systems, approximately 30% of the classes participate in one or more instances of GoF patterns [42–44]. This picture adds up to a growing concern and interest in the research community. In this context, if a pattern instance is not the optimal design solution, an alternative (non-pattern) design solution can be applied. Several authors (including GoF design pattern advocates) have proposed such alternatives [38, 45–49].

In efforts to investigate the aforementioned effect, Litke et al. [50] studied the energy consumption of five design patterns[8] through six toy examples and were able to detect a negligible consumption overhead for the Factory Method and Adapter pattern. Sahin et al. [51] investigated 15 design patterns[9]; however, there were some inconclusive results, as they could observe both an increase and a decrease in energy consumption. To shed further light on the matter, Noureddine and Rajan [52] examined in detail two design patterns for which they identified a significant overhead, namely Observer and Decorator patterns. The comparison involved not only pattern and alternative non-pattern solutions but also a transformed pattern solution that optimizes the number of object creations and method calls. Although the pattern solution showed overheads between 15% and 30%, the optimized solution reduced these observations by up to 25%.

The preceding work shows that there is indeed a potential systematic effect of GoF patterns in energy consumption and that negative effects may be countered on certain cases. Such knowledge is relevant for both greenfield projects (i.e., fresh development), where it can support an energy-smart application of patterns, and brownfield projects (e.g., refactoring of a system to a new purpose), where it can inform decisions on what parts of the system to refactor. However, to fulfill these goals, more insights and guidelines are necessary to fully understand what influences the energy consumption of GoF patterns.

To that end, one of the authors was the lead researcher in a study to investigate the effect of Template Method and State/Strategy patterns on energy consumption [53]. In particular, an experiment was set up to compare the energy consumption of pattern and alternative (non-pattern) solutions and, more importantly, to examine factors that influenced the observed results. To improve accuracy, the energy measurements were collected at both system and method level. The energy efficiency of pattern instances was analyzed at the method level, from which both the size (measured in source lines of code—SLOC) and the number of foreign calls (measured via the message passing coupling metric—MPC[10]) were assessed.

The results of the study showed that the non-pattern solutions consume less energy than their pattern counterpart. However, as in other studies, there were cases in which the pattern solution had a similar or marginally lower energy consumption. One of the main contributions of this work is the investigation of the related factors. Upon examining the SLOC and MPC metrics, it was possible to establish that instances of GoF patterns tend to provide an equitable or more energy-efficient solution when used to implement logic with longer methods and multiple calls to external classes, i.e., complex behaviors. These findings are illustrated in Fig. 5.2, which compares the energy consumption of pattern (y-axis, left chart) and non-pattern (x-axis, left chart) solutions for all assessed methods. These data points

[8]Factory Method, Adapter, Observer, Bridge, and Composite.

[9]Abstract Factory, Bridge, Builder, Command, Composite, Decorator, Factory Method, Flyweight, Mediator Observer, Prototype, Proxy, Singleton, Strategy, and Visitor.

[10]Number of invocations to methods that are not owned or inherited by the class being measured.

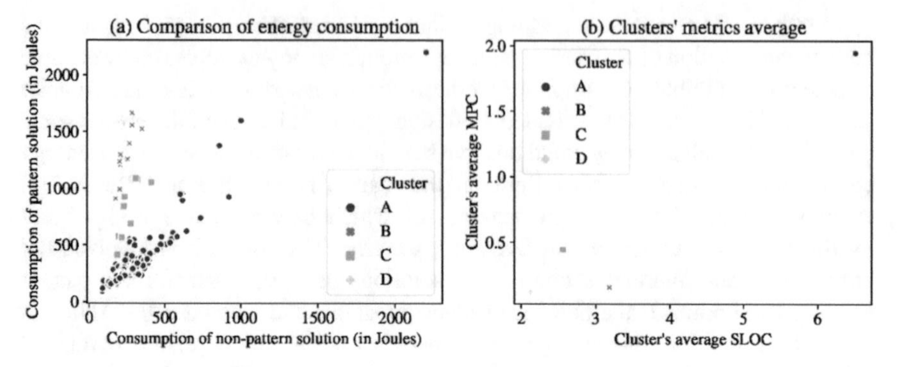

Fig. 5.2 Comparison of energy consumption and associated factors

are clustered by energy efficiency (distinguished by shape and color) and the average SLOC and MPC of each cluster are depicted in the right-hand chart.

These findings serve to reiterate and discuss a set of recurring concerns around the use of GoF patterns. First, they should only be applied if the extra (design) complexity that they introduce is lower than the one that they resolve. In other words, if the context or logical complexity is trivial, the design solution should also be trivial. Otherwise, quality attributes, including energy efficiency, are likely to deteriorate [40, 41]. For example, longer methods reduce the ratio between localization time of the overall computation (i.e., logic) and thus also the overall overhead caused by the polymorphic mechanism.

Finally, note that as patterns promote improved structuring of the source code, energy efficiency may also be achieved through more efficient bytecode. For example, we observed that the Java Virtual Machine applies internal optimizations when pattern-related methods comprise a set of external invocations (i.e., to methods that are not owned or inherited by the pattern class). Such optimizations might not be triggered in a non-pattern alternative, as the structure is altered.

5.5 Patterns in Context

In this section, we present a series of cases describing situations in which patterns can help improve the energy consumption of software-intensive systems. These cases were extracted from real projects or created based on scenarios that practitioners may regularly encounter. As the cases comprise the application of patterns, we resort to a well-known template for capturing design decisions related to patterns described by Harrison et al. [54]. Each case is described according to the fields presented in Table 5.1. We clarify that there are additional fields available in the template by Harrison et al., e.g., related patterns and related requirements. However, we restricted our analyses to the parts of the systems on which we report, and thus we do not establish links between decisions within a project.

Table 5.1 Template for documenting pattern-related decisions

Field	Description
Context	Scenario (incl. constraints) in which the pattern is (or would be) applied.
Problem	Stakeholders' concern that must be addressed.
Alternatives	Alternatives (according to forces) that have been considered to tackle the issue.
Solution	Generic solution (provided by the pattern) to the design problem.
Rationale	Rationale of applying the pattern's solution in relation to the forces.
Pattern	Pattern name.
Consequences	Context and implications of applying the pattern.
Notes	Relevant points that do not fit in another field.
Source	Origin of the case, or description of the fictional context.

Case: Android Token

Fig. 5.3 The main activity of *Android Token*

Android Token is an application suited for generating and managing One-Time Password (OTP) tokens, to be used in software requiring Open Authentication (OATH). It is completely free and open source, and is available in the F-Droid application catalog.

Context The main purpose of this application is to provide information regarding the properties of the generated tokens, such as their value, where are they being used, and how much time is left until the token expires. As such, the application's main view (which is managed by the main *Activity*, depicted in Fig. 5.3) shows a list of all tokens, with all of the aforementioned properties displayed. Since the information per token is the same, it is expected that there will be several identical view components displayed (such as labels or progress bars).

Problem Drawing the same type of view components for each token means repeating almost the exact same task, but with different values. Once an application is created, the Android system puts all the metadata of all view components within the application inside the same wrapper class. Each *Activity* is then responsible for fetching the required components to be drawn in their associated layouts. The fetching process is available in Android only through an API call already known to be energy greedy [19]. Moreover, due to how Android internally handles the process of swapping between activities, moving to a new activity or going back to a previously visited one means redrawing all the components. As expected, this has a huge impact on the amount of work performed by both the CPU and the GPU.

Essentially, this problem creates two optimization challenges. The first one is the excessive number of component fetching and redrawing tasks, which should be reduced. Second, since the actual component's *draw* operation itself is repeated several times, it should be focused only on the component-drawing process, and not on tasks such as setting up of any kind, or creating objects.

Solution For the first problem, the solution requires a caching strategy, to avoid unnecessary fetching, and to optimize the redrawing process. Therefore, the *Activity* responsible for fetching and drawing the view components should internally keep a copied reference of each one, collected the first time they are drawn.

The second problem can be tackled by reducing to a minimum the number of instructions not related to the drawing process. As such, creating new objects should be avoided in the *onDraw* method of a view component, as described in the Android documentation.[11]

Rationale Caching view components means reducing the effort required by the CPU to traverse through (potentially) all existing components, and avoiding unnecessary calls to an energy-greedy Android API. It also means reducing the effort required by the GPU to redraw the same components. Avoiding object allocation inside the *onDraw* method is also a CPU effort reduction optimization, since many objects require an expensive initialization procedure.

Ultimately, reducing the effort on these tasks translates to reducing the energy consumed by the application, and consequently increasing the device's battery uptime.

Pattern The patterns that provide the solution to the aforementioned problems are commonly known as *ViewHolder* and *DrawAllocation*, respectively.

Consequences Implementing both the patterns has a significant impact on code readability and maintainability, especially for *ViewHolder*. It requires including an inner class inside the *Activity* to hold the view components, and to increase the complexity of the fetching/drawing method. As for *DrawAllocation*, developers should preallocate objects (by using class variables), which, depending on the type of object, may require additional effort and reduce the code readability. When applying both patterns on an existing application, it also means restructuring code, critical to the application, with a new concept, which can be a delicate and costly task.

[11]Android View documentation: https://developer.android.com/training/custom-views/custom-drawing#createobject

Case: Nextcloud Android app

Nextcloud is a file hosting service client-server solution for file hosting services. Anyone can install it on their own private server. It is distributed under the General Public License v2.0 open-source license, which also means that anyone can contribute to the project. It provides a software suite with a cloud server and client apps for different desktop and mobile platforms. In this particular case, we are looking at their Android app.

Context As in most mobile apps for cloud services, data exchanging is a recurrent task in their feature set. In the case of Nextcloud, all the files need to be synchronized with the different user devices. Thus, whenever a new file is added or updated, it needs to be uploaded to the cloud server.

Problem Uploading files is a resource-intensive task that may take a few minutes to execute. This may considerably reduce battery level. However, there are cases in which the user is not so interested in having all the files immediately uploaded to the server. Depending on the user context, the trade-off between file consistency and battery level may be different.

Solution Allow the user to define when the app should prioritize energy efficiency above other features. Typically, mobile operating systems already provide a power save mode that can be activated manually or when the battery reaches a critical level (e.g., 20% of full capacity). All the apps have access to this setting and can change their behavior accordingly. In the example of Nextcloud, developers decided to deactivate any file upload during this mode.

Rationale The power save mode is a deliberate user action that expresses that the user is prioritizing battery life above other features. Thus, it is important that energy-intensive features, such as file transfers, are avoided.

Pattern Power Save Mode.

Alternatives The patterns Inform Users (i.e., warn users of energy-intensive actions) and Power Awareness (e.g., change behavior according to the battery level) can also be used in this context.

Consequences This strategy can have a big impact on the user experience. It is important that users understand that during this mode their files are not going to be uploaded to the server. Thus, this behavior should be properly flagged in the user interface, so that users are well informed of it. In this particular case, the Nextcloud app allows users to override the *Power Save Mode* behavior by clicking on a button that manually triggers a synchronization with the server. Finally, some studies have found evidence that, when not coded properly, this pattern may hinder the maintainability of the project.

Notes This pattern is usually supported by any modern mobile operating system. It is always a good practice to implement this pattern in a mobile app.

Source This case is reported in the Nextcloud app's GitHub project: https://github.com/nextcloud/android/commit/8bc432027e0d33e8043cf401922

Case: K-9 Mail

K9-Mail is a free and open-source e-mail client for Android. It was first written in 2008 and it is still under active development, being one of the oldest Android apps. Like any mobile application, K-9 Mail runs under limited energy resources. Battery life needs to be optimized to prevent hindering user experience. Thus, along the history of its project, we encounter a number of code changes that were made to improve energy efficiency.

Context An important activity done by an e-mail client app is synchronizing data and communicating with e-mail providers. For example, when new emails appear in the user's inbox, the app needs to communicate with the server and download this new data.

Problem Servers do not always work as intended. There are many reasons for servers being unreachable: slow or no internet connection, too many users accessing the server, server is down for maintenance, and so on. This means that the features requiring server communication will fail until the required server can be reached again. Typically, the communication can be established after a few unsuccessful attempts. Thus, it is common that for asynchronous tasks the app will try the communication again after some delay. However, in some cases the server may be unreachable for hours or days. This means that the app will silently be draining the battery while continuously attempting to establish a connection with the server. Debugging this behavior is not trivial since the app will not necessarily fail but the task keeps running in the background. In this particular case, K9-Mail is trying to communicate with the server to set up the synchronization mechanism IMAP IDLE protocol.[12]

Solution The typical fix for this situation is creating a threshold for the maximum number of times a communication can fail. After this defined threshold, the app should permanently stop trying to reach the server. In addition, it is a good practice to increase the delay between attempts. For example, while the initial attempts can be made within a few seconds, the following delays should be subsequently increased.

Rationale Often when a server is not reachable within seconds, it is due to a more severe communication problem. Thus, it is unwise to continuously attempt new

[12]IMAP IDLE is a feature defined by the standard RFC 2177 that allows a client to indicate to the server that it is ready to accept real-time notifications.

connections. It is better to kill the task and wait for the user to trigger a new attempt later. This approach gives more control to the user to define whether (1) the task is indeed critical and battery life is not so important or (2) the other way around.

Pattern This pattern is commonly known as *Dynamic Retry Delay*.

Alternatives Alternatives (according to forces) that have been considered to tackle the issue.

Consequences The main consequence of this approach is that new code needs to be added to accomplish this behavior. It is always a good practice to use existing APIs to schedule this kind of task in the background.

Notes The same problem can be found in other features of a mobile app, for example syncing with a wearable device, getting location data, and accessing.

Source This issue was found by K-9 Mail developers and their solution can be found on GitHub: https://github.com/k9mail/k-9/commit/86f3b28f79509d1a4d

Case: WebAssembly design

WebAssembly is an assembly-like language that can be executed in modern web browsers.[13] With this context in focus, the language was designed to produce a compact binary that can be executed with near-native performance, i.e., comparable to binaries compiled for native platforms (e.g., x86, ARM).[14]

The WebAssembly project has a repository dedicated to its design[15] and the bug tracking system is used to discuss issues related to it. Among the discussed issues are matters related to energy efficiency.

Context The WebAssembly group aims at providing a Just-in-Time (JIT) interface part of its specification.[16] However, the level of detail provided in the specification dictates the level of flexibility that library implementations would have. For example, depending on the level of detail in the specification, a library could allow for more undefined behaviors, e.g., at what moment a function definition is evaluated and how deep the checking goes.

Problem The specification of the moment in which a function is evaluated also requires the specification of when errors are reported. This concern was brought up and discussed in an issue opened on the aforementioned GitHub repository.[17] In

[13] https://developer.mozilla.org/en-US/docs/WebAssembly

[14] https://webassembly.org/

[15] https://github.com/WebAssembly/design

[16] https://webassembly.org/docs/jit-library/

[17] https://github.com/WebAssembly/design/pull/719

short, developers argued about the proper moment for a JIT compiler to flag a malformed or not fully implemented feature (e.g., function or module) as an error.

Alternatives The main alternatives discussed by the developers were threefold:

- *Ahead of time*. Maintain the current situation and enforce the validation of features as early as possible. This option provides a more deterministic solution but also may result in waste of resources.
- *Lazy loading*. Modify the expected behavior to validate features at call time. This option allows potential savings w.r.t. resources as modules and functions will only be validated and loaded if used, which may oftentimes not be the case.
- *Mixed approach*. Use lazy loading by default, but provide a compiler setting (*WebAssembly.validate*) that allows compilation ahead of time. This option will require library developers to maintain the two behaviors.

Solution Although the aforementioned issue is still open at the time of writing this chapter, the current solution is to partially abandon WebAssembly.

Rationale A specification that allows for a greater degree of lazy loading gives library developers the freedom to define the level of aggressiveness of the JIT compiler and balance responsiveness with other aspects, notably startup performance, battery, and memory. Furthermore, some stakeholders expected that WebAssembly code would be mainly generated by tools, which provides less room for true positives (i.e., actually malformed or defective features).

Pattern Lazy loading.

Consequences There are three main side effects raised by those involved in the discussion. First, the JIT compilation is abstracted from developers, who lose some control over optimization (e.g., for parallelizing loading tasks). However, it is expected that the benefits outweigh the optimizations that could be manually implemented. Second, although validation is performed at call time, the time at which errors are thrown is non-deterministic. This behavior may change entirely if a variable is set to enforce validation ahead of time. Finally, it is possible that non-deterministic aspects of the compiler may make testing more complicated. However, foreseeable problems can be averted by enforcing feature validation ahead of time (manually or by setting).

Source This issue was found by WebAssembly developers and their solution can be found at the aforementioned link.

5.6 Conclusions

In this chapter, we addressed energy efficiency as a pattern-related problem, where issues are not unique and reoccur systematically in a variety of software systems. In particular, we looked at two levels of abstraction, namely code and design, to

Fig. 5.4 Word cloud of chapter content

identify recurrent issues and solutions. Furthermore, we acknowledge that parts of a system are rarely islands, isolated from each other, and rather comprise a network of interconnected components, in which other patterns may be in play. Thus, we also considered and discussed energy efficiency from two perspectives: as a main concern of patterns and as a side effect of applying patterns.

To consolidate the concepts in this chapter, we showed how the different patterns were used in four real scenarios. These use cases emphasize the importance of being aware of energy-efficiency strategies to make informed decisions when developing sustainable software systems. In Fig. 5.4, we depict the most recurrent words in this chapter and, in light of the presented knowledge, we provide the following takeaway messages and advice.

There exists a consolidated list of refactorings for code-level patterns that can consistently be explored to improve the energy efficiency of Android mobile applications. Along these lines, we should, however, note that we have previously shown that combining as many individual refactorings as possible most often, but not always, increases energy savings. The interested reader may consult all the details on the magnitude and realization of the expected savings in [14].

On a different level of abstraction, design patterns have been used to improve energy efficiency. These patterns ought to be considered when designing software with critical energy requirements, such as mobile applications. By gaining knowledge about these patterns, developers can learn from the vast experiences of different developers across different platforms.

Finally, even if a pattern is not intended to address energy-related issues, it may still have a substantial effect on energy consumption. Thus, it is paramount to not only be aware of the patterns applied in the system but also how to harvest their benefits while avoiding detriments to the overall energy consumption of the system. As a rule of thumb for OO systems, we suggest avoiding the application of patterns to encapsulate trivial functionality, e.g., small in size or that do not communicate with other classes.

References

1. Buschmann F, Meunier R, Rohnert H, Sommerlad P, Stal M (1996) Pattern-oriented software architecture: a system of patterns, vol 1. Wiley
2. Andrae A, Edler T (2015) On global electricity usage of communication technology: trends to 2030. Challenges 6(1):117–157. https://doi.org/10.3390/challe6010117
3. Power consumption in data centers is a global problem. https://www.datacenterdynamics.com/en/opinions/power-consumption-data-centers-global-problem/. Accessed 10 Jun 2020
4. Pinto G, Castor F (2017) Energy efficiency: a new concern for application software developers. Commun ACM 60(12):68–75. https://doi.org/10.1145/3154384
5. Thorwart A, O'Neill D (2017) Camera and battery features continue to drive consumer satisfaction of smartphones in US. https://www.prnewswire.com/news-releases/camera-and-battery-features-continue-to-drive-consumer-satisfaction-of-smartphones-in-us-300466220.html. Accessed 06 Feb 2019
6. The most wanted smartphone features. https://www.statista.com/chart/5995/the-most-wanted-smartphone-features. Accessed 24 Jan 2018
7. Mickle T (2018) Your phone is almost out of battery. Remain calm. Call a doctor. https://www.wsj.com/articles/your-phone-is-almost-out-of-battery-remain-calm-call-a-doctor-1525449283. Accessed 05 Feb 2019
8. Bragazzi NL, Del Puente G (2014) A proposal for including nomophobia in the new dsm-v. Psychol Res Behav Manag 7:155. https://doi.org/10.2147/PRBM.S41386
9. Fu B, Lin J, Li L, Faloutsos C, Hong J, Sadeh N (2013) Why people hate your app: making sense of user feedback in a mobile app store. In: Proc. ACM SIGKDD 19th Int. Conf. Knowledge Discovery and Data Mining (KDD '13). ACM, Chicago, IL, pp 1276–1284. https://doi.org/10.1145/2487575.2488202
10. Khalid H, Shihab E, Nagappan M, Hassan AE (2015) What do mobile app users complain about? IEEE Softw 32(3):70–77. https://doi.org/10.1109/MS.2014.50
11. Manotas I, Bird C, Zhang R, Shepherd D, Jaspan C, Sadowski C, Pollock L, Clause J (2016) An empirical study of practitioners' perspectives on green software engineering. In: Proc. IEEE/ACM 38th Int. Conf. Software Engineering (ICSE '16), pp. 237–248. IEEE, Austin, TX. https://doi.org/10.1145/2884781.2884810
12. Pang C, Hindle A, Adams B, Hassan AE (2016) What do programmers know about software energy consumption? IEEE Softw 33(3):83–89. https://doi.org/10.1109/MS.2015.83
13. Pinto G, Castor F, Liu YD (2014) Mining questions about software energy consumption. In: Proc. 11th Working Conf. Mining Software Repositories (MSR '14). ACM, Hyderabad, pp 22–31. https://doi.org/10.1145/2597073.2597110
14. Couto M, Saraiva J, Fernandes JP (2020) Energy refactorings for android in the large and in the wild. In: Proc. IEEE 27th Int. Conf. Software Analysis, Evolution and Reengineering (SANER '20). London, ON, pp 217–228. https://doi.org/10.1109/SANER48275.2020.9054858
15. Cruz L, Abreu R (2017) Performance-based guidelines for energy efficient mobile applications. In: Proc. IEEE/ACM 4th Int. Conf. Mobile Software Engineering and Systems (MobileSoft '17). IEEE, Buenos Aires, pp 46–57. https://doi.org/10.1109/MOBILESoft.2017.19

16. Cruz L, Abreu R (2018) Using automatic refactoring to improve energy efficiency of android apps. In: Proc. XXI Ibero-American Conf. Software Engineering (CIbSE '18). Bogota, Colombia, pp 1–14
17. Palomba F, Di Nucci D, Panichella A, Zaidman A, De Lucia A (2019) On the impact of code smells on the energy consumption of mobile applications. Inf Softw Technol 105:43–55. https://doi.org/10.1016/j.infsof.2018.08.004
18. Vekris P, Jhala R, Lerner S, Agarwal Y (2012) Towards verifying Android apps for the absence of no-sleep energy bugs. In: Proc. USENIX 5th Conf. Power-Aware Computing and Systems (HotPower '12). USENIX Association, Hollywood, CA
19. Linares-Vásquez M, Bavota G, Bernal-Cárdenas C, Oliveto R, Di Penta M, Poshyvanyk D (2014) Mining energy-greedy API usage patterns in android apps: an empirical study. In: Proc. 11th Working Conf. Mining Software Repositories (MSR '14). ACM, Hyderabad, pp 2–11. https://doi.org/10.1145/2597073.2597085
20. Carette A, Younes MAA, Hecht G, Moha N, Rouvoy R (2017) Investigating the energy impact of Android smells. In: Proc. IEEE 24th Int. Conf. Software Analysis, Evolution and Reengineering (SANER '17). Klagenfurt, Austria, pp 115–126. https://doi.org/10.1109/SANER.2017.7884614
21. Morales R, Saborido R, Khomh F, Chicano F, Antoniol G (2018) EARMO: an energy-aware refactoring approach for mobile apps. IEEE Trans Softw Eng 44(12):1176–1206. https://doi.org/10.1109/TSE.2017.2757486
22. Saborido R, Morales R, Khomh F, Guéhéneuc YG, Antoniol G (2018) Getting the most from map data structures in Android. Empir Softw Eng 23(5):2829–2864. https://doi.org/10.1007/s10664-018-9607-8
23. Li D, Halfond WG (2014) An investigation into energy-saving programming practices for android smartphone app development. In: Proc. 3rd Int. Workshop on Green and Sustainable Software (GREENS '14). ACM, Hyderabad, pp 46–53. https://doi.org/10.1145/2593743.2593750
24. Cruz L, Abreu R (2019) Catalog of energy patterns for mobile applications. Empir Softw Eng 24(4):2209–2235. https://doi.org/10.1007/s10664-019-09682-0
25. Agolli T, Pollock L, Clause J (2017) Investigating decreasing energy usage in mobile apps via indistinguishable color changes. In: Proc. IEEE/ACM 4th Int. Conf. Mobile Software Engineering and Systems (MOBILESoft '17). IEEE, Buenos Aires, pp 30–34. https://doi.org/10.1109/MOBILESoft.2017.17
26. Li D, Tran AH, Halfond WG (2014) Making web applications more energy efficient for old smartphones. In: Proc. 36th Int. Conf. Software Engineering (ICSE '14). ACM, Hyderabad, pp 527–538. https://doi.org/10.1145/2568225.2568321
27. Li D, Tran AH, Halfond WG (2015) Nyx: a display energy optimizer for mobile web apps. In: Proc. 10th Joint Meeting on Foundations of Software Engineering (ESEC/FSE '15). ACM, Bergamo, Italy, pp 958–961. https://doi.org/10.1145/2786805.2803190
28. Linares-Vásquez M, Bernal-Cárdenas C, Bavota G, Oliveto R, Di Penta M, Poshyvanyk D (2017) Gemma: multi-objective optimization of energy consumption of guis in android apps. In: Proc. 39th Int. Conf. Software Engineering Companion (ICSE-C '17). IEEE, Buenos Aires, pp 11–14. https://doi.org/10.1109/ICSE-C.2017.10
29. Banerjee A, Roychoudhury A (2016) Automated re-factoring of android apps to enhance energy-efficiency. In: Proc. IEEE/ACM 3rd Int. Conf. Mobile Software Engineering and Systems (MOBILESoft '16). ACM, Austin, TX, pp 139–150
30. Liu Y, Xu C, Cheung SC, Terragni V (2016) Understanding and detecting wake lock misuses for android applications. In: Proc. ACM SIGSOFT 24th Int. Symposium on Foundations of Software Engineering (FSE '16). ACM, Seattle, WA, pp 396–409. https://doi.org/10.1145/2950290.2950297
31. Pathak A, Jindal A, Hu YC, Midkiff SP (2012) What is keeping my phone awake?: Characterizing and detecting no-sleep energy bugs in smartphone apps. In: Proc. 10th Int. Conf. Mobile Systems, Applications, and Services (MobiSys '12). ACM, Windermere, pp 267–280. https://doi.org/10.1145/2307636.2307661

32. Metri G, Agrawal A, Peri R, Shi W (2012) What is eating up battery life on my smartphone: a case study. In: Proc. 2nd Int. Conf. Energy Aware Computing (ICEAC '12). IEEE, Morphou, Cyprus, pp 1–6. https://doi.org/10.1109/ICEAC.2012.6471003
33. Chowdhury S, Di Nardo S, Hindle A, Jiang ZMJ (2018) An exploratory study on assessing the energy impact of logging on android applications. Empir Softw Eng 23(3):1422–1456. https://doi.org/10.1007/s10664-017-9545-x
34. Corral L, Georgiev AB, Janes A, Kofler S (2015) Energy-aware performance evaluation of android custom kernels. In: Proc. IEEE/ACM 4th Int. Workshop on Green and Sustainable Software (GREENS '15). IEEE, Florence, pp 1–7. https://doi.org/10.5555/2820158.2820160
35. Huang G, Cai H, Swiech M, Zhang Y, Liu X, Dinda P (2017) DelayDroid: an instrumented approach to reducing tail-time energy of Android apps. SCIENCE CHINA Inf Sci 60 (1):012106. https://doi.org/10.1007/s11432-015-1026-y
36. Gottschalk M, Jelschen J, Winter A (2014) Saving energy on mobile devices by refactoring. In: Proc. 28th Conf. Environmental Informatics (EnviroInfo '14). BIS-Verlag, Oldenburg, Germany, pp 437–444
37. Kim D, Jung N, Chon Y, Cha H (2016) Content-centric energy management of mobile displays. IEEE Trans Mob Comput 15(8):1925–1938. https://doi.org/10.1109/TMC.2015.2467393
38. Gamma E, Helm R, Johnson R, Vlissides JM (1994) Design patterns: elements of reusable object-oriented software, 1st edn. Addison-Wesley Professional
39. Ampatzoglou A, Charalampidou S, Stamelos I (2013) Research state of the art on GoF design patterns: a mapping study. J Syst Softw 86(7):1945–1964. https://doi.org/10.1016/j.jss.2013.03.063
40. Hsueh NL, Chu PH, Chu W (2008) A quantitative approach for evaluating the quality of design patterns. J Syst Softw 81(8):1430–1439. https://doi.org/10.1016/j.jss.2007.11.724
41. Huston B (2001) The effects of design pattern application on metric scores. J Syst Softw 58 (3):261–269. https://doi.org/10.1016/s0164-1212(01)00043-7
42. Ampatzoglou A, Chatzigeorgiou A, Charalampidou S, Avgeriou P (2015) The effect of GoF design patterns on stability: a case study. IEEE Trans Softw Eng 41(8):781–802. https://doi.org/10.1109/tse.2015.2414917
43. Feitosa D, Ampatzoglou A, Avgeriou P, Chatzigeorgiou A, Nakagawa E (2019) What can violations of good practices tell about the relationship between GoF patterns and run-time quality attributes? Inf Softw Technol 105:1–16. https://doi.org/10.1016/j.infsof.2018.07.014
44. Khomh F, Gueheneuc YG, Antoniol G (2009) Playing roles in design patterns: An empirical descriptive and analytic study. In: Proc. IEEE 25th Int. Conf. Software Maintenance (ICSM '09). IEEE, Timişoara, Romania. https://doi.org/10.1109/icsm.2009.5306327
45. Adamczyk P (2004) Selected patterns for implementing finite state machines. In: Proc. 11th Conf. Pattern Languages of Programs (PLoP '04). Monticello, IL, pp 1–41
46. Ampatzoglou A, Charalampidou S, Stamelos I (2013) Design pattern alternatives. In: Proc. 17th Panhellenic Conf. Informatics (PCI '13). ACM, Thessaloniki. https://doi.org/10.1145/2491845.2491857
47. Fowler M, Beck K, Brant J, Opdyke W, Roberts D (1999) Refactoring: improving the design of existing code. Object technology series. Addison-Wesley
48. Lyardet FD (1997) The dynamic template pattern. In: Proc. 4th Conf. Pattern Languages of Programs (PLoP '97). Monticello, IL, pp 1–8. https://hillside.net/plop/plop/plop97/Proceedings/chai.pdf
49. Saúde AV, Victório RASS, Coutinho GCA (2010) Persistent state pattern. In: Proc. 17th Conf. Pattern Languages of Programs (PLoP '10). ACM, Reno, NV. https://doi.org/10.1145/2493288.2493293
50. Litke A, Zotos K, Chatzigeorgiou A, Stephanides G (2005) Energy consumption analysis of design patterns. Proc World Acad Sci Eng Technol 6:86–90

51. Sahin C, Cayci F, Gutiérrez ILM, Clause J, Kiamilev F, Pollock L, Winbladh K (2012) Initial explorations on design pattern energy usage. In: Proc. 1st Int. Workshop on Green and Sustainable Software (GREENS '12). IEEE, Zurich, pp 55–61. https://doi.org/10.1109/GREENS.2012.6224257
52. Noureddine A, Rajan A (2015) Optimising energy consumption of design patterns. In: Proc. 37th Int. Conf. Software Engineering (ICSE '15). IEEE, pp 623–626
53. Feitosa D, Alders R, Ampatzoglou A, Avgeriou P, Nakagawa EY (2017) Investigating the effect of design patterns on energy consumption. J Softw Evol Process 29(2):e1851. https://doi.org/10.1002/smr.1851
54. Harrison NB, Avgeriou P, Zdun U (2007) Using patterns to capture architectural decisions. IEEE Softw 24(4):38–45. https://doi.org/10.1109/MS.2007.124

Chapter 6
Small Changes, Big Impacts: Leveraging Diversity to Improve Energy Efficiency

Wellington Oliveira, Hugo Matalonga, Gustavo Pinto, Fernando Castor, and João Paulo Fernandes

Abstract In this chapter, we advocate that developers should leverage software diversity to make software systems more energy efficient. Our main goal is to show that non-specialists can build software that consumes less energy by alternating at development time between readily available, diversely designed pieces of software implemented by third parties. By revisiting the main findings of research work we conducted in the past few years, we noticed that they share a common observation: small changes can make a big difference in terms of energy consumption. These changes can usually be implemented by very simple modifications, sometimes amounting to a single line of code. Based on experimental results, one small change that could make a big difference is to replace most of the uses of a Hashtable class with uses of the ConcurrentHashMap class. In most of the cases, it was only necessary to modify the line where the Hashtable object was created. This simple reengineering effort promoted a reduction of up to 17.8% in the energy consumption of Xalan and up to 9.32% for Tomcat, when using the workloads of the DaCapo benchmark suite.

Conclusions: The main insight we draw is that small changes can make a big contribution to reducing energy consumption, especially in mobile devices. We have also witnessed in practice that the huge variability of devices in the market and the vast number of factors influencing energy consumption is a real problem when experimenting with energy consumption. To try to minimize this problem, we finally

W. Oliveira · F. Castor
Federal University of Pernambuco, Recife, Brazil
e-mail: fjclf@cin.ufpe.br

H. Matalonga
Minho University, Braga, Portugal
e-mail: hugo@hmatalonga.com

G. Pinto (✉)
Federal University of Pará, Belém, Brazil
e-mail: gpinto@ufpa.br

J. P. Fernandes
CISUC and University of Coimbra, Coimbra, Portugal
e-mail: jpf@dei.uc.pt

C. Calero et al. (eds.), *Software Sustainability*,
https://doi.org/10.1007/978-3-030-69970-3_6

present an initiative that aims to collect real-world usage information about thousands of mobile devices and make it publicly available to researchers and companies interested in energy efficiency.

6.1 Introduction

In 2012, information and communication technology was estimated to be responsible for 4.7% of the world's electrical energy consumption [1]. Although that energy is to a large extent used to reduce energy consumption in other productive sectors [1, 2], it is still a considerable percentage. Moreover, that figure is estimated to grow to between 8 and 21% of the global demand for energy by 2030 [3]. In addition, energy has a high cost for many organizations [4]. Reducing that cost, even by a small percentage, can mean savings in the order of millions of dollars.

High energy consumption also has a direct impact on our daily lives, especially when we consider mobile devices. Long battery life is considered one of the most important smartphone features by users [5, 6]. In addition, from a sustainability standpoint, batteries that last longer need to be recharged less often, which also increases the lifespan of mobile devices. Making the battery last longer with a single charge involves a combination of energy-efficient hardware, infrastructure software, and applications.

In the last few years, a growing body of research has proposed methods, techniques, and tools to support developers in the construction of software that consumes less energy. These solutions leverage diverse approaches such as version history mining [7], analytical models [8], identifying energy-efficient color schemes [9], and optimizing the packaging of HTTP requests [10].

In this chapter, we present a complementary approach. We advocate that developers should leverage software diversity to make software systems more energy efficient. Our main insight is that non-specialists can build software that consumes less energy by alternating at development time between readily available, diversely designed pieces of software implemented by third parties. These pieces of software can vary in nature, granularity, and quality attributes. Examples include data structures and constructs for thread management and synchronization.

Diversity can be leveraged in a number of different situations to improve the quality of both software systems and the processes through which they are built. According to the Merriam-Webster dictionary, diversity is "the quality or state of having many different forms, types, ideas, etc." In the context of fault-tolerant software, *design diversity* has been employed since the 1970s [11, 12]. The idea is that different implementations built from the same specification are likely to fail independently and thus can be combined to build more reliable software. Another flavor of design diversity aiming to improve reliability can be observed when developers write detailed behavioral contracts for functions [13]. A contract can be seen as a diverse implementation written in a declarative language that is close to mathematics. Design diversity is also important for the construction of software systems that have dependencies on external libraries, components, or frameworks. In 2016, the unpublication of a small npm Javascript package[1] broke thousands of client projects. Availability of diverse packages with similar functionality can help reduce the impact of this kind of problem. Diversity is applicable beyond software design, in other software-related situations. Not long ago, Google discussed [14] one of its approaches to reducing latency: to have multiple servers serve the same request. In this scenario, we have *latency (or timing) diversity*, since a multitude of factors can affect the response time of each server at any given moment.

In this chapter we discuss the use of software diversity as a tool in the developers' toolbox to build more energy-efficient software. Diversity, in this case, expands the design and implementation options [15] available for developers. To assess the impact of these options, throughout this chapter we revisit the main findings of research work we conducted in the past few years (e.g., [16–21]). Although these works target different programming languages, execution environments, and programming constructs, they share a common observation: small changes can make a big difference in terms of energy consumption. These changes can usually be implemented by very simple modifications, sometimes amounting to a single line of code. Nonetheless, the results can be significant.

In our work we have, for example, refactored two Java systems, the `TOMCAT` web server and the `XALAN` library for XML processing. Based on experimental results [20], we replaced most of the uses of the `Hashtable` class, which implements the `Map` interface, with uses of the `ConcurrentHashMap` class, which implements the same interface. In most of the cases, it was only necessary to modify the line where the `Hashtable` object was created. This simple reengineering effort promoted a reduction of up to 17.8% in the energy consumption of `XALAN` and up to 9.32% for `TOMCAT`, when using the workloads of the DaCapo [22] benchmark suite.

This chapter first introduces some of the aforementioned studies (Sect. 6.3). It then proceeds to present an automated approach to help developers to select potentially more energy-efficient options in situations where diversity is available (Sect. 6.4). On the one hand, this approach works statically, and experiments conducted show that it is able to improve the energy efficiency of real-world systems. On the other hand, for mobile devices, results vary widely (particularly due to the fragmentation of Android devices and their versions), which requires additional information and experimentation on their usage profiles. Based on this, we present a more recent initiative that aims to collect real-world usage information about thousands of mobile devices and make it publicly available to researchers and companies interested in energy efficiency (Sect. 6.5).

[1] https://www.theregister.co.uk/2016/03/23/npm_left_pad_chaos/

6.2 Software Energy Consumption

Although software systems do not consume energy themselves, they affect hardware utilization, leading to indirect energy consumption. Energy consumption E is an accumulation of power dissipation P over time t, that is, $E = P \ t$. Power P is measured in watts, whereas energy E is measured in joules. As an example, if one operation takes 10 seconds to complete and dissipates 5 watts, it consumes 50 joules of energy $E = 5 \ 10$. In particular, when talking about software energy consumption, one should pay attention to:

- The hardware platform.
- The context of the computation.
- The time spent.

To understand the importance of a *hardware platform*, consider an application that communicates through the network. Any commodity smartphone supports, at least, WiFi, 3G, and 4G. Some researchers observed that 3G can consume about 70% more energy than WiFi, whereas 4G can consume about 30% more energy than 3G, while performing the same task, on the same hardware platform [23].

Context is relevant because the way in which software is built and used has a critical influence on energy consumption. A program may impact the energy consumption of different parts of a device, for instance, the CPU, when performing CPU-intensive computations [24], the DRAM, when performing intensive accesses to data structures [25], the network, when sending and receiving HTTP requests [26], or on OLED displays, when using lighter-colored backgrounds [9].

Finally, *time* plays a key role in this equation. A common misconception among developers is that reducing execution time also reduces energy consumption [27, 28], the t of the energy equation. However, chances are that this reduction in execution time might increase energy consumption by imposing a heavier burden on the device, e.g., by using multiple CPUs [16]. This in turn can increase the number of context switches and, as a consequence, might also increase the P of the equation, impacting the overall energy consumption.

6.2.1 Gauging Energy Consumption

Power Measurement and Energy Estimation are high-level approaches encompassing multiple techniques to gauge energy consumption at different levels of granularity. The first group of techniques makes use of power measurement hardware to obtain power samples. The main advantage of this technique is its ability to capture actual power use, possibly with high precision. Its main disadvantage, however, is that it is only possible to attribute the measured power to specific hardware or software elements indirectly. This usually requires software-based techniques and energy estimation (see below). Many different power meters are currently available in the market. Different power meters have different characteristics. Among these characteristics, one of the most important is the sampling rate, that is, the number of samples obtained per second. The sample is often measured in watts, *P* (power). Depending on the power meter used, the sampling rate can vary from 1 sample per second to more than 10,000 samples per second. The higher the sampling rate, the more accurate the power curve will be.

The second area, energy estimation, assumes that developers do not have access to power measurement hardware and uses software-based techniques to predict how much energy an application will consume at run time. These predictions are based on mathematical models of how the different aspects of the hardware under examination consume energy, while accounting for their workloads. One example of this approach is the powertop[2] utility. This tool takes one sample per second and generates a log with these measurements. It analyzes the programs, device drivers, and kernel options running on a computer based on the Linux and Solaris operating systems, and estimates the power consumption resulting from their use. Powertop can also instrument laptop battery features in order to estimate power usage (in watts) and battery life.

The Running Average Power Limit (RAPL) interface [29], originally designed by Intel to enable chip-level power management, is widely supported in today's Intel architectures, including Xeon server-level CPUs and the popular i5 and i7. RAPL-enabled architectures monitor performance counters in a machine and estimate the energy consumption, storing the estimates in Machine-Specific Registers (MSRs). Such MSRs can be accessed by the OS, e.g., by means of the msr kernel module in Linux. RAPL is an appealing design, particularly because it allows energy/power consumption to be reported at a fine-grained level, e.g., monitoring CPU core, CPU uncore (caches, on-chip GPUs, and interconnects), and DRAM separately. Previous work has shown that RAPL estimates are precise when compared to measurements obtained by power measurement equipment [8]. One drawback of this approach is the fact that programmers need a deep knowledge on how to use these low-level registers, which is not straightforward.

Liu and colleagues [25] introduced jRAPL, a library for profiling Java programs running on CPUs with RAPL support. This library can be viewed as a software wrapper to access the MSRs. Since the user interface for jRAPL is simple, the programmer can focus her efforts on the high-level application design. For any block of code in the application whose energy/performance information is of interest to the user, she just needs to enclose the code block with a pair of `statCheck` invocations. For example, the following code snippet attempts to measure the energy consumption of the `doWork()` method, whose value is the difference between the `beginning` and `end` variables:

[2]https://01.org/powertop

```
double beginning = EnergyCheck.statCheck();
doWork();
double end = EnergyCheck.statCheck();
```

A shortcoming of the jRAPL library is that it can only be used on desktop computers that leverage Intel CPUs. Thus, it provides little help for measuring energy consumption of mobile apps in tablets, smartphones, or smartwatches.

In November, 2014, as part of Android 5.0, Google released the Android Power Profiler, which queries battery information from Android devices. It is currently available on every Android device since version 5.0 (which corresponds to over 85% of all Android devices[3]). The Android Power Profiler has many advantages over similar libraries. First, it requires no extra instrumentation. As the profiler is natively executed, no external applications are needed either. Second, it provides a straightforward interface to gather battery information and it does not require any setup. Third, the profiler distinguishes battery usage in terms of the different components used on the device (e.g., WiFi, CPU, GPS). The Android Power Profiler, similarly to RAPL, is based on energy estimation.

To use the Android Power Profiler tools it is necessary to use the Android Debug Bridge (ADB).[4] ADB is a command-line tool that works like a communication interface, using the client-server model, where the device being used is the client and the development machine is the server. ADB allows one to install and debug apps, collect data about the device, execute automated tests, etc. For instance, the `adb shell dumpsys batterystats` command collects battery information and may save it onto an output file. The exported file could be an input to other programs to manipulate and analyze data. A growing number of research works are taking advantage of the Android Power Profiler (e.g., [8, 18, 30, 31]).

6.3 Design Decisions

In this section we explore three different approaches that share the same observation that small design decisions can greatly impact energy consumption. More specifically, we discuss how it is possible to reduce the energy footprint of software systems by leveraging diversity in IO primitives (Sect. 6.3.1), collection implementations (Sect. 6.3.2), and concurrent programming constructs (Sect. 6.3.3).

[3]https://www.statista.com/statistics/271774/share-of-android-platforms-on-mobile-devices-with-android-os/

[4]https://developer.android.com/studio/command-line/adb

6.3.1 I/O Constructs

I/O programming constructs are not only the building blocks of several low-level communication channels such as sockets or database drivers, but also the bedrock of high-level software applications that have anything to do with data storage or transmission. Despite their widespread use, the energy consumption of I/O programming constructs is not well understood. This is particularly unfortunate since related work suggests that I/O APIs could severely impact energy consumption. For instance, Lyu and colleagues [32] indicated that about 10% of the energy consumption of mobile applications is spent in I/O operations. Similarly, Liu and colleagues [25] pointed out that it was possible to save 4.29% of energy consumption by changing I/O programming constructs. A comprehensive energy characterization of I/O programming constructs could help practitioners to further improve the energy behaviors of their software applications.

In the study by Rocha and colleagues [21], we presented a comprehensive characterization of Java I/O APIs. In this work we conducted a broad experimental exploration of 22 Java I/O APIs, aiming to answer two research questions: (RQ1) *What is the energy consumption behavior of the Java I/O APIs*? and (RQ2) *Can we improve the energy consumption of non-trivial benchmarks by refactoring their use of Java I/O APIs*?

To answer these research questions, we employ what we consider to be three types of benchmarks. For the first research question, we created and instrumented 22 *micro-benchmarks*. The *micro-benchmarks* are small programs (around 200 lines of code) that perform a single task (e.g., reading a file from the disk), each one using a different Java I/O API. These Java I/O APIs have been introduced in the Java programming language in its very early versions and are in widespread use. For instance, the `FileInputStream` Java I/O API is used in 2823 open-source projects in BOA [33] (we ran this query in April 2019). Each one of the studied Java I/O APIs implements at least one method for input operations or at least one method for output operations.

For the second research question, we performed refactorings in the code base of *optimized benchmarks* and *macro-benchmarks*. On the one hand, *optimized benchmarks* are similar to micro-benchmarks in size, but are optimized for performance, while *macro-benchmarks* are full-fledged working software systems comprising thousands of lines of code. The *macro-benchmarks* used are as follows: `XALAN` (an XSLT processor that translates XML documents into HTML files, or other types of documents), `FOP` (an XSLT processor that translates XML documents into HTML files, or other types of documents), `BATIK` (a toolkit for applications that want to use images in the Scalable Vector Graphics (SVG) format), `COMMONS-IO` (a utility library used to provide high-level I/O abstractions to third-party software applications), and `PGJDBC` (the official PostgreSQL driver for the Java programming language).

To avoid non-working solutions, we focused on refactorings that do not require extensive code changes (e.g., changes between Java I/O APIs that extend the same

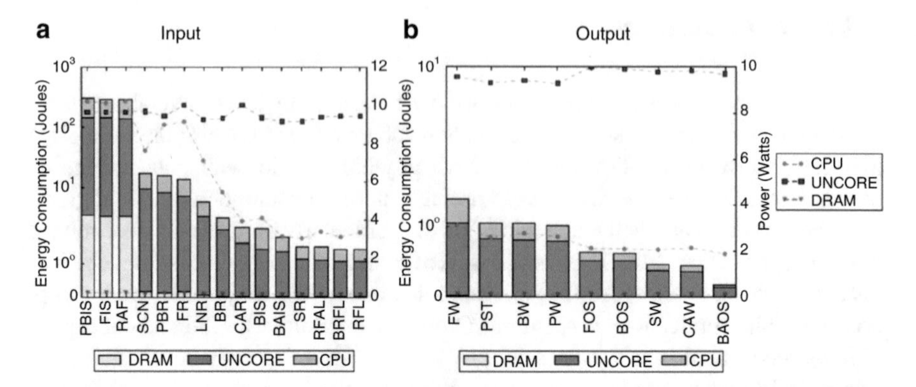

Fig. 6.1 Energy consumption behavior of Java I/O APIs. Energy data is presented in a logarithmic scale. For the figure on the left, PBIS stands for `PushbackInputStream`, FIS stands for `FileInputStream`, RAF stands for `RadomAccessFile`, SCN stands for Scanner, PBR stands for `PushbackReader`, FR stands for `FileReader`, LNR stands for `LineNumberReader`, BR stands for `BufferedReader`, CAR stands for `CharArrayReader`, BIS stands for `BufferedInputStream`, BAIS stands for `ByteArrayInputStream`, SR stands for `StringReader`, RFAL stands for `Files.readAllLines`, BRFL stands for `Files.newBufferedReader`, and RFL stands for `Files.lines`. For the figure on the right, FW stands for `FileWriter`, PST stands for `PrintStream`, BW stands for `BufferedWriter`, PW stands for `PrintWriter`, FOS stands for `FileOutputStream`, BOS stands for `BufferedOutputStream`, SW stands for `StringWriter`, CAW stands for `CharArrayWriter`, and BAOS stands for `ByteArrayOutputStream`

interface). These refactorings could also be easily automated by a general purpose tool. To conduct the experimentation process, we executed each benchmark 10 times. Since it requires some time for the Just-In-Time (JIT) compiler to identify the hot code and perform optimizations, we discarded the first three executions of the benchmarks. We report the average of the seven remaining executions. We also fixed the garbage collector and the heap size accordingly: we used the parallel garbage collector (`-XX:+UseParallelGC`), and the heap size was fixed at 261 MB, minimum (−`Xms`), and 4183 MB, maximum (−`Xmx`). No other JVM options were employed.

After conducting this process, we observed many interesting findings. Figure 6.1 shows an overview of the energy behavior of Java I/O APIs for the micro-benchmarks. First, we found that input operations consume more energy than output operations (on average: 96 joules vs 0.80 joules, respectively). The `PushbackInputStream` Java I/O API is the most energy-consuming one (492 joules consumed), followed by `FileInputStream` (474 joules). Analyzing the `PushbackInputStream` implementation, we perceived that this Java I/O API adds a flag in the `InputStream` that marks bytes as "not read." Such bytes are included back in the buffer to be read again. However, before reading the bytes, this Java I/O API also checks whether the stream is still open using the `ensureOpen()` method. This repetitive operation could be the source of this high energy consumption. The `Files` Java I/O API, however, which could act as a potential replacement for `FileInputStream`, is the one with the least energy consumption, when executing its `lines` method (1.86 joules).

When considering the macro- and optimized benchmarks, it was not possible to use all the Java I/O APIs mentioned in Fig. 6.1. This happened because there is a semantic gap between the Java I/O APIs that do not inherit from the same parent, and we opted not to bridge this gap. We then only refactored instances of Java I/O APIs that share the same parent class. This problem did not occur for the micro-benchmarks because of the more straightforward way in which they use the APIs. In the end, we had 21 refactored versions of these benchmarks.

Overall, when refactoring the macro- and optimized benchmarks, we observed energy improvements in 8 out of the 21 refactored versions of the benchmarks. In particular, we observed that one optimized benchmark and one macro-benchmark improved their overall energy consumption when changing their use of Java I/O APIs to the `Files` class. With very minor modifications, we were able to improve up to 17% of the energy consumption of these benchmarks. These initial results provide evidence that small changes in Java I/O APIs might have the potential for improving the energy consumption of benchmarks already optimized for performance.

6.3.2 *Collections Constructs*

Collections provide easy access to reliable implementations that can reduce the complexity of developing applications. In Java, each collection's API has multiple implementations. Collection implementations that can be safely used by several concurrent threads are considered "thread-safe." This safety usually comes with extra complexity or inferior performance, which might favor the use of "thread-unsafe" collections. This is expected, since there are a number of different algorithms and data structures that can implement the abstract concept of lists, sets, and maps. There are a number of different ways in which a collection can be implemented, and these diverse implementations can have a non-negligible impact on energy consumption.

In the last few years, a number of researchers have attempted to address the problem of helping developers to understand collections energy usage [16, 19, 20, 34–36]. These works conducted extensive exploration of collection usage. While some papers focused on the energy usage of collection implementations that are part of the Java Development Kit [20], others were broader in scope and covered not only the official implementations but also third-party libraries [19]. Similarly, while some works performed the experiments on commodity devices [36], others conducted experiments on servers [20], while others also experimented with mobile devices [19, 34]. Generally speaking, these works followed a similar approach for collecting data: they created small and large benchmarks, and executed these benchmarks 10 or more times, reporting the averages as the results.

The work of Oliveira and colleagues [19] followed a slightly different approach because they created the so-called energy profiles, inspired by previous work [34], and attempted to make recommendations by leveraging these profiles. We devote Sect. 6.4 to providing a more comprehensive overview of this work. An energy profile is a number that can be used to compare similar constructs under the same circumstances. Energy profiles for collections can be produced by executing several micro-benchmarks on different collection operations, aiming to gather information about the energy behavior of these programming constructs in an application-independent way. For instance, an energy profile for the operation `ArrayList.add(Object o)` could be 10. After we create the profiles, we perform static analysis to estimate in which ways and how intensively a system employs these collections. If we know that the program under investigation uses exclusively `ArrayList.add(Object o)` 100 times, its energy consumption could be (roughly) inferred as 100 10 (their energy profile). Since the work of Oliveira and colleagues [19] focuses on code recommendation, a collection is more likely to be recommended if its energy profiles are low.

Since these works performed computations in very different environments, the results cannot be easily merged together. However, some interesting findings seem to emerge. For instance, these papers explored the energy consumption of the most commonly used methods. In the case of `ArrayList`, they investigated the `add(Object o)` method. For example, in the work of Pinto and colleagues [20], the authors observed that the method `ArrayList.add(Object o)` consumes the least energy, when compared to the thread-safe implementations. On the other hand, both Pinto et al. [20] and Pereira et al. [36] observed that the most energy-consuming implementation among the thread-safe collections is `CopyOnWriteArrayList`. In particular, Pinto et al. [20] noted that insertion operations over `CopyOnWriteArrayList` consumed about 152$\times$ more energy than `Vector` (which consumes 14$\times$ more than `ArrayList`). In terms of `Map` implementations, it was found that the concurrent implementation `ConcurrentHashMap` had a similar performance when compared to the non-thread safe implementation, `LinkedHashMap`, on both insertion and removal operations. Indeed, `ConcurrentHashMap` performed around three times better than `Hashtable`, one of the most common `Map` implementations.

It is important to note that these findings were observed in small benchmarks, that is, $\sim$100 lines of code programs that perform one collection operation a number of times. Given these observations in a controlled setting, Pinto et al. [20] also manually refactored two large-scale open-source programs: XALAN (a program that transforms XML documents into HTML, which had 170 k lines of code in the version we studied) and TOMCAT (an open-source web server, which had 188 k lines of code in the version we studied). In both programs, the authors changed 100+ uses of `Hashtable` to `ConcurrentHashMap`. After applying these modifications, it was observed that there was an energy saving of 12% for XALAN and 17% TOMCAT, considering the workloads of the DaCapo benchmark suite [22]. Oliveira and colleagues [19] also employed a similar approach for alternating between collection implementations. In their work, they found that by refactoring from `ArrayList` to

`FastList` (a third-party `List` implementation) it was possible to save 17% in energy consumption of one mobile app, PASSWORDGEN. These findings share a common trend: with no prior knowledge of the application domains or the system implementations, it was possible to reduce the energy consumption of a software system by means of simple changes in collection usage.

6.3.3 Concurrent Programming Constructs

Concurrency control and thread management are additional software features where it is possible to reap the benefits of software diversity. Early work by Trefethen and Thiyagalingam [37] observed that, for parallel applications in the area of scientific computing, performance is often not a proxy for energy consumption. A subsequent study [24] investigated the impact of different approaches to manage concurrent and parallel execution in Java programs. This study found that different thread management approaches, e.g., percore threads, thread pools, and work-stealing, have diverse, significant, and hard to predict impacts on energy consumption. It also observed that performance is not a good proxy for energy efficiency in the studied benchmarks, which comprised both small programs and real-world, high-performance Java applications.

These studies inspired us to investigate how thread management constructs affect energy consumption in a different setting, namely, programs written in Haskell, a lazy, purely functional programming language. Haskell programs can create lightweight threads that may be associated with a specific physical core or operating system thread, or managed entirely by the Haskell scheduler. Furthermore, the language has multiple primitives for data sharing between threads which act as concurrency control primitives, including a lock-based approach, a fully featured implementation of software transactional memory [38] (STM), and an STM-based solution that simulates locks.

We conducted a study with nine Haskell benchmarks. The benchmarks were selected from multiple sources, such as the Computer Language Benchmarks Game[5] and Rosetta Code.[6] We selected the benchmarks based on their diversity. For instance, two of them are synchronization-intensive programs, two are CPU-intensive and scale up well on a multicore machine, two are CPU- and memory-intensive, one is I/O-intensive, one is CPU- and I/O-intensive, and one is peculiar in that it is CPU-, memory-, synchronization-, and I/O-intensive. We implemented and ran different variants of these benchmarks considering the nine possible combinations of thread management constructs and data sharing/concurrency control primitives. Not every possible variant could be used, e.g., because

[5]https://benchmarksgame-team.pages.debian.net/benchmarksgame/index.html

[6]http://www.rosettacode.org

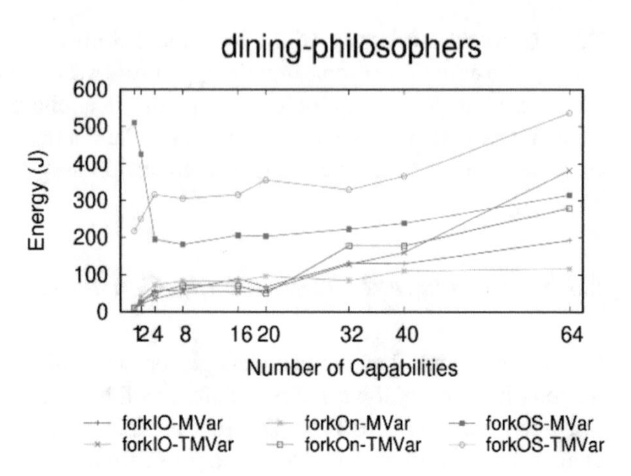

Fig. 6.2 Energy measurements for the dining philosophers benchmark, considering six combinations of thread management constructs (`forkIO`, `forkOn`, and `forkOS`) and concurrency control primitives (`MVar` and `TMVar`) [16]

some benchmarks do not leverage concurrency control. The details of the methodology of this study are presented elsewhere [16].

We ran all the experiments on a server machine with 2x10-core Intel Xeon E5-2660 v2 processors (Ivy Bridge microarchitecture, 2-node NUMA) and 256GB of DDR3 1600 MHz memory. This machine runs the Ubuntu Server 14.04.3 LTS (kernel 3.19.0-25) OS. The compiler was GHC 7.10.2. The benchmarks were exercised by Criterion [39], a benchmarking library to measure the performance of Haskell code. To collect information about energy usage, we had to modify the implementations of Criterion and the Haskell profiler to make them energy-aware. We executed the benchmarks with 1, 2, 4, 8, 16, 20, 32, 40, and 64 capabilities. The number of capabilities of the Haskell run time determines how many Haskell threads can run truly simultaneously at any given time.

Once again, we found that small changes can make a big difference in terms of energy consumption. For example, in one of our benchmarks, under a specific configuration, choosing one data sharing primitive (`MVar`) over another (`TMVar`) can yield 60% energy savings. However, there is no universal winner. The results vary depending on the characteristics of each program. In another benchmark, `TMVars` can yield up to 30% energy savings over `MVars`.

Figure 6.2 illustrates an extreme case. When considering 20 capabilities, the `forkOS-TMVar` variant of the dining philosophers benchmark consumed 268% more energy than the `forkIO-MVar` variant. These results indicate that it is also possible to exploit software diversity in Haskell in order to improve energy efficiency.

Similar to previous studies [24, 37], we found that the relationship between energy consumption and performance is not always clear. High performance is usually a proxy for low energy consumption. Nonetheless, we found scenarios where the configuration with the best performance (30% faster than the one with the worst performance) also exhibited the second worst energy consumption (used 133% more energy than the one with the lowest usage). The scatterplots in Fig. 6.3 illustrate how energy and time are imperfectly aligned. This is different from what

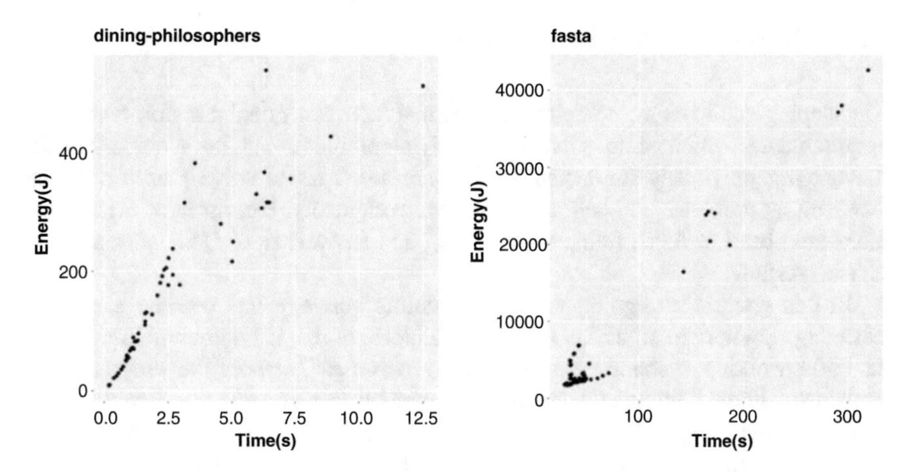

Fig. 6.3 Scatterplots for the relationship between time (*x*-axis) and energy (*y*-axis) for two of the analyzed Haskell benchmarks [16]

we would observe, for example, in sequential Haskell collections [16], where the points would be almost perfectly aligned along the diagonal.

In addition to these results, we propose some guidelines that Haskell developers can follow to make applications more energy efficient, based on our empirical results. First, CPU-bound applications should avoid setting more capabilities than physical cores, since these applications will not benefit from the enhanced thread switching afforded by hyperthreading[7] and similar approaches. Second, they should use the `forkOn` function, which attempts to pin threads to specific physical cores, to create threads in embarrassingly parallel applications. This reduces thread migration overhead in applications where workloads are evenly distributed among threads and threads do not have data dependencies. Third, they should avoid using the `forkOS` function to spawn new threads. Since this function binds a Haskell thread to an OS thread, it results in thread switching involving OS threads whenever new Haskell threads must be executed. Fourth, if energy matters, only use STM if transaction conflicts are rare. Although transactional memory may improve performance due to optimistic concurrency, the large number of conflicts can have a strong impact on energy consumption, even if they do not hinder performance. As mentioned earlier, for one of the analyzed benchmarks, the variant with the best performance exhibited more than twice the energy consumption of the most energy-efficient variant.

[7]https://www.intel.com/content/www/us/en/architecture-and-technology/hyper-threading/hyper-threading-technology.html

6.4 Recommending Java Collections

Developing applications, complex systems that have functional and non-functional requirements combined to solve a non-trivial problem, can be a difficult task. Leveraging previously implemented software solutions to solve pieces of these challenging problems can help to reduce that complexity. Examples of these solutions are libraries, APIs, frameworks, gists,[8] and answers from Q&A sites such as StackOverflow.

In this context, designing and implementing software has become a task of selecting appropriate solutions among multiple options [15] and combining them to build working systems. We call *energy variation hotspots* the programming constructs, idioms, libraries, components, and tools in a system for which there are multiple, interchangeable, readily available solutions that have potentially different energy footprints. A number of previous papers have measured and analyzed different types of energy variation hotspots, such as programming languages [18, 40, 41], API usage [9, 21, 42], thread management constructs [16, 24], data structures [16, 20, 34, 36], color schemes [43, 44], and machine learning approaches [45], among many others. Having to choose the most energy-efficient solution for an energy variation hotspot can be difficult for developers, as the energy consumption of these constructs is usually not easily measurable. Furthermore, information on how to execute tests to measure the energy impact of different solutions can be hard to find.

In this section, we present our solution to reduce the energy consumption of software applications, making it easier for non-specialist developers to exploit energy variation hotspots. This solution can be separated into three different steps. In the first step, we exercise the available alternative solutions, aiming to build energy consumption profiles [34]. In the second step, we analyze the application, collecting and organizing the usage of the selected energy variation hotspots, in particular, to estimate how intensively the system uses them. Finally, in the third step, we combine the energy profile and the results of analyzing the system to make potentially energy-saving recommendations specific to the application-device pair. This approach is instantiated in an energy-saving tool called CT+. Using this tool, non-specialist developers can optimize the energy efficiency of Java collections.

While experimenting with CT+, we selected collections from three different sources: Java Collections Framework (JCF),[9] Apache Commons Collections,[10] and Eclipse Collections.[11] These sources are widely used on Java projects, with a query on GitHub projects[12] showing 1,276,939 occurrences for Apache Commons, 537,956 occurrences for Eclipse Collections, and 85,865,270 occurrences for the

[8]https://gist.github.com

[9]https://docs.oracle.com/javase/8/docs/technotes/guides/collections/

[10]https://commons.apache.org/proper/commons-collections/

[11]https://www.eclipse.org/collections/

[12]These queries were executed in April 2020.

most widely used collection implementations from the JCF. All sources have thread-safe and thread-unsafe collections.

We implemented CT+ following our approach step by step. In the first step, it automatically runs multiple micro-benchmarks (i.e., executing specific collection operations such as `List.add(Object o))` for 39 distinct Java collection implementations in an application-independent manner and builds their energy profiles. `List`, `Map`, and `Set` are the three collection APIs targeted by our tool. A varying number of operations was exercised for each API, with `List` having 12 operations, `Map` 4, and `Set` 3. Our collection pool comprises implementations from the Java Collections Framework (25 implementations), Apache Commons Collections (5 implementations), and Eclipse Collections (9 implementations). The energy consumption profile is built with the data from these micro-benchmarks.

For the second step, an inter-procedural static analysis using WALA[13] is performed on the application source code. This analysis collects and organizes data on how the application uses collection implementations on its source code, such as frequency and location of use, which operations were used, method and variable names, and calling context, among others.

The third and final step consists of combining these two pieces of information, that is, the energy profile and the analysis of the application. CT+ identifies the most energy-efficient collection implementations across the whole program and automatically applies these recommendations to the source code. We evaluated CT+ in two distinct studies, analyzing the impact of different devices and different energy profiles, aiming to answer the following four research questions: (RQ1) *To what extent can we improve the energy efficiency of an application by statically replacing Java collections implementations?* (RQ2) *Are the recommendations device-independent?* (RQ3) *To what extent does workload size impact the energy efficiency of a Java collection implementation?* and (RQ4) *Are the recommendations profile-independent?*

6.4.1 Evaluation

The main objective of our evaluation was to compare the energy consumption of the original versions of software systems with the versions where the recommendations made by CT+ were applied. This was made across two different studies. Overall, our evaluation comprises two different execution environments, **desktop** and **mobile**, and six distinct devices.

In the first part of our experiment, our main goal was to evaluate the collection implementation recommendations made by CT+ (RQ1 and RQ2). On the desktop environment we executed CT+ across two machines, a notebook and a high-end server. We labeled the notebook as **dell** (Dell Inspiron 7000) and the server as **server**

[13]http://wala.sourceforge.net/wiki/index.php/Main_Page

(the same machine described in Sect. 6.3.3). On the mobile environment, we executed our tool on three smartphones and a tablet: Samsung Galaxy J7 (**J7**), Samsung Galaxy S8 (**S8**), Motorola G2 (**G2**), and Samsung Galaxy Table 4 (**Tab4**). In this experiment, we analyzed seven desktop-based software systems, BARBECUE, BATTLECRY, JODATIME, version 6.0.20 of TOMCAT, TWFBPLAYER, XALAN, and XISEMELE; two mobile-based software systems, FASTSEARCH and PASSWORDGEN; and three that work on both environments: APACHE COMMONS MATH 3.4 (COMMONS MATH for short), GOOGLE GSON, and XSTREAM.

In the second part of the experiment, we have analyzed the energy impact of three different strategies to build energy profiles (RQ3 and RQ4) For this study, a single device was used: a notebook **asus** (ASUS X555UB). To explore the impact of different energy profiles on the energy-efficiency behavior of Java collection implementations, we created three different profiles for **asus**: `small`, `medium`, and `big`. These profiles were created to simulate three different scenarios of usage intensity of the collection implementations: `small`, to be used for applications that have a light usage of collections; `medium`, an intermediate profile, for general purpose usage; and `big` to be used on applications that have a very intense usage of collections. We used as targets systems six applications from the latest version of Dacapo: BIOJAVA, CASSANDRA, GRAPHCHI, KAFKA, ZXING, and version 9.0.2 of TOMCAT, the latter with two different types of workload: LARGE and HUGE.

To measure the energy consumption of the devices in both studies, we employ jRAPL to collect the energy data in the desktop environment and the Android Energy Profiler in the mobile environment. While running our experiments, for some systems the difference between original and modified versions was not statistically significant. That was the case for TWFBPLAYER, XISEMELE, CASSANDRA, and KAFKA and for the device **Tab4**. To better focus on more relevant data, these results are not presented here. However, all data from every system used in our experiments can be found at the companion website, https://energycollections.github.io/.

Across both studies, CT+ performed 1454 changes that impacted the energy consumption across 17 software systems, 12 targeting a desktop environment, 2 targeting a mobile environment, and 3 that work in both scenarios, for a total of 46 modified versions. The analyzed applications were, for the most part, mature systems comprising thousands of lines of code (LoC), such as BIOJAVA with 914kLoC, CASSANDRA with 466kLoC, and TOMCAT with 433kLoC. Even without any prior knowledge of the application domains, CT+ reduced the energy consumption of 13 out of the 17 systems.

Analyzing Different Devices Figure 6.4 summarizes the results of our modifications on the desktop and mobile devices. For the desktop environment, CT+ made 477 recommendations, with all the modified systems consuming less energy than the original versions. Among the software systems that only ran on the **dell** machine, JODATIME exhibited the greatest improvement, with the modified version consuming 7.13% less than the original one. The modified version of TOMCAT v6 was

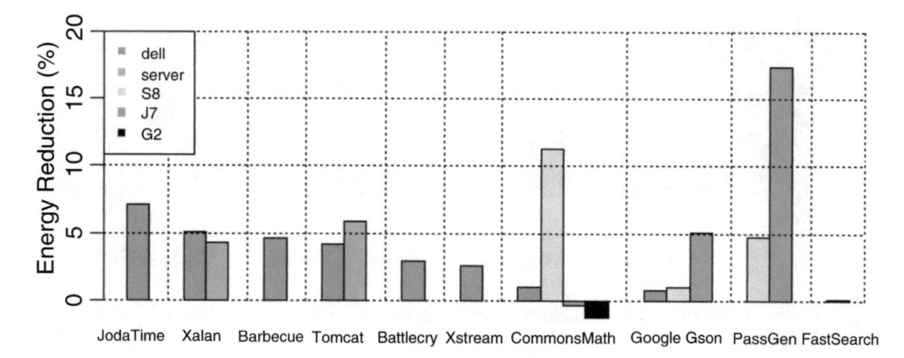

Fig. 6.4 Percentage of energy reduction in the study on different devices. Greater is better

energy efficient on **dell** and **server**, consuming 4.12% and 4.3% less energy, respectively. We found that, for the same workload, systems running on **server** consumed more than twice the energy they consumed on **dell**.

For both devices we can observe a trend of recommendations to replace well-known collections from the JCF (`Vector`, `ArrayList`, and `HashMap`) with alternative sources. For the specific case of XALAN, among the 119 recommendations across the two desktop machines, just three suggested the use of implementations from the Java Collections Framework.

On the mobile environment, CT+ made 107 recommendations among the analyzed devices, with an expressive variation in their effectiveness. Modified versions of PASSWORDGEN on **S8** and **J7** devices exhibited significant improvements over the original versions, consuming 4.7% and 17.34% less energy, but on **G2** the modifications did not have a significant impact. The modified version of GOOGLE GSON exhibited an improvement of 5.03% on **J7**; however, the modifications yielded a small 0.95% improvement on **S8**. COMMONS MATH had more inconsistent results. Although the modified version consumed 11.31% less energy than the original version on **S8**, this was not the case on **G2** and **J7**, where the modified versions consumed 1.2% and 0.33% **more** energy, respectively. Finally, FASTSEARCH showed statistically significant results only on **S8**, with the modified version having a very slight reduction of 0.09% in energy consumption.

Analyzing Different Profiles Figure 6.5 summarizes the results of our modifications on the three different profiles. Among the modified systems, GRAPHCHI was the one which presented the best results, with a reduction of 12.73% on `small`, 11.09% on `medium`, and 5.30% on `big`. While executing TOMCAT v9 using the LARGE workload, CT+ recommendations resulted in a reduction of energy consumed across all profiles. On the other hand, while using the HUGE workload, CT+ modifications did not provide a statistically significant result, with the exception of the profile `small`, where the modified version consumed more energy than the original version, making it less energy efficient.

Among the profiles, there was no overall winner. Each profile had at least one application with the best energy efficiency. TOMCAT-LARGE consumes less energy

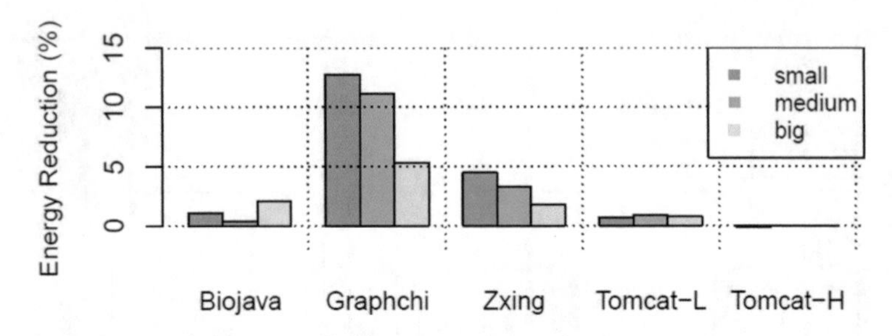

Fig. 6.5 Percentage of energy reduction in the study on different profiles. Greater is better

using the recommendations made using the `small` profile; `GRAPHCHI` and `ZXING` the `medium` profile; and `BIOJAVA` the `big` profile.

There was an expressive variation in the number of implementation changes across the profiles, with `medium` having the most with 456 changes, followed by `small` with 352, and finally `big` with 62, for a total of 870 recommendations that wielded reduced energy consumption.

These changes were not evenly distributed, with some cases having a noticeable change in the number of recommendations made for an application based on the profile (e.g., `BIOJAVA` had 13 times more changes on the profile `medium` than on the profile `big`). The changed collections were also different. As an example, the number of list implementations changed on `big` are 5% of the number of list implementations changed on the other profiles. In addition, the changed collections were different. As an example, the number of list implementations changed on `big` are 5% of the number of list implementations changed on the other profiles. Most of the time, these recommendations changed an implementation from JCF to an implementation from one of our alternative sources, i.e., Eclipse Collections or Apache Commons Collections. More specifically, this was the case for 95% of the recommendations when using the `small` profile, 98% for `medium`, and 60% for `big`.

6.4.2 Findings

Analyzing more in depth the results from both studies produced some interesting lessons.

Java Collections Framework is not the most energy efficient. The majority of the CT+ recommendations were for collection implementations not in the JCF. In the desktop environment, only 5.8% of the changes recommended by CT+ used JCF implementations, while in the mobile environment CT+ recommended JCF collection implementations in one third of the cases. Across the two different environments, 91.9% of the recommendations originated from the Apache Common Collections and Eclipse Collections.

Collection popularity does not reflect energy efficiency. Looking at the most widely used collections in Java projects, i.e., `Hashtable`, `HashMap`, `HashSet`, `Vector`, and `ArrayList`, CT+ changes to them comprise the best part of all changes made (94.2% on the desktop environment and 94.39% on the mobile environment). In the specific case of `ArrayList`, the overall most popular Java collection by far [19], this happened for two reasons. First, two common operations (namely, `insert(value)` and `iteration(random)`) usually have a worse performance in `ArrayList` when compared to other implementations. Second, in the cases where it was the best implementation, due to its widespread use, it is already being employed and thus no benefits could be achieved.

The energy behavior varied heavily across devices, even when executing the same application. Using XALAN as an example, while being analyzed on **dell**, CT+ recommended ten `ArrayList` instances to be changed to `FastList` and one to `NodeCachingLinkedList`. This was not the case on **server**, where CT+ recommended only two instances of `ArrayList` and suggested the use of `TreeList`. Nevertheless, there was an improvement in energy efficiency in both machines. Another example is XSTREAM. None of the mobile modified versions, even while consuming less energy, differed statistically from their original version. This was not the case on **dell**, where the modified version consumed less energy and exhibited a statistically significant difference.

The profiles heavily influenced the energy savings. Using the wrong profile can result in the energy consumption of the application rising instead of dropping, as in the case of the modified version of TOMCAT, on the profile `small` and using the workload with size HUGE. This happened because `small` was created to use small-sized collections, and thus it is optimized to that case while HUGE represents exactly the opposite. This illustrates that even though profile creation is an application-independent step of the proposed approach, knowledge about actual usage profiles can be leveraged to produce more useful energy profiles. A better use of the profiles can be seen in the recommendations applied to TOMCAT using the workload LARGE, resulting in a positive impact on energy efficiency, with statistical significance in the three different profiles.

The best implementation is workload-dependent. Among our recommendations on **asus**, 95% of list modifications on `small` and `medium` were changes from `ArrayList` to a different implementation. In BIOJAVA, the system representing 60% of all `ArrayList` modifications, two operations were the most intensively used: `insert(value)` and `iteration(iterator)`. This is reflected in the collection implementations that most often replace `ArrayList` in the profiles `small` and `medium`, `NodeCachingLinkedList` and `FastList`, respectively, consuming less energy than `ArrayList` for these two operations. On the other hand, the profile `big` did not have a single implementation that had lower consumption for these operations, with `ArrayList` outperforming all the other implementations. Nevertheless, the changes made by CT+ on BIOJAVA resulted in an improvement in energy efficiency across all profiles.

There is dominance between collections implementations Out of the 39 possible implementations available to CT+ only 20 were recommended. When trying to understand this behavior, we observed that some collection implementations consistently dominate [46] others. An implementation C_1 dominates implementation C_2 when every operation in the former consumes less energy, on average, than the same operation in the latter. In this case, the dominated collection is never recommended by CT+. Among all implementations, `ConcurrentHashMap` shows a particular behavior that is worth mentioning. That implementation was changed 26 times on the desktop environment, 25 out of 26 cases for `ConcurrentHashMap(EC)`. Nevertheless, `ConcurrentHashMap` was also recommended 47 times, always replacing `Hashtable`. This illustrates that if an implementation is not dominated by another, there will be cases where it may still perform better.

Energy profile creation is not trivial During our experiments, we noticed that some factors could make it infeasible to create profiles at a larger scale. Due to the enormous variance in the execution times of operations, the original process [19] of creating the energy profiles can take a long time, i.e., hours for desktop devices and days for mobile devices. To reduce this time we used two approaches. First, we executed each operation three times, and measured and collected the energy consumption of those operations. In the cases where a relation of dominance was found, the dominated collection was not included as an option for recommendation. Second, we delimited a threshold based on how long each operation could run. This threshold was based on the fastest operation among all implementations in a specific group (e.g., `insert(start)` for thread-safe `Lists`). Very expensive operations were discarded if they spent more time than our threshold. In our experiments, this threshold was set at two orders of magnitude, i.e., 100 times the average time of the fastest alternative.

6.5 Energy Profiling in the Wild

Addressing energy efficiency within mobile devices is particularly relevant as these devices have become one of our most used gadgets, and most often run powered by batteries. As a consequence, battery life is a high-priority concern for users and one of the major factors influencing consumer satisfaction [5, 6]. On the other hand, battery life is also important for app developers, as excessive battery consumption is one of the most common causes for bad app reviews in app stores [47, 48].

As we have already witnessed in the previous section regarding the choice of data structures, developer decisions can directly impact the energy consumption of a mobile application (or simply “app”). In general, when considering other factors such as location services [49], programming languages [18], color Schemes [9, 50], or code refactorings [51], the use of one available solution over another can have a non-negligible effect on energy consumption.

Keeping energy usage to a minimum is so important for app developers that IDEs for the most popular smartphone platforms include energy profilers. However, profiling for energy within mobile environments is a particularly difficult problem, and especially within Android, the mobile platform with the largest market share, and by a big margin.[14] Indeed, Android is a highly heterogeneous platform: in 2015 there were already more than 24,000 Android device models available,[15] and a recent study found that there are more than 2.5 million apps in the Google Play Store.[16] In addition, the Android operating system is currently in its tenth major release, with multiple minor releases throughout the years. These numbers combined with the different ways in which apps and devices are used produce a virtually infinite number of potential usage scenarios.

In this context, profiling for energy consumption has only limited applicability. Alternatively, one needs to obtain large-scale information about energy use in real usage scenarios to make informed, effective decisions about energy optimization. In this section, we describe how we leverage crowdsourcing to collect information about energy in real-world usage scenarios. We introduce the GreenHub initiative, https://greenhubproject.org/, which aims to promote collaboration as a path to produce the best energy-saving solutions. The most visible outcome of the initiative is a large dataset, called Farmer, that reflects *in the wild*, real-world usage of Android devices [17].

The entries in Farmer include multiple pieces of information such as active sensors, memory usage, battery voltage and temperature, running applications, model and manufacturer, and network details. This raw data was obtained by continuous crowdsourcing through a mobile application called BatteryHub. The collected data is strictly, and by construction, anonymous so as to ensure the privacy of all the app users. Indeed, it is impossible to associate any data with the user who originated it. The data collected by BatteryHub is then uploaded to a remote server, where it is made publicly available to be used by third parties in research on improving the energy efficiency of apps, infrastructure software, and devices.

In order to foster the involvement of the community, Farmer is available for download in raw format and can be accessed by means of a backend web app that provides an overview of the data and makes it available through a REST API. The dataset can also be queried by means of Lumberjack, a command-line tool for interacting with the REST API.

Within the GreenHub initiative, we have so far been able to collect a dataset which is *sizable*. Thus far it comprises 48+ million unique samples. The dataset is also *diverse*. It includes data stemming from 2.5 k + different brands, 15 k + smartphone models, from over 73 Android versions, across 211 countries.

In the remainder of this section, we describe in detail the alternatives that we have implemented to allow the community to access the data within our dataset. The main motivation for the section is to foster the engagement of the community in exploring the data we are providing, in this way contributing to increasing the knowledge on how energy-saving strategies can be realized within Android devices. We therefore

[14] https://gs.statcounter.com/os-market-share/mobile/worldwide

[15] https://www.zdnet.com/article/android-fragmentation-there-are-now-24000-devices-from-1300-brands/

[16] https://www.statista.com/statistics/276623/number-of-apps-available-in-leading-app-stores/

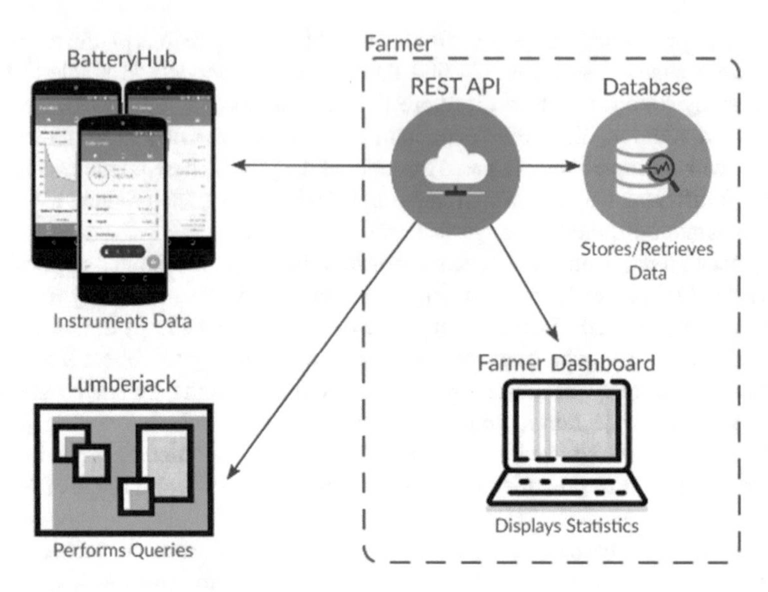

Fig. 6.6 GreenHub platform architecture

expect that this section will be particularly interesting for researchers and/or practitioners focusing on mobile development, and who have a data mining mindset.

6.5.1 A Collaborative Approach to Android Energy Consumption Optimization

The success of GreenHub is dependent on its data, and to keep such data coming in we plan to give back to the community in concrete and valuable ways. In this section, we focus on the ways that the community can access the data we are collecting.

The initiative relies on an open-source technological platform,[17] whose architecture overview is shown in Fig. 6.6. This platform includes our data collection Android app called BatteryHub, a command-line application interface called Lumberjack, and the Farmer REST API for prototyping queries, dashboard interface, and database for storing data. These components are further defined in the following sub-sections.

Data Collection is provided by BatteryHub, an Android app whose development was inspired by Carat [52]. Initially, we forked Carat's open-source code to take advantage of the data collection and storage mechanisms. On top of that, we updated its data model to consider more details on modern devices, such as NFC and Flashlight usage. In the same spirit as Carat, BatteryHub is entirely open-source.

[17]https://github.com/greenhub-project

In contrast with Carat, however, all our collected data is permanently and publicly available, so as to strongly encourage and help others in collaborating, inspecting, and/or reusing any artifact that we have developed or collected.

BatteryHub is available at Google's Play Store,[18] and tracks the broadcast of system events, such as changes to the battery's state, and, when such an event occurs, obtains a sample of the device's current state. BatteryHub either uses the official Android SDK or custom implementations for universal device compatibility support, and periodically communicates with the server application (over HTTP) to upload, and afterwards remove, the locally stored samples. Each sample characterizes a wide range of aspects that may affect battery usage, such as sensor usage, temperature, and the list of running applications.

It is important to mention that the data collected from each user is made anonymous by design. Each installation of BatteryHub is associated with a random unique identifier and no personal information, such as phone number, location, or IMEI, is collected. This means that it is (strictly) not possible to identify any BatteryHub user, nor is it possible to associate any data with the user from whom it originates.

In regard to sample collection frequency, a new data measurement is collected (to be sent to our server) when the battery's state changes. In most cases, this translates to a sample being sent at each 1% battery change (which accounts for 95% of the time according to our data). The app allows for configurable alerts, e.g., when the battery reaches a certain temperature. Our overarching goal is to use BatteryHub to give suggestions to users, based on their usage profiles, on how to reduce the energy consumption of their device.

Besides BatteryHub, our infrastructure includes four additional components, as depicted in Fig. 6.6. We envision that they can be used in different stages of mining our dataset, which are described in the remainder of this section. Finally, our infrastructure also includes a web dashboard interface[19] that provides access to up-to-date statistics about the collected samples.

Fast prototyping of queries can be made by using Farmer's REST API, which was designed as a means to quickly interface with and explore the dataset. As every request made to the API must be authenticated, users must first obtain an API key in order to access the data in this fashion.[20] The API provides real-time, selective access to the dataset and one may query, e.g., all samples for a given brand or OS version. Since the API is designed according to the REST methodology, this allows us to incrementally add new data models to be reflected within the API itself as the data protocol evolves over time. After an API key has been successfully generated, one may request his/her own user profile from the API:

```
farmer.greenhubproject.org/api/v1/me?api_token=yourTokenHere
```

[18]https://play.google.com/store/apps/details?id=com.hmatalonga.greenhub

[19]https://farmer.greenhubproject.org/

[20]https://docs.greenhubproject.org/api/getting-started.html

Every successful API response is a JSON formatted document, and in this case the server will reply with the user details, as shown next.


```
{ "data": {
 "id": XX,
 "name": "Your Name",
 "email": "your@email.com",
 "email_verified_at": "YYYY-MM-DD HH:MM:SS",
 "created_at": "Mon. DD, YYYY",
 "updated_at": "YYYY-MM-DD HH:MM:SS",
 "roles": [ ... ] } }
```


It is now possible to use the API, for example, to list devices:

```
farmer.greenhubproject.org/api/v1/devices?api_token=yourToken
```

This request can take additional parameters for *page* and devices *per page*. A full description of all the available parameters for each request can be found in the API Reference.[21]

The expected response from the request above is as follows:


```
{ "data": [
  { "brand": "asus",
    "created_at": "2017-10-28 02:51:09",
    "id": 2518,
    "is_root": false,
    "kernel_version": "3.1835+",
    "manufacturer": "asus",
    "model": "ASUS_X008D",
    "product": "WW_Phone",
    "os_version": "7.0",
    "updated_at": "2017-10-28 02:51:09"
  }, ... ],
 "links": { ... },
"meta": { ... } }
```


To obtain more detailed information, e.g., about a particular device whose identifier is 123, it is possible to request its samples:

```
farmer.greenhubproject.org/api/v1/devices/123/samples?
api_token=yourToken
```

A complementary approach to interface with the API is to use its command-line application interface, Lumberjack. Using this tool, users can perform flexible, on-demand queries to the data repository, to support quick prototyping of data queries applying different filters and parameters. Furthermore, users can quickly

[21] https://docs.greenhubproject.org/api-reference/

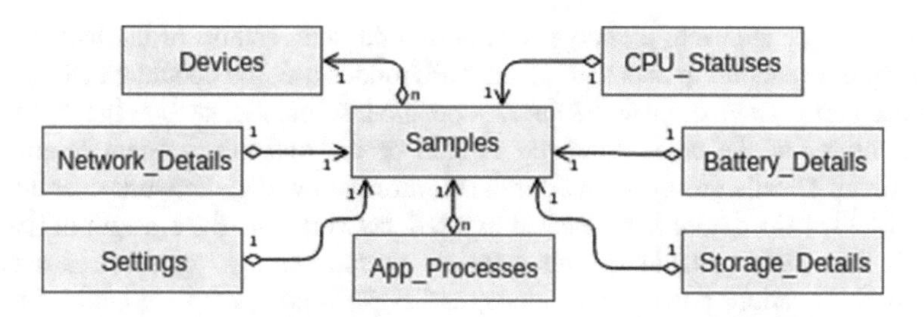

Fig. 6.7 Entity relationship diagram of the dataset

fetch subsets of the data without the need to download a snapshot of the entire dataset. The following is an example of a Lumberjack query to obtain the list of Google brand devices:

```
$ greenhub lumberjack devices brand:google -o googleDevices.json
```

The following example queries the dataset for samples whose model is nexus and that were uploaded before May 31, 2018:

```
$ greenhub lumberjack samples model:nexus -R ..2018-05-31
```

Extensive Mining can be conducted from the samples we collected, which are accessible through Farmer. The dataset is available as a zip archive file, in CSV format,[22] and also in Parquet[23] binary format,[24] which can be analyzed more efficiently than a plain text dump. The dataset is also available as a MariaDB relational database. The samples sent by BatteryHub are queued to be processed by a PHP server application built using the Laravel framework.[25] Each sample is received as a JSON formatted string that is deconstructed and correctly mapped within the database.

The (simplified) data model that we employ is shown in Fig. 6.7, where each box represents a table (or a CSV file) in the dataset. Samples is the most important of them, including multiple features of varied nature, e.g., the unique sample id, the timestamp for each sample, the state of the battery (charging or discharging), the level of charge of the battery, whether the screen was on or not, and the free memory on the device.

App_Processes is the largest among the tables of the dataset, containing information about each running process in the device at the time the sample was collected, e.g., whether it was a service or an app running in the foreground, its name, and version. Battery Details provides battery-related information such as whether the

[22]https://farmer.greenhubproject.org/storage/dataset.7z

[23]http://parquet.apache.org/

[24]https://farmer.greenhubproject.org/storage/dataset.parquet.7z

[25]https://laravel.com/

device was plugged into a charger or not and the temperature of the battery. Cpu Statuses indicates the percentage of the CPU under use, the accumulated up time, and sleep time. Devices provides device-specific information, such as the model and manufacturer of the device and the version of the operating system running on it. Network Details groups network-related information, e.g., network operator and type, whether the device is connected to a wifi network, and the strength of the wifi signal. The Settings table records multiple yes/no settings for services such as Bluetooth, location, power saver mode, and NFC, among others. Finally, Storage Details provides multiple features related to the secondary storage of the device.

6.6 Conclusion

Although developers have recently become more aware of the importance of creating energy-efficient software systems, they are still missing important knowledge and tools to help them achieve their goals. In summary, with this chapter we hope to show the following relevant aspects of green development.

Small changes can make a big difference in terms of energy consumption, especially in mobile devices. Throughout this chapter we discussed the importance of diversity and the energy consumption of different pieces of software, such as I/O APIs, concurrent mechanisms, and Java collections. For every one of these constructs, performing modifications to employ diversely designed, more energy-efficient versions resulted in a reduction in energy consumption, with some cases only requiring changing a single line of code. Focusing on these energy variation hotspots can greatly reduce the complexity of improving the energy efficiency of an application, making it viable for non-specialists to enhance their systems. Knowing this, developers and researchers can focus on easily exchangeable, power-hungry aspects of the software to reduce the energy consumption of an application with minimum effort.

Device variability is a real problem when experimenting on energy consumption. Similar to performance, energy efficiency can be impacted by a number of factors, at different abstraction levels, including factors that are not obvious for non-specialists. Factors such as a device's battery's age or the room temperature can have a significant influence on the energy consumption. Even worse, when dealing with different devices, every decision made by the manufacturer can change the energy efficiency of the device and turn a seemingly sound experiment into conclusions that only work in specific cases. Knowing this, developers and researchers can try to mitigate this factor by executing their experiments on a bigger pool of devices, greatly reducing this bias factor in their results. In case only a single device is available, its characteristics should be fully presented so other researchers/developers can have a clear understanding of the results and in which devices they would be relevant.

Crowdsourcing can be used to see the big picture of energy consumption. Although very important for several factors, such as studying particular cases and learning about energy consumption behavior, controlled experiments on mobile devices hardly extrapolate to the whole environment. Even when dealing with just the Android OS, there is so much variability (e.g., OS versions, manufacturers, device versions) that no controlled experiment will be truly general enough to cover a significant number of devices and thus have widespread applicably. Crowdsourcing can be a way to mitigate this by using the data provided by the users, leveraging thousands of different devices to achieve a truly panoramic view of the ecosystem as a whole. The biggest problem with the crowdsourcing solution is that it depends on end-users for the energy samples. Convincing them to provide their data and, even more importantly, getting enough of it to be statistically relevant can be arduous. We propose that developers and researchers who want to investigate the energy behavior of the mobile devices in the Android environment should use the GreenHub initiative, an already well-established database with millions of data-points.

References

1. Gelenbe E, Caseau Y (2015) The impact of information technology on energy consumption and carbon emissions. Ubiquity, 2015 (June)
2. Coroama V, Hilty LM (2009) Energy consumed vs. energy saved by ICT – a closer look. In: Wohlgemuth V, Page B, Voigt K (eds) Environmental informatics and industrial environmental protection: concepts, methods and tools. Shaker Verlag, Aachen
3. Andrae A, Edler T (2015) On global electricity usage of communication technology: trends to 2030. Challenges 6(1):117–157
4. Andrews RNL, Johnson E (2016) Energy use, behavioral change, and business organizations: Reviewing recent findings and proposing a future research agenda. Energy Res Soc Sci 11:195–208
5. Richter F. The most wanted smartphone features. https://www.statista.com/chart/5995/the-most-wanted-smartphone-features. Accessed 24 Jan 2018
6. Thorwart A, O'Neill D (2017) Camera and battery features continue to drive consumer satisfaction of smartphones in US. https://www.prnewswire.com/news-releases/camera-and-battery-features-continue-to-drive-consumer-satisfaction-of-smartphones-in-us-300466220.html Last visit: 2019-02-06
7. Hindle A (2012) Green mining: a methodology of relating software change to power consumption. In: 9th IEEE working conference on Mining Software Repositories (MSR), June 2012, pp 78–87
8. Di Nucci D, Palomba F, Prota A, Panichella A, Zaidman A, De Lucia A (2017) Software-based energy profiling of Android apps: simple, efficient and reliable? In 2017 IEEE 24th international conference on software analysis, evolution and reengineering (SANER), pp 103–114
9. Linares-Vásquez M, Bavota G, Bernal-Cárdenas C, Di Penta M, Oliveto R, Poshyvanyk D (2018) Multi-objective optimization of energy consumption of GUIs in android apps. ACM Trans Softw Eng Methodol 27(3):14:1–14:47
10. Li D, Lyu Y, Gui J, Halfond WGJ (2016) Automated energy optimization of HTTP requests for mobile applications. In Dillon LK, Visser W, Williams L (eds) Proceedings of the 38th international conference on software engineering, ICSE 2016, ACM, Austin, TX, May 14–22, 2016, pp 249–260
11. Avizienis A, Kelly JPJ (1984) Fault tolerance by design diversity: concepts and experiments. IEEE Comp 17(8):67–80

12. Randell B (1975) System structure for software fault tolerance. IEEE Trans Softw Eng 1 (2):221–232
13. Rustan Leino K (2017) Accessible software verification with Dafny. IEEE Softw 34(6):94–97
14. Dean J, Barroso LA (2013) The tail at scale. Commun ACM 56(2):74–80
15. Baldwin CY, Clark KB (2000) Design rules, vol 1: the power of modularity. MIT Press
16. Lima LG, Soares-Neto F, Lieuthier P, Castor F, Melfe G, Fernandes JP (2019) On haskell and energy efficiency. J Syst Softw 149:554–580
17. Matalonga H, Cabral B, Castor F, Couto M, Pereira R, de Sousa SM, Fernandes JP (2019) GreenHub farmer: real-world data for android energy mining. In 2019 IEEE/ACM 16th international conference on mining software repositories (MSR), pp 171–175. IEEE
18. Oliveira W, Oliveira R, Castor F (2017) A study on the energy consumption of android app development approaches. In 2017 IEEE/ACM 14th international conference on mining software repositories (MSR)
19. Oliveira W, Oliveira R, Castor F, Fernandes B, Pinto G (2019) Recommending energy-efficient java collections. In 2019 16th international conference on mining software repositories (MSR), pp 160–170
20. Pinto G, Liu K, Castor F, Liu YD (2016) A comprehensive study on the energy efficiency of java thread-safe collections. In ICSME, 2016
21. Rocha G, Castor F, Pinto G (2019) Comprehending energy behaviors of Java I/O APIs. In 2019 ACM/IEEE international symposium on empirical software engineering and measurement (ESEM), pp 1–12. IEEE
22. Blackburn SM, Garner R, Hoffmann C, Khang AM, McKinley KS, Bentzur R, Diwan A, Feinberg D, Frampton D, Guyer SZ, Hirzel M, Hosking A, Jump M, Lee H, Moss JEB, Phansalkar A, Stefanovic D, VanDrunen T, von Dincklage D, Wiedermann B (2006) The dacapo benchmarks: Java benchmarking development and analysis. In Proceedings of the 21st annual ACM SIGPLAN conference on object-oriented programming systems, languages, and applications, OOPSLA '06, ACM, New York, NY, pp 169–190
23. Kwon Y-W, Tilevich E (2013) Reducing the energy consumption of mobile applications behind the scenes. In 2013 IEEE international conference on software maintenance, IEEE Computer Society, Eindhoven, September 22–28, pp 170–179
24. Pinto G, Castor F, Liu YD (2014) Understanding energy behaviors of thread management constructs. In Proceedings of the 2014 ACM international conference on object oriented programming systems languages and applications, OOPSLA '14, pp 345–360
25. Liu K, Pinto G, Liu D (2015) Data-oriented characterization of application-level energy optimization. In Proceedings of the 18th international conference on fundamental approaches to software engineering, FASE'15
26. Chowdhury SA, Sapra V, Hindle A (2016) Client-side energy efficiency of HTTP/2 for web and mobile app developers. In IEEE 23rd international conference on software analysis, evolution, and reengineering, SANER 2016, Suita, Osaka, March 14–18, 2016, vol 1. IEEE Computer Society, pp 529–540
27. Manotas I, Bird C, Zhang R, Shepherd DC, Jaspan C, Sadowski C, Pollock LL, Clause J (2016) An empirical study of practitioners' perspectives on green software engineering. In Proceedings of the 38th international conference on software engineering, ICSE 2016, Austin, TX, May 14–22, 2016, pp 237–248. ACM
28. Pinto G, Castor F, Liu YD (2014) Mining questions about software energy consumption. In Proceedings of the 11th working conference on mining software repositories, MSR 2014, pp 22–31
29. David H, Gorbatov E, Hanebutte UR, Khanna R, Le C (2010) Rapl: memory power estimation and capping. In 2010 ACM/IEEE international symposium on low-power electronics and design (ISLPED), pp 189–194
30. Di Nucci D, Palomba F, Prota A, Panichella A, Zaidman A, De Lucia A (2017) Petra: a software-based tool for estimating the energy profile of android applications. In 2017 IEEE/ACM 39th international conference on software engineering companion (ICSE-C), pp 3–6

31. Gao X, Liu D, Liu D, Wang H, Stavrou A (2017) E-Android: a new energy profiling tool for smartphones. In 2017 IEEE 37th international conference on distributed computing systems (ICDCS), pp 492–502
32. Lyu Y, Gui J, Wan M, Halfond WGJ (2017) An empirical study of local database usage in android applications. In Proceedings of the international conference on software maintenance and evolution (ICSME), Sept 2017
33. Dyer R, Nguyen HA, Rajan H, Nguyen TN (2015) Boa: Ultralarge-scale software repository and source-code mining. ACM Trans Softw Eng Methodol 25(1):7:1–7:34
34. Hasan S, King Z, Hafiz M, Sayagh M, Adams B, Hindle A (2016) Energy profiles of java collections classes. In Proceedings of the 38th international conference on software engineering, New York, NY, pp 225–236
35. Manotas I, Pollock L, Clause J (2014) Seeds: a software engineer's energy-optimization decision support framework. In Proceedings of the 36th international conference on software engineering, ICSE 2014, pp 503–514
36. Pereira R, Couto M, Saraiva J, Cunha J, Fernandes JP (2016) The influence of the java collection framework on overall energy consumption. In Proceedings of the 5th international workshop on green and sustainable software, GREENS '16, pp 15–21, ACM, New York, NY
37. Trefethen AE, Thiyagalingam J (2013) Energy-aware software: challenges, opportunities and strategies. J Comput Sci 4(6):444–449
38. Shavit N, Touitou D (1997) Software transactional memory. Distributed Comput 10(2):99–116
39. O'Sullivan B (2009) Criterion: robust, reliable performance measurement and analysis. http://www.serpentine.com/criterion/. Last access 22 Jan 2019
40. Georgiou S, Spinellis D (2020) Energy-delay investigation of remote inter-process communication technologies. J Syst Softw 162:110506
41. Pereira R, Couto M, Ribeiro F, Rua R, Cunha J, Fernandes JP, Saraiva J (2017) Energy efficiency across programming languages: How do energy, time, and memory relate? In Proceedings of the 10th ACM SIGPLAN international conference on software language engineering, SLE 2017, pp 256–267, ACM, New York, NY
42. Aggarwal K, Zhang C, Campbell JC, Hindle A, Stroulia E (2014) The power of system call traces: predicting the software energy consumption impact of changes. In Proceedings of 24th annual international conference on computer science and software engineering, CASCON 2014, pp 219–233. IBM/ACM
43. Li D, Tran AH, Halfond WGJ (2014) Making web applications more energy efficient for OLED smartphones. In 36th international conference on software engineering (ICSE '2014), ACM, pp 527–538
44. Linares-Vásquez M, Bavota G, Bernal Cárdenas CE, Oliveto R, Di Penta M, Poshyvanyk D (2015) Optimizing energy consumption of GUIs in android apps: a multi-objective approach. In Proceedings of the 2015 10th joint meeting on foundations of software engineering, ESEC/FSE 2015, pp 143–154, ACM, New York, NY
45. Mcintosh A, Hassan S, Hindle A (2019) What can android mobile app developers do about the energy consumption of machine learning? Empirical Softw Eng 24(2):562–601
46. Peterson M (2009) Decisions under ignorance, pp 40–63. Cambridge introductions to philosophy. Cambridge University Press
47. Fu B, Lin J, Li L, Faloutsos C, Hong J, Sadeh N (2013) Why people hate your app: Making sense of user feedback in a mobile app store. In Proceedings of the 19th ACM SIGKDD international conference on knowledge discovery and data mining. ACM, pp 1276–1284
48. Khalid H, Shihab E, Nagappan M, Hassan AE (2015) What do mobile app users complain about? IEEE Softw 32(3):70–77

49. Lin K, Kansal A, Lymberopoulos D, Zhao F (2010) Energy-accuracy trade-off for continuous mobile device location. In Proceedings of the 8th international conference on Mobile systems, applications, and services. ACM, pp 285–298
50. Wan M, Jin Y, Li D, Gui J, Mahajan S, Halfond WGJ (2017) Detecting display energy hotspots in android apps. Softw Test Verification Reliab 27(6):16–35
51. Couto M, Saraiva J, Fernandes JP (2020) Energy refactorings for android in the large and in the wild. In Proceedings of the IEEE 27th international conference on software analysis, evolution and reengineering (SANER '20), pp 217–228
52. Oliner AJ, Iyer AP, Stoica I, Lagerspetz E, Tarkoma S (2013) Carat: collaborative energy diagnosis for mobile devices. In Proceedings of the 11th ACM conference on embedded networked sensor systems, SenSys '13, Roma, November 11–15, 2013, pp 10:1–10:14. ACM

Chapter 7
Tool Support for Green Android Development

Hina Anwar, Iffat Fatima, Dietmar Pfahl, and Usman Qamar

Abstract Mobile applications are developed with limited battery resources in mind. To build energy-efficient mobile apps, many support tools have been developed which aid developers during the development and maintenance phases. To understand what is already available and what is still needed to support green Android development, we conducted a systematic mapping study to overview the state of the art and to identify further research opportunities. After applying inclusion/exclusion and quality criteria, we identified tools for detecting/refactoring code smells/energy bugs, and for detecting/migrating third-party libraries in Android applications. The main contributions of this study are: (1) classification of identified tools based on the support they offer to aid green Android development, (2) classification of the identified tools based on techniques used to offer support to developers, and (3) characterization of the identified tools based on the user interface, IDE integration, and availability. The most important finding is that the tools for detecting/migrating third-party libraries in Android development do not provide support to developers to optimize code w.r.t. energy consumption, which merits further research.

7.1 Introduction

Global warming due to CO_2 emissions has been one of the most prominent environmental issues in the past decade. A part of these CO_2 emissions is contributed by the information and communication technology (ICT) industry [1]. Therefore, producing green or sustainable products and practices has been the focus of many researchers in the ICT community. Recently, however, the focus of research in the

H. Anwar (✉) · D. Pfahl
Institute of Computer Science, University of Tartu, Tartu, Estonia
e-mail: hina.anwar@ut.ee; dietmar.pfahl@ut.ee

I. Fatima · U. Qamar
College of Electrical and Mechanical Engineering, National University of Sciences and Technology, Islamabad, Pakistan
e-mail: iffat.fatima@ce.ceme.edu.pk; usmanq@ceme.nust.edu.pk

C. Calero et al. (eds.), *Software Sustainability*,
https://doi.org/10.1007/978-3-030-69970-3_7

ICT community has shifted from optimizing the energy consumption of hardware to optimizing the energy consumption of software [2–8], as software indirectly consumes energy by controlling the equipment. An efficiently designed software might use resources optimally, thus reducing energy consumption [9–11]. Among portable devices, mobile phones are the most commonly used. Statistics show that the usage of mobile devices will grow in the coming years [12], indicating an increase in the carbon footprint.

Green software development encompasses green by software and green in software. Green by software means using software products to make other domains of life more sustainable. Green in software refers to the study and practice of designing, developing, maintaining, and disposing of software products in such a way that they have a minimal negative impact on the environment, community, economy, individuals, and technology [13, 14].

This chapter mostly focuses on green in software and summarizes the tool support available to improve the green-ability of Android apps in the development and maintenance phases. The term "Android development" refers to the development of applications that are developed to operate on devices running the Android operating system. These applications can be developed in various languages; however, in this chapter, we focus on Android development in Java. Android development differs from traditional software development in terms of context, user experience, and a touch-based interface. Android applications are designed for portable devices, which have limited resources such as memory or battery. A common struggle during Android application development is how to make the applications efficient in terms of resource usage. Banarjee et al. summarize the problem nicely as follows: "High computational power coupled with small battery capacity and the application development in an energy-oblivious fashion can only lead to one situation: short battery life and an unsatisfied user base" [15].

Previous studies have explored applications in app stores in order to define procedures to optimize their energy consumption [16–23]. Some studies have focused on profiling energy [24–28] consumed by applications, while others have developed support tools [29, 30]. As compared to desktop or web applications, Android applications contain multiple components that have user-driven workflows. A typical Android application consists of activities, fragments, services, content providers, and broadcast receivers. Due to the difference in architecture, the support tools used in the development of traditional Java-based applications are not so useful in Android application development and maintenance. Android application code can be roughly divided into two part: custom code and reusable code. While custom code is unique to each app, reusable code includes third-party libraries that are included in apps to speed up the development process.

In the domain of Android application development, research has been focused on development activities related to energy efficiency, memory usage, performance, etc., and maintenance activities related to code smell detection and correction, energy bug detection and correction, detection/migration of third-party libraries, etc.

Code smells are an indication of possible problems in source code or design of the applications. Such problems can be avoided by refactoring the code [31]. However, object-oriented code smells are different from Android-specific code smells. In Android development, code smell can appear due to frequent development and update cycles of applications. Some studies [32–34] have focused on identifying, cataloguing, and profiling the energy consumption of Android-specific code smells. Energy bugs are scenarios which cause unexpected energy drains such as preventing the mobile device from going into the idle state even after the application execution has completed. Such malfunctioning can cause battery drain and should be avoided [15]. To build an energy-efficient Android application developers need to identify and refactor code smells/energy bugs.

Third-party libraries are reusable components available to implement various functionalities in the app, such as billing, advertisement, and networking. Up until June 2020, the online Maven repository[1] contained 344,869 unique libraries. Such a huge supply of third-party libraries is linked to the demands and needs of developers [35]. Almost 60% of code in Android applications is related to third-party libraries [36]. However, these libraries could introduce various security-, privacy-, permission-, and resource usage-related issues in applications [37]. The research on the detection/migration of third-party libraries has many uses. Some studies have used third-party library detection techniques for finding security vulnerabilities [38–41] in Android apps, while others have focused on privacy leaks [42–46]. Third-party libraries have been detected and removed as noise in clone, app repackage, and malicious app detection studies [47–51]. Third-party libraries are detected and removed from these studies in order to improve the accuracy of the analysis. Studies related to the energy impact of third-party Android libraries are limited [52].

In order to build effective Android-specific support tools to aid green Android development, we first need to understand what is already available, what is still needed, and how the problems in existing tools can be overcome. Based on published literature we outlined an explorative analysis of support tools available to (1) optimize code in Android applications through code smell detection/refactoring, and (2) optimize reusable code in Android applications through detection/migration of third-party libraries. This study extends our previous work [53] comparing 21 tools in the following ways: we have improved search string and extended our analysis for one more year, which gave us 30 more tools and also additional results. We provide further information about the interface, availability, and integrated development environment (IDE) integration of all 51 tools.

The remainder of this chapter is organized as follows. Section 7.2 presents related work. Section 7.3 describes the methodology used to analyze the literature. Section 7.4 presents the result of screening the publications and classification and analysis. Section 7.5 provides a discussion to identify future research directions. Section 7.6 provides possible threats to the validity of this study. Section 7.7 concludes the chapter by summarizing the main findings.

[1]https://search.maven.org/stats, statistics for central repository

7.2 Related Work

Secondary studies related to energy efficiency in Android development are scarce. Some [54–56] have reviewed tools and techniques for improving the quality of Java projects in the object-oriented paradigm (with regard to performance or maintainability).

Most Android projects use Java as the programing language; however, the support tools and techniques used for Java projects reviewed by previous secondary studies [54–56] cannot be effectively applied to Android projects. Therefore, many specialized support tools have been developed to improve the quality of Android apps with regard to maintainability, performance, security, or energy. Li et al. [57] performed a systematic literature review to analyze static source code analysis techniques and tools proposed for Android to assess issues related to security, performance, or energy. The authors have reviewed work published between 2011 and 2015, consisting of 124 studies. The review concluded that the majority of static analysis techniques only uncover security flaws in Android apps. Degu A [58]. performed a systematic literature review to classify primary studies with a focus on resource usage, energy consumption, and performance in Android apps. The classification is high level based on the main research focus, type of contribution, and type of evaluation method adopted in selected studies. Their results did not provide an in-depth review of support tools in green Android development.

Another group of studies has compared the state-of-the-art tools through experiments in order to benchmark their performance, accuracy, and reporting capabilities. Qiu et al. [59] provide a comparison between three static analysis tools: FlowDroid/IccTA, Amandroid, and Droidsafe. They evaluated these tools using a common configuration setup and the same set of benchmark applications. Results were compared to those of previous studies in order to identify reasons for inaccuracy in existing tools. Corrodi et al. [60] review the state-of-the-art in Android data leak detection tools. Out of 87 state-of-the-art tools, they executed five based on availability. They compared these five tools against a set of known vulnerabilities and discussed the overall performance of the tools. Ndagi and Alhassan [61] provide a comparison of machine classifiers for detecting phishing attacks in Adware in Android applications. This study concluded that many existing machine classifiers, if adequately explored, could yield more accurate results for phishing detection.

Another group of studies has focused on reviewing the technique related to security, malware, similarity, and repackaging in Android apps. Cooper et al. [62] provide an overview of security threats posed by Android malware. They also survey some common defense techniques to mitigate the impact of malware applications and characteristics that are commonly found in malware applications that could enable detection techniques. Li et al. [63] provide a literature review that summarizes the challenges and current solutions for detecting repackaged apps. They concluded that many existing solutions merit further research as they are tested on closed datasets and might not be as efficient or accurate as they claim to be. Roy et al.

[64] provide a qualitative comparison of clone detection techniques and tools. They classify, compare, and evaluate these tools.

We found some studies that have conducted controlled experiments to measure the energy consumption of third-party libraries. For example, Wang et al. [65] presented an algorithmic solution to model the energy minimization problem for ad prefetching in Android apps. Rasmussen et al. [66] conducted a study to compare the power efficiency of various methods of blocking advertisements on an Android platform. They found many cases where ad-blocking software or methods resulted in increased power usage. In Android applications, there could be many reasons for long-running operations in the background that continuously consume resources. Such operations could cause battery drain and performance degradation. Shao et al. [67] demonstrated through an experiment that sometimes such behavior is not intentional and is caused by third-party libraries.

However, we could not find any secondary study that provides an overview of the state of the art w.r.t. to support tools available for detecting/migrating third-party libraries in Android apps. To the best of our knowledge, none of the previous secondary studies has reviewed the literature from the point of view of support tools developed to aid green Android development. Most of the secondary studies discussed above have covered published work until 2015 or 2017 in the object-oriented paradigm, and many of the reviewed tools in these studies are now outdated/obsolete. Therefore, in this study we provide a different view of the literature by analyzing recently developed support tools for energy profiling, code optimization, refactoring, and third-party library detection or migration in Android development to improve energy efficiency in apps. We explore whether these support tools aid green Android development. We also provide an overview of the techniques used in these support tools.

7.3 Methodology

We conducted a systematic mapping study following the method described in [68]. First, we formulated research questions, and then based on those research questions we formulated two general search queries and conducted the search in the following online repositories for primary publications: IEEE Xplore, ACM digital library, Science Direct, and Springer. In this study, we cover publications from 2014 to June 2020, as from 2014 onwards the focus of many publications has been Android and energy-efficient app development, indicating a shift in research focus.

7.3.1 Research Questions

As the objective of this study is to analyze the current support tools available to improve custom code through detection/refactoring of code smell/energy bugs and to improve reusable code through detection/migration of third-party libraries in Android applications, we formulated the following research questions.

RQ1: *What state-of-the-art support tools have been developed to aid software practitioners in detecting/refactoring code smells/energy bugs in Android apps?*
RQ2: *What state-of-the-art support tools have been developed to aid software practitioners in detecting/migrating third-party libraries in Android apps?*
RQ3: *How do existing support tools compare to one another in terms of techniques they use for offering the support?*
RQ4: *How do existing support tools compare to one another in terms of the support they offer to practitioners for improving energy efficiency in Android apps?*

RQ1and RQ2 aim to classify publications based on the tools they offer. RQ3 aims to classify and analyze publications based on techniques used in the tool to offer support to developers. RQ4 deals with the characterization of all the identified tools in terms of the support (such as output or interface or availability) they offer to developers to aid green Android development.

7.3.2 Search Query

We derived search terms to use in our search query from the research questions of this study. We looked for alternatives to the search terms in publications we already knew and refined our search terms to return the most relevant publications. We used the "*" operator to cover possible variations on the selected search terms in the search query. The keyword "OR" was used to improve search coverage.

Based on our previous work [53], we improved our search query and extended our search in terms of publication years to include one more year. The first search query is designed to retrieve publications that provide a support tool to detect/refactor code smells/energy bugs in Android apps. The second search query is designed to retrieve publications that provide a support tool to detect/migrate third-party libraries in Android apps.

We did not use the search terms "mobile development," "apps," "optimization," "green," "sustainability," and "recommendation" in isolation as they were too high level and produced quite a large corpus consisting of a high number of irrelevant publications, while the search terms "resource leaks," "API," "tool," "framework," and "technique" were eliminated to avoid being too specific. The search queries were applied to popular online repositories (IEEE Xplore, ACM digital library, Science Direct, and Springer) to find a dataset of relevant primary publications. In each repository, based on available advanced search options, filters were applied to refine

Table 7.1 Search query filter

Filter	Value
Publication year	2014–2020 (up until June)
Content-type	Journal Article, Conference Paper

the query results. Applied filters are shown in Table 7.1. The search queries were applied to the titles, abstracts, and keywords of the publications.

Search Query 1
Android AND (energy OR code smell OR bug OR refactor OR correct* OR detect* OR optimiz* OR efficien*) AND NOT (environ* OR iot OR edu* OR hardware OR home)*

Search Query 2
(Android) AND ("third-party libr" OR "third-party Android lib*" OR "libr*") AND NOT (environ* OR iot OR indus* OR edu* OR hardware OR home)*

7.3.3 Screening of Publications

We first removed duplicate results and then defined inclusion, exclusion, and quality criteria for further screening of search results.

7.3.3.1 Duplicate Removal

The search results from online repositories were first loaded in Zotero[2] (an open-source reference management system) to create a dataset of relevant publications. Using the feature in Zotero, duplicate publications were removed from the dataset. Next, we manually applied inclusion, exclusion, and quality criteria to the remaining publications.

7.3.3.2 Inclusion Criteria

For inclusion, the selected publication should be a primary study generally related to the software engineering domain with a focus on third-party libraries or code smells or energy bugs in Android apps. A tool/automated technique for third-party library/code smell/energy bug detection, modification, or replacement was presented in the publication to support Android development. We considered only conference and journal articles published in English.

[2]https://www.zotero.org/

Table 7.2 Quality assessment criteria

ID	Description	Rating
1	Does the publication clearly state contributions that are directly related to third-party libraries/code smells/energy bugs in Android apps?	0.5
2	Is the contributions related to green in Android development?	0.5
3	Is the contributions a tool/automated technique that could be used in Android development/maintenance?	1
4	Is the research method adequately explained?	0.5
5	Are threats to validity and future research directions discussed separately?	0.5
	Total	3

7.3.3.3 Exclusion Criteria

Publications that were unrelated to Android development or third-party library/code smell/energy bugs in Android apps were excluded. The publications that focused only on hardware, environmental, security, privacy, networks, malware, clones, repackaging of apps, obfuscation issues, iOS, or present secondary data were also excluded. Work presented in a thesis or a book chapter is usually published in relevant journals or conferences as well. Therefore, doctoral symposium papers, magazine articles, book chapters, work-in-progress papers, and papers that were not in English were excluded as well.

7.3.3.4 Quality Criteria

The quality criteria applied to selected publications are shown in Table 7.2. Abstracts of the publications and structure of the publication were inspected for further quality assessment. If a quality rule was true for a publication, it was awarded full points; otherwise, no points were awarded. In case a rule partially applied to a publication, half points were awarded. After applying all five quality rules, the points were added to get a final quality score for a publication. A maximum quality score of 3 could be assigned to a publication. If a publication was below a total quality score of 2, it was removed from the results.

7.3.4 Classification and Analysis

To answer RQ1–RQ3, we identified the main keywords of the selected publications along with the commonly used terms in the abstracts to define categories of support tools. Research methodology and results of selected publications were additionally studied when needed. We kept extracted data in Excel spreadsheets for further processing. During data extraction, if there was a conflict of opinion, it was discussed among the authors until a consensus was reached. To answer RQ1 and

Table 7.3 Categories of support tools (RQ1)

ID	Category	Description
CP	Profiler	A software program that measures the energy consumption of an Android app or parts of apps
CD	Detector	A software program that only identifies and detects energy bugs/code smells in an Android app
CO	Optimizer	A software program that identifies energy bugs/code smells as well as refactor source code of an Android app to improve energy consumption

Table 7.4 Categories of support tools (RQ2)

ID	Category	Description
CI	Identifier	A software program that only identifies and detects third-party libraries in an Android app
CM	Migrator	A software program that identifies third-party libraries as well as helping in updating or migrating the third-party libraries (to an alternative library or version) in the source code of an Android app
CC	Controller	A software program that identifies third-party libraries to control, isolate, or de-escalate privileges and permissions granted to third-party libraries in an Android app

RQ2, a bottom-up merging technique was adopted to build our own classification schemes (see Tables 7.3 and 7.4). Once classification schemes were established, we extracted data from each selected publication to identify its main contribution and assigned the tool mentioned in the publication to a category based on the classification scheme. To answer RQ3, a classification scheme was needed to classify techniques used in support tools for offering support to aid Android development. We used the bottom-up approach to build this classification scheme by combining the specialized analysis methods/techniques into more generic higher-level techniques. The identified generic techniques along with their definitions are described in Tables 7.5 and 7.6. Once we had established the classification schemes, we extracted data from the abstract and research methodology of each selected publication and assigned it to a category defined in the classification schemes.

To answer RQ4, we extracted data form each selected publication to gather information about the kind of support the identified tool offers. We compare these tools based on the inputs of the tool, outputs of the tool, code smells/energy bugs/third-party libraries coverage, interface type, integrated development environment (IDE) support, and availability. In general, a code smell is defined as "a surface indication that usually corresponds to a deeper problem in the system" [69] and an energy bug is defined as an "error in the system (application, OS, hardware, firmware, external conditions or combination) that causes an unexpected amount of high energy consumption by the system as a whole" [70]. A third-party library is a reusable component related to specific functionality that can be integrated into the application to speed up the development process. A third-party library could be for advertising, analytics, Image, Network, Social Media, Utility, etc. [71]. In the light

Table 7.5 Categories of techniques used in support tools for code smell/energy bugs (RQ3)

ID	Technique	Definition
T1	Byte Code Manipulation	A technique that injects code in the Smali files of the app under test. The injected code is either a log statement or an energy evaluation function. These statements help determine the part of the source code that consumes a specific amount of energy at runtime.
T2	Code Instrumentation	A technique that instruments the app, using instrumented test cases that are capable of running specific parts of the app, in such a way that it is run in a specific environment while calling known methods/classes of the app under test. It uses finite state machines and device-specific power consumption details to measure energy.
T3	Logcat Analysis	A technique that uses system-level log files to obtain energy consumption information provided by the OS for the app under test. These logs are compared with application-level logs to give graphical information about the energy consumption of the app.
T4	Static Source Code Analysis	A technique that uses the source code of the app and analyzes it using one or a combination of the following methods: control flow graphs analysis, point-to-analysis, inter-procedural, intra-procedural, component call analysis, abstract syntax tree traversal, or taint analysis.
T5	Search-Based Algorithms	A technique that uses a multi-objective search algorithm to find multiple refactoring solutions and the most optimal solution is selected as final refactoring output by iteratively comparing the quality of design and energy usage.
T6	Dynamic Analysis	A technique based on the identification of information flow between objects at runtime for the detection of vulnerabilities in the app under test. It monitors the spread of sensory data during different app states.

of these definitions, we looked for Android-specific code smells, energy bugs, and third-party libraries in the studies.

7.4 Results

In this section, we present the result of the mapping study. The list of selected publications and additional details about code smells/energy bugs covered by support tools are shown in a separate file (additional materials).[3]

[3]Additional material: https://figshare.com/s/da429977adc4e928fd64

Table 7.6 Categories of techniques used in support tools for third-party libraries (RQ3)

ID	Technique	Description
T7	Feature Similarity	A technique that uses machine learning to extract code clusters or train classifiers by using feature hashing or similarity metrics or pattern digest or similarity digest on apps and third-party libraries code in order to identify and classify third-party libraries.
T8	Whitelist Comparison	A technique that compares third-party library names/versions/ package information to whitelist in order to detect third-party libraries.
T9	API Hooking	A technique that intercepts or redirects API calls at various levels in order to regulate permission or policy-related operations.
T10	Module Decoupling	A technique to divide code into modules and extract code features such as package name, package structure, and inheritance relationships for clustering/classification to detect library.
T11	Process Isolation	A technique to isolate untrusted components in the operating system. This technique requires system-level modification.
T12	Class Profile Similarity	A technique to extract (strict or relaxed) profiles from libraries and apps code based on structural hierarchies. Based on similarity (exact or fuzzy) between these profiles library is detected.
T13	Collaborative Filtering	A technique to predict or recommend third-party libraries based on feature vectors and their similarity against a set of similar apps or neighborhood apps. It includes model-based approaches (such as matrix factorization), memory, and item-based approaches.
T14	Natural Language Processing	A technique used to identify or recommend third-party libraries based on textual descriptions. It includes techniques such as word embedding, skip-gram model, continuous bag-of-words model, domain-specific relational and categorical tag embedding, and topic modeling.

7.4.1 *Results of Screening*

Search Query 1 (Support Tools for Code Smell/Energy Bugs)
As a result of running search query 1 and applying filters (see Table 7.1) to search results, 2334 publications were found from the selected online repositories. These publications were loaded into the Zotero software for the screening and removal of duplicates, and the total number of publications was reduced to 2241 after duplicate removal. Inclusion and exclusion criteria were applied to the remaining publications, and the number was reduced to 575. We read abstracts of these publications and looked at the structure to assign them a quality score based on quality criteria. After applying the quality criteria, the number of selected publications was reduced to 24 (see Tables 7.7, 7.8, and 7.9).

Search Query 2 (Support Tools for Third-Party Libraries)
As a result of running search query 2 and applying filters (see Table 7.1) to search results, 545 publications were found from the selected online repositories. These publications were loaded into the Zotero software for the screening and removal of duplicates, and the total number of publications was reduced to 521 after duplicate

Table 7.7 Number of studies extracted per online repository (search query 1)

Sr.	Repo.	# of papers	Conference papers	Journal articles
1	IEEE Xplore	1170	910	260
2	ACM Digital library	483	459	24
3	Springer	595	362	231
4	Science Direct	86	4	82

Table 7.8 Number of articles per screening step (search query 1)

Sr.	Step in the screening of publications	# of publications
1	Search string results after applying filters	2334
2	Remove duplicates	2241
3	Apply inclusion and exclusion criteria	575
4	Apply quality criteria	24

Table 7.9 Quality score assigned to each selected publication (search query 1)

Publication ID	Quality score
P2, P11, P12, P14, P16, P17, P19, P20	2
P5, P6, P7, P13, P21, P24	2.25
P1, P4, P8, P9, P10, P15, P22, P23	2.5
P3, P18	2.75

Table 7.10 Number of studies extracted per online repository (search query 2)

Sr.	Repo.	# of papers	Conference papers	Journal articles
1	IEEE Xplore	312	296	12
2	ACM Digital library	177	157	20
3	Springer	28	22	6
4	Science Direct	28	0	28

Table 7.11 Number of articles after applying filters and screening steps (search query 2)

Sr.	Step in the screening of publications	# of publications
1	Search string results after applying filters	545
2	Remove duplicates	521
3	Apply inclusion and exclusion criteria	131
4	Apply quality criteria	27

removal. Inclusion and exclusion criteria were applied to the remaining publications and the number was reduced to 131. We read abstracts of these publications and looked at the structure to assign them a quality score based on quality criteria. After applying the quality criteria, the number of selected publications was reduced to 27 (see Tables 7.10, 7.11, and 7.12).

Table 7.12 Quality score assigned to each selected publication (search query 2)

Publication ID	Quality score
P31, P43	2
P26, P29, P33, P34, P35, P36, P39, P40, P48, P49	2.25
P25, P27, P28, P30, P32, P37, P38, P41, P42, P44, P45, P46, P47, P50, P51	2.5

Table 7.13 Distribution of studies in each category (search query 1)

ID	Selected publications	# Tools
CP	P6, P14, P16, P12, P13, P20, P19	7
CD	P1, P3, P4, P5, P8, P9, P7, P17	8
CO	P10, P11, P15, P18, P2, P21, P22, P23, P24	9

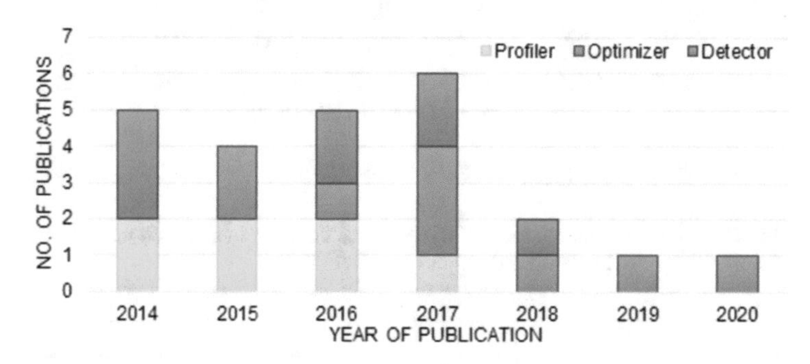

Fig. 7.1 Publications per year per category (search query 1)

7.4.2 Classification and Analysis

RQ1: What State-of-the-Art Support Tools Have Been Developed to Aid Software Practitioners in Detecting/Refactoring Code Smells/Energy Bugs in Android Apps?

To answer RQ1, the classification scheme defined in Table 7.3 (cf. Sect. 7.3.4) was used and the selected publications were divided into three categories, i.e., (1) "Profiler," (2) "Detector," and (3) "Optimizer," based on the support tool they offer to aid green Android development. Table 7.13 gives an overview of the distribution of selected publications in each category, along with the total number of tools in each category. Figure 7.1 shows the number of publications each year. The colors in the bars indicate the number of tools in each category each year from 2014 to 2020. We can see a decrease in the number of "Profiler" tools while there is an increase in the number of "Optimizer" tools. In 2019 and 2020 (until June), no new "Detector" tool was published.

Table 7.14 Distribution of publications in each category (search query 2)

ID	Selected publications	# Tools
CI	P26, P27, P29, P30, P31, P32, P33, P37, P40, P41, P42, P44, P47, P48, P49	16
CM	P35, P45, P50, P51	4
CC	P25, P28, P34, P36, P38, P39, P43, P46	7

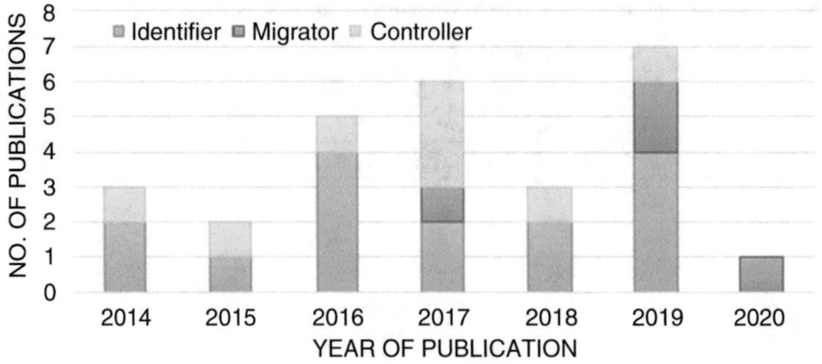

Fig. 7.2 Publications per year per category (search query 2)

Table 7.15 Overview of support tools (for code smell/energy bug detection and refactoring) showing the technique used for offering support to developers

Techniques						
Ct.	T1	T2	T3	T4	T5	T6
CP	P6, P16	P12, ‘P13, P19	P14, P20	–	–	P17
CD	–		–	P1, P3, P7, P8, P9, P4, P5	–	–
CO	–	–	–	P11, P21, P2, P10, P22, P23, P24	P15	P18

RQ2: What State-of-the-Art Support Tools Have Been Developed to Aid Software Practitioners in Detecting/Migrating Third-Party Libraries in Android Apps?
To answer RQ2, the classification scheme defined in Table 7.4 (cf. Sect. 7.3.4) was used and the selected publications were divided into the categories (1) "Identifier," (2) "Migrator," and (3) "Controller," based on the support tool they offer to aid Android development. Table 7.14 gives an overview of the distribution of selected publications in each category, along with the total number of tools in each category. Figure 7.2 shows the number of publications each year. The colors in the bars indicate the number of tools in each category each year from 2014 to 2020 (until June). We can see at least one "Identifier" and "Controller" tool each year. In addition, we can see an increase in the number of "Migrator" tools in 2019 and 2020.

RQ3: How Do Existing Support Tools Compare to One Another in Terms of Techniques They Use for Offering the Support?
To answer RQ3, we identify techniques used in each tool for improving the energy efficiency of apps. Tables 7.15 and 7.16 give an overview of tools and techniques

Table 7.16 Overview of support tools (for third-party library detection and migration) showing the technique used for offering support to developers

Techniques								
Ct.	T7	T8	T9	T10	T11	T12	T13	T14
CI	P26, P42, P49, P33, P37, P40	P31, P47		P27, P29, P32, P44, P33, P37, P40, P47		P30, P41, P48		
CM				P45			P35, P51	P35, P50
CC			P34, P36, P38, P39, P43, P46	P28	P25			

along with reference to selected publications. Based on Table 7.15, we observed that no tool in any category used a combination of techniques. Each tool could be easily classified into exactly one category of techniques (defined in Sect. 7.3.4). However, in Table 7.16, many tools used a combination of techniques such as module decoupling and feature similarity, or collaborative filtering and natural language processing.

As a result of fine-tuning search query 1, we were able to identify three new "Optimizer" tools [P22, P23, and P24] which used static source code analysis to refactor and optimize the application code. See additional materials[3] for more details on techniques used in the "Profiler," "Detector," and "Optimizer" categories.

Identifier "Identifier" tools mostly used feature similarity or module decoupling or both techniques to detect third-party libraries. The authors of [P26] used similarity digests (which are similar to standard hashes) and compared them against a database consisting of original compiled code of third-party libraries. The authors of [P42] also used similarity digests to measure the similarity between data objects. The authors of [P49] used design pattern digests, fuzzy signatures, and fuzzy hash to match design patterns from app and library code. The authors of [P27 and P29] identified third-party libraries by decoupling an app into modules using package hierarchy clustering and clustering based on locality sensitive hashing, respectively. The authors of [P32 and P44] decoupled apps into modules to extract package dependencies for identifying third-party libraries. The authors of [P33, P37 and P40] used a combination of module decoupling and feature hashing/digests to provide a list of detected third-party libraries. The authors of [P47] used whitelist-based detection for non-obfuscated[4] apps and used motifs subgraph-based detection for obfuscated apps. The authors of [P31] used whitelist-based detection by comparing library name and package information against a list of commonly used third-party libraries. The authors of [P31, P41, and P48] extracted method signatures and

[4]Code obfuscation is used to conceal or obscure the code in order to avoid tempering.

package hierarchy structures from libraries to build profiles per library and used these profiles for third-party library identification.

Migrator "Migrator" tools mostly used a combination of collaborative filtering and natural language processing techniques. The authors of [P35] used collaborative filtering in combination with topic modeling (applied to the textual description in readme files). Based on results of topic modeling, similar apps were identified, and the set of third-party libraries extracted from these similar apps were then used to recommend libraries to developers. The authors of [P50] applied word embedding and domain-specific relational and categorical knowledge on stack overflow questions to recommend alternative libraries. The authors of [P51] used collaborative filtering and applied the matrix factorization approach to neutralize bias while recommending libraries. The authors of [P45] used the "LibScout" tool to extract library profiles. These profiles were then used to determine if a library version should be updated or not.

Controller "Controller" tools mostly used API hooking techniques to provide control over library privileges based on policy. The authors of [P34] intercepted and controlled framework APIs. The authors of [P36] intercepted system APIs to extract runtime library sequence information. The authors of [P38] tracked the execution entry of the module and all related asynchronous executions at thread level. The authors of [P39] used the tool "Soot Spark" to get call graphs in order to identify Android APIs that leak data (based on a given policy). The authors of [P43] used binder hooking, in-VM API hooking, and GOT (global offset table) hooking to regulate permission and file-related operation of third-party libraries. The authors of [P46] intercepted permissions protected calls and checked them against a compiled list of third-party libraries in order to regulate privileges. The authors of [P28] extracted code features and package information to train a classifier to detect libraries and grant them privileges. The authors of [P25] used system-level process isolation in order to separate third-party library privileges.

Techniques used to provide support by the various categories of support tools for detecting and refactoring code smells/energy bugs are as follows:

> "Profiler" tools typically use a variety of techniques to measure energy consumption but none of the tools in this category uses static source code analysis.
>
> Almost all "Detector" and "Optimizer" tools use static source code analysis of APK/SC based on a predefined set of rules.

Techniques used to provide support by the various categories of support tools for detecting and migrating third-party libraries are as follows:

> "Identifier" tools use a variety of techniques for detecting third-party libraries. However, feature similarity and/or module decoupling techniques are more frequent.
>
> Almost all "Migrator" tools used collaborative filtering and/or natural language processing techniques to recommend library migration.
>
> Almost all "Controller" tools used API hooking techniques to control privileges/permissions related to third-party libraries.

RQ4: How Do Existing Support Tools Compare to One Another in Terms of the Support They Offer to Practitioners for Improving Energy Efficiency in Android Apps?

To answer RQ4, we first list all the support tools for code smell/energy bug detection/correction (see Table 7.17) and compare them in terms of input, output, user interface, integrated development environment (IDE) integration, availability, and code smell/energy bug coverage. Second, we list all the support tools for detecting/migrating third-party libraries (see Table 7.18) and compare them in terms of input, output, library coverage, user interface, availability, and IDE integration support.

In Tables 7.17 and 7.18, the "input" column provides information about the input for each tool. The "output" column provides information about the support the tool offers based on the input. The "UI" column provides information about the user interface of the tool. The "open source" column provides information about tool availability for usage/extension. The "IDE" column (in Table 7.17) provides information about the IDE integration capability of tools. The "TPL Type" column (in Table 7.18) provides information about the third-party library (TPL) coverage of the tool.

Support Tools for Code Smell/Energy Bug Detection and Refactoring

In Table 7.17, we provide a list of all the tools identified in the "Profiler," "Detector," and "Optimizer" categories. As a result of fine-tuning search query 1, we were able to identify three new "Optimizer" tools [P22, P23, P24] that were not included in our previous work [53]. For all the 24 tools listed in Table 7.17 we provide additional information related to interface, availability, and IDE integration that was not included in previous work [53].

Studies in the category "Profiler" offer support to the practitioners by providing tools that can measure the energy consumed by the whole/parts of an app or device sensors used in the apps. The measured information is usually presented to practitioners as graphs for energy consumption over time. Studies in the "Profiler" category do not recommend when, where, and how practitioners can use the information from these graphs during development to improve the energy consumption of their apps. Studies in the category "Detector" offer support to practitioners by developing tools that present as output lists of energy bugs/code smells causing a change in energy consumption of apps. Studies in the category "Optimizer" offer support to practitioners by developing tools that present as output refactored source code of apps optimized for energy. The studies in this category do not explicitly give the recommendation to the developers about how to optimize the source code for energy efficiency as the tools automatically refactor the code.

Out of the 24 tools listed in Table 7.17, only seven are open source. Out of the seven open-source tools, three are "Detector" tools, and four are "Optimizer" tools. Most of the tools do not offer IDE integration. Four tools in "Optimizer" category support integration with Eclipse IDE [P11, P18, P21, P24] while one tool [P22] supports integration with Android Studio IDE. Out of 24 tools, 12 offer command-line interface (CMD) [P1, P3, P5, P9, P7, P10, P13, P15, P17, P20, P22, P23], eight

Table 7.17 List of support tools in "Profiler," "Detector," and "Optimizer" categories along with information about their inputs and outputs, user interface, IDE support, and availability

Ct.	Tool	Input	Output	UI	IDE	Open source	ID
CP	Orka	APK	ECG	GUI	No	No	P6
	SEPIA	AE	ECG	GUI	No	No	P12
	Mantis	PBC	Program CRC predictors	CMD	No	No	P13
	AEP*	SL, PID via ADB	ECG	GUI	No	No	P14
	E-Spector	SL, AL via ADB	ECG	GUI	No	No	P16
	SEMA	PID, MVC	Log of EC	CMD	No	No	P20
	Keong et. al	SC	ECG	GUI + CMD	No	No	P19
CD	Wu et al.	SC	List of energy bugs	CMD	No	No	P1
	Kim et al.	PBC	List of energy bugs	CMD	No	No	P3
	Statedroid	APK	List of energy bugs	CMD	No	No	P5
	PatBugs	SC	List of detected warnings	NS	No	No	P8
	SAAD	APK	List of energy bugs	CMD	No	No	P9
	aDoctor	SC	List of code smells	GUI + CMD	No	Yes	P4
	GreenDroid	PBC, CF	List of energy bugs + severity level	CMD	No	Yes	P17
	Paprika	APK, PM	List of code smells	CMD	No	Yes	P7
CO	DelayDroid	APK	Refactored APK	NS	No	No	P2
	HOT-PEPPER	APK	Most energy efficient APK, Refactored SC, and List of refactoring	CMD	No	Yes	P10
	Asyncdroid	SC	Refactored SC	GUI	Eclipse	No	P11
	EARMO	APK	Refactored APK	CMD	No	Yes	P15
	EnergyPatch	APK	Refactored APK	GUI	Eclipse	No	P18
	Nguyen et al.	SC	Refactored SC	GUI	Eclipse	No	P21
	Chimera	SC	Refactored APK	CMD	Android Studio	No	P22
	ServDroid	APK	Refactored APK	CMD	No	Yes	P23
	Leafactor	SC	Refactored APK file	GUI	Eclipse	Yes	P24

Ct. category, *SC* source code, *APK* android package kit, *PBC* program byte code, *SL* system log files, *AL* application log files, *PID* process ID, *ADB* android debug bridge, *CRC* computational resource consumption, *AE* application events, *CF* configuration files, *MVC* measurements of voltage and current, *ECG* energy consumption graph, *SM* software metrics values, *PM* playstore metadata, *GUI* graphical user interface, *CMD* command line, *EC* energy consumption

Table 7.18 List of support tools in "Identifier," "Migrator," and "Controller" categories along with information about their inputs and outputs, library coverage, UI, and availability

Ct.	Tool	Input	Output	TPL Type	UI	Open source	Ref
CI	Duet	APK	Library integrity pass/fail ratio	Java	NS	No	P26
	AdDetect	APK	List of detected TPLs	Java-Ad	NS	No	P27
	AnDarwin	APK	Detect and exclude TPLs + Set clone or rebranded apps	Java	NS	No	P29
	LibScout	TPL .jar/.aar + APK	Presence of given TPL based on similarity score	Java	CMD	Yes	P30
	DeGuard	APK	De-obfuscated APK (containing detected TPLs)	Java	GUI[a]	Yes	P31
	LibSift	APK	List of detected TPLs	Java	NS	No	P32
	LibRadar	APK	List of detected TPLs sorted by popularity + info about TPLs	Java	GUI[a]	Yes	P33
	LibD	APK	List of detected TPLs	Java	CMD	Yes	P37
	Ordol	APK	List of detected TPL versions + similarity score.	Java	NS	No	P40
	LibPecker	TPL name + APK	Presence of given TPL based on the similarity score	Java	NS	No	P41
	Orlis	APK	List of detected TPLs	Java	NS	Yes	P42
	PanGuard	APK	List of detected TPLs	Java	GUI[a]	No	P44
	He et al.	APK	List of detected TPLs + risk assessment	Java	NS	No	P47
	Feichtner et al.	APK/TPL	List of detected TPLs and versions + similarity score	Java	CMD[b]	Yes	P48
	DPAK	APK/ Android jar	List of detected TPLs	Java	CMD[b]	No	P49
CM	AppLibRec	SC	List of recommended TPLs	Java	NS	No	P35
	Appcommune	APK	Tailored app without TPLs and updated/ customized TPLs	Java	GUI[c]	No	P45
	SimilarTech	TPL name	List of recommended TPLs + information about usage	Java	GUI[a]	No	P50
	LibSeek	APK	List of recommended TPLs	Java	NS	No	P51

(continued)

Table 7.18 (continued)

Ct.	Tool	Input	Output	TPL Type	UI	Open source	Ref
CC	NativeGuard	APK	Split original APK into Service APK and Client APK	Native	CMD	No	P25
	Pedal	APK	Repackaged APK with privilege de-escalated for detected TPLs	Java-Ad	GUI[c]	No	P28
	LibCage	SC + list of permissions required by TPLs	Deny unnecessary TPL permission on runtime	Java+ Native	NS	No	P34
	Zhan et al.	SC + Policy	Grant or deny permissions to TPLs based on policy	Java	NS	No	P36
	Perman	APK	Grant or deny permissions to TPLs based on policy	Java	GUI[c]	No	P38
	SurgeScan	TPL bytecode + Android.jar + policy	Dex and jar files of TPL with the policy implemented	Java	NS	No	P39
	AdCapsule	SC + policy	Grant or deny permissions to TPLs based on policy	Java-Ad	NS	No	P43
	Reaper	APK	Grant or deny permissions to TPLs based on user preference	Java + Native	GUI[c]	Yes	P46

Ct. category, *UI* user interface, *SC* source code, *APK* android package kit, *TPL* third-party libraries, *GUI* graphical user interface, *CMD* command-line interface, *NS* not specified in publication
[a]Web service
[b]Executable jar
[c]App on Android device

tools offer graphical user interface (GUI) [P6, P11, P12, P14, P16, P18, P21, P24], and two tools offer both [P4, P19], while for the rest of them information about interface is not specified in the publications.

See additional material[3] for details about definitions of code smells/energy bugs covered by tools in the "Detector" and "Optimizer" categories. Figure 7.3 shows the Android energy bug coverage of tools in the "Detector" and "Optimizer" categories. The Android energy bugs are shown on the horizontal axis. The percentage of tools in the "Detector" and "Optimizer" categories covering Android energy bugs is shown on the vertical axis. We can see that Android energy bugs "TMV," "TDL," "UL," "UP," and "VBS" are detected by 13% of the tools, whereas "RL," "WB," and "NCD" are detected by 75%, 50%, and 38% of the tools in the "Detector" category respectively. None of the tools in the "Optimizer" category covers. "TMV," "TDL,"

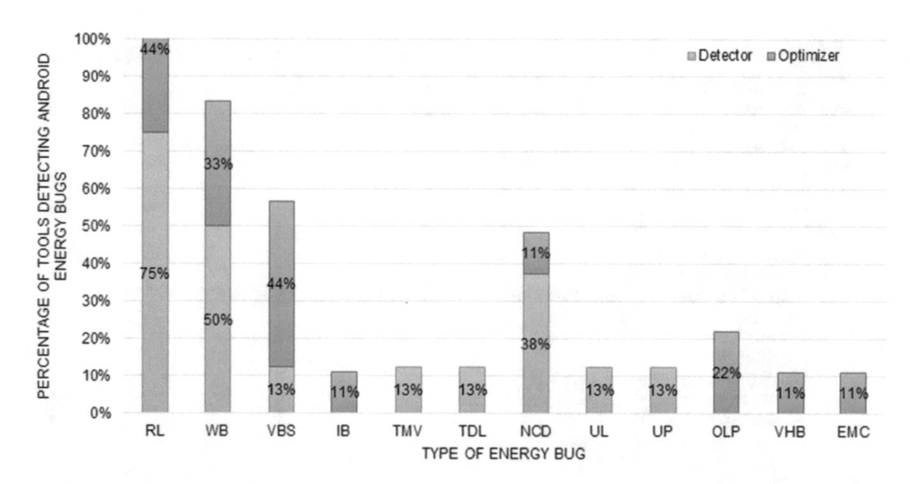

Fig. 7.3 Percentage of the tools in "Detector" and "Optimizer" categories that can detect Android energy bugs (*RL* resource leak, *WB* wake-lock bug, *VBS* vacuous background services, *IB* immortality bug, *TMV* too many views, *TDL* too deep layout, *NCD* not using compound drawables, *UL* useless leaf, *UP* useless parent, *OLP* obsolete layout parameter, *VHB* view holder bug, *EMC* excessive method calls)

"UL," and "UP" energy bugs. On the other hand, energy bugs "IB," "OLP," "VHB," and "EMC" are covered by tools in the "Optimizer" category, whereas none of the tools in the "Detector" category covers them. "RL" and "VBS" energy bugs are detected by 44% of the tools in the "Optimizer" category.

Figure 7.4 shows the Android code smell coverage of tools in the "Detector" and "Optimizer" categories. The Android code smells are shown on the vertical axis. The percentage of tools in the "Detector" and "Optimizer" categories covering Android code smells is shown on the horizontal axis. We can see that Android code smells "ERB" and "VHP" are not detected by any tool in the "Detector" category, whereas "LWS," "LC," "RAM," "PD," "ISQLQ," "IDFP," "DW," "DR," and "DTWC" are not detected by any of the tools in the "Optimizer" category. Android code smells such as "IOD," "HBR," "HSS," "HAT," "IWR," "UIO," "BFU," "UHA," "LWS," "LC," "SL," "RAM," "PD," "NLMR," "MIM," "LT," "IDS," "IDFP," "DW," "DR," and "DTWC" are detected by 13–25% of the tools in the "Detectors" category.

Typical support given by the various categories of support tools for detecting and refactoring code smells/energy bugs are as follows:

"Profiler" tools support developers by visualizing the energy consumption of the whole app or parts of it.

"Detector" tools support developers with lists of energy bugs and code smells to be manually fixed by the developer for energy improvement.

"Optimizer" tools support developers by automatically refactoring APK/SC versions based on predefined rules.

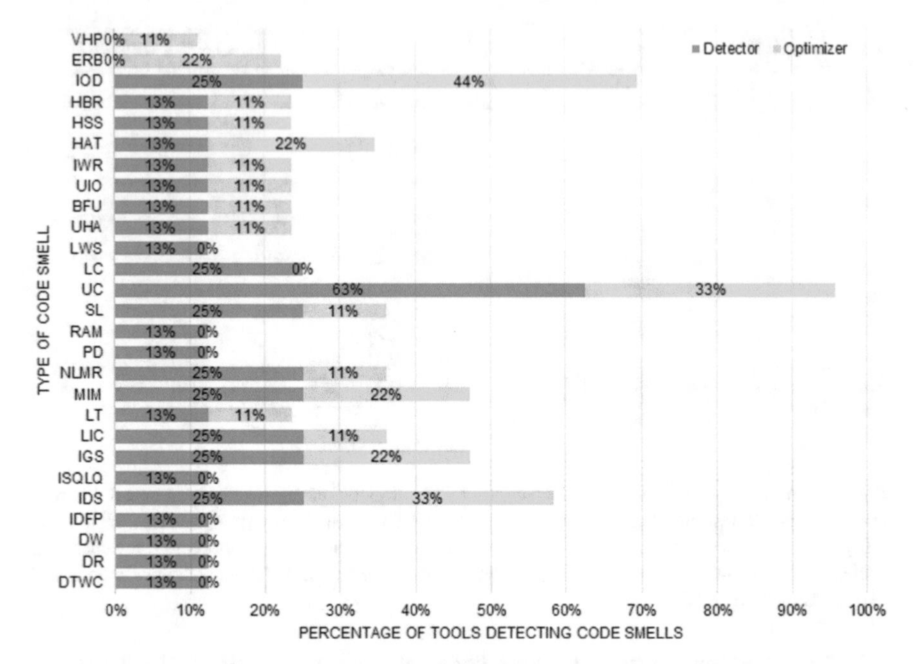

Fig. 7.4 Percentage of code smells detected by each tool in "Detector" and "Optimizer" categories. *DTWC* data transmission without compression, *DR* debuggable release, *DW* durable wake-lock, *IDFP* inefficient data format and parser, *IDS* inefficient data structure, *ISQLQ* inefficient SQL query, *IGS* internal getter and setter, *LIC* leaking inner class, *LT* leaking thread, *MIM* member ignoring method, *NLMR* no low memory resolver, *PD* public data, *RAM* rigid alarm manager, *SL* slow loop, *UC* unclosed closeable, *LC* lifetime containment, *LWS* long wait state, *UHA* unsupported hardware acceleration, *BFU* bitmap format usage, *UIO* UI overdraw, *IWR* invalidate without rect, *HAT* heavy AsyncTask, *HSS* heavy service start, *HBR* heavy broadcast receiver, *IOD* init ONDraw, *ERB* early resource binding, *VHP* view holder pattern

Support Tools for Third-Party Library Detection and Migration

In Table 7.18, we provide a list of all the tools identified in the "Identifier," "Migrator," and "Controller" categories. Publications in the category "Identifier" offer support to practitioners by providing tools that detect third-party libraries present in the apps. The information is usually presented to practitioners as a list of detected libraries along with their version/similarity scores. Publications in the category "Migrator" offer support to practitioners by developing tools that present as output lists of recommended third-party libraries. Publications in the category "Controller" offer support to practitioners by developing tools that present as output policy-based privilege/permission control over third-party libraries. Most tools in the "Identifier," "Migrator," and "Controller" categories provide coverage for all types (advertisement, social, network, billing, analytics, etc.) of Java-based third-party libraries. Some tools such as AdDetect (CI) or Pedal (CC) cover only the advertisement-related third-party libraries. NativeGuard (CC) provides coverage for only native third-party libraries. Reaper (CC) and LibCage (CC) provide coverage for native and Java-based third-party libraries. For many tools listed in

Table 7.18, interface type was not specified in publications, while others provide either a command-line interface (CMD) or a graphical user interface (GUI). Out of the 27 tools listed in Table 7.18, only seven tools are open source. Out of the seven open-source tools, six tools [P30, P31, P33, P37, P42, P48] are "Identifier" tools and one tool [P46] is a "Controller" tool. None of the tools listed in Table 7.18 provides IDE integration support.

The tools in the "Identifier," "Migrator," and "Controller" categories do not detect/update/control/migrate third-party libraries to optimize the source code of Android applications for energy efficiency.

Typical support given by the various categories of support tools for detecting and migrating third-party libraries are as follows:

"Identifier" tools support developers by detecting third-party libraries present in apps.

"Migrator" tools support developers with lists of recommended third-party libraries along with the mapping information of these libraries for updating/migrating them.

"Controller" tools support developers by separating third-party library privileges from the app privileges based on policy defined by developers.

7.5 Discussion

In this section, we discuss the results of the mapping study to identify future research opportunities.

7.5.1 Support Tools for Code Smell/Energy Bug Detection and Refactoring

We observed that most of the support tools in the "Profiler," "Detector," and "Optimizer" categories are not open source, making them inaccessible to many developers. On top of that, most of these support tools do not support IDE integration. Due to the rapid development process of Android applications, developers are more likely to use tools that are integrated with the IDEs and share the same interface design. The current state-of-the-art tools could be extended to integrate with other industrially famous code analyzers like Android Lint, Check Style, Find Bugs and PMD. Each tool in "Detector" and "Optimizer" category provided a limited coverage over Android-specific code smells/energy bugs. The industry relevance of the current state-of-the-art support tools might not be obvious because they are not evaluated in industrial settings. In principle, if developers spent time and effort to learn one such tool they still might not be able to identify many code smells/energy bugs in their code, unless they use a combination of these tools to get complete coverage. Most tools in the "Detector" and "Optimizer" categories used static source code analysis, which indicates that dynamic issues such as those related to asynchronous tasks are not covered by these tools. For the development of better support tools, hybrid techniques encompassing both dynamic and static analysis could be used. In addition, non-intrusive techniques could be used to collect software metrics for identifying code smells/energy bugs. The results from the selected publications could be expanded to include cross-project predictions and corrections for energy bugs. Analysis and inclusion of multi-threaded programming approaches in the experiments could be another direction for future researchers.

7.5.2 Support Tools for Third-Party Library Detection and Migration

We observed that none of the support tools in the “Identifier,” “Migrator,” and “Controller” categories provides support for IDE integration and many of these tools are also not open source, making them inaccessible to developers. We also observed the none of the support tools in these categories offers any support to developers to aid green Android development. One possible reason could be that so far research related to third-party library identification is mostly used in clone detection, detection of rebranded/similar/malicious apps, and detection of issues related to security, privacy, or data leaks (see related work). However, there is a gap in the literature regarding support tools that identify/update/recommend third-party libraries to aid green Android development. Anwar et al. [52] have investigated the energy consumption of third-party libraries in Android applications, indicating that the energy consumption of alternative third-party libraries varies significantly in various use cases. Rasmussen et al. [66] showed that blocking advertisements in Android apps reduces energy consumption. However, these studies have only focused on a small subset of network- and advertisement-related libraries. Energy consumption of other types of libraries such as social, analytical, or utility has not yet been explored, and merits further research. Data from such studies could be used by tool developers to recommend energy-efficient libraries to developers during development. Support tools in the “Migrator” category are good candidates for this type of research as the collaborative filtering and natural language processing techniques could supplement the data gathered from energy reading of third-party libraries. Such information could be useful in mapping the function of one library to another alternative library for a smooth migration. Support tools in the “Identifier” category generally use two techniques: (a) whitelist-based and (b) similarity-based. Tools that used whitelist-based approaches are fast due to a smaller feature set, and thus could perform better in large-scale analysis. However, this technique cannot identify third-party libraries without prior knowledge. On the other hand, tools that use similarity-based approaches such as feature hashing use a larger feature set and can identify third-party libraries without prior knowledge. Due to the extended feature set, these tools might be more accurate but time-consuming. Many tools in the “Identifier” category (such as “LibD,” “LibScout,” “LibRadar,” or “AdDetect”) consider code obfuscation during library detections in order to give accurate results. However, not many tools are resilient against code shrinking as they rely on package hierarchies. Support tools in the "Controller" category rely on API hooking techniques which separate libraries from app code. Such tools could also benefit from using an access control list to split privileges. Because current techniques require system-level changes, this makes the deployment of "Controller" tools difficult.

7.6 Threats to Validity

The search queries and classification of selected publications could be biased by the researcher's knowledge. We mitigated this threat by defining the inclusion, exclusion, and quality criteria for the selection of the publications. Conflicting opinions were discussed among authors of this study until a consensus was reached. In order to avoid false-positives and false-negatives in the search results, we used the wildcard character (*) to maximize coverage and the keyword "AND NOT" to remove irrelevant studies. We did not use the terms "energy" or "efficiency" in combination with "Android" in the second search query, as we had already executed this combination in search query 1. The results of the search strings were manually checked and further refined by the authors. Online repositories continuously update their databases to include new publications, and therefore executing the same queries might yield some additional results that were not included in this study. We already knew about many relevant studies and we recaptured almost 90% of them when we executed the search queries. On each online repository the search mechanism is slightly different and we tried to keep the queries as consistent as possible, but there might be a slight difference due to the difference in search mechanism provided by different online repositories. Some selected publications use the terms code smells and energy bugs interchangeably, which could affect the classification. To mitigate this threat, we used the selected definitions (cf. Sect. 7.3.4) for code smells and energy bugs to correctly classify the studies in the right category.

We have excluded publications that did not focus on Android development yet still contributed a tool for detecting/recommending third-party libraries. Maven central repository contains a huge quantity of Java-based third-party libraries that can be used in any Java-based application. However, in this study, we focused particularly on the support tools for energy profiling, code optimization and refactoring of code smells/energy bugs, and detection/migration of third-party libraries to help aid green Android development. Other types of support tools, such as tools for style checking, interface optimization, test generation, requirement engineering, and code obfuscation, were not in the scope of this study. Therefore, while applying inclusion/exclusion criteria, we filtered support tools such as "LibFinder," LibCPU, CrossRec, and RAPIM [72–75]. These tools could identify/recommend third-party libraries but they were not designed to be used specifically with Android applications. We plan to cover such tools in future work.

7.7 Conclusions

We conducted a mapping study to give an overview of the state of the art and to find research opportunities with respect to support tools available for green Android development. Based on our analysis we identified tools for detecting/refactoring code smells/energy bugs, which were classified into three categories: (1) "Profiler," (2) "Detector," and (3) "Optimizer." Additionally, we identified tools for detecting/migrating third-party libraries in Android applications, which were classified into (1) Identifier, (2) Migrator, and (3) Controller categories. The main findings of this study are that most "Profiler" tools provide a graphical representation of energy consumption over time. Most "Detector" tools provide a list of energy bugs/code smells to be manually corrected by a developer for the improvement of energy. Most "Optimizer" tools automatically convert original APK/SC into a refactored version of APK/SC. Tools in the "Identifier," "Migrator," and "Controller" categories do not provide support to developers to optimize code w.r.t. energy consumption. The most typical technique in the "Detector" and "Optimizer" categories was static source code analysis using a predefined set of code smells and rules. The most typical techniques in the "Identifier" category were module decoupling and feature similarity, while in the "Migrator" and "Controller" categories, API hooking and collaborative filtering in combination with natural language processing were used, respectively.

Acknowledgments This work was supported by the Estonian Center of Excellence in ICT research (EXCITE), the group grant PRG887 funded by the Estonian Research Council, and the Estonian state stipend for doctoral studies.

References

1. GeSI (2015) #SMARTer2030 ICT solutions for 21st century challenges. Accessed 06 Jun 2020. http://smarter2030.gesi.org/downloads/Full_report.pdf
2. Acar H (2017) Software development methodology in a Green IT environment. Université de Lyon
3. Calero C, Piattini M (2015) Introduction to green in software engineering. In: Calero C, Piattini M (eds) Green in software engineering. Springer International Publishing, Cham, pp 3–27
4. Chauhan NS, Saxena A (2013) A green software development life cycle for cloud computing. IT Prof 15(1):28–34. https://doi.org/10.1109/MITP.2013.6
5. Federal Ministry for Economic Affairs and Energy (2014) Energy-efficient ICT in practice: planning and implementation of GreenIT measures in data centres and the office
6. Jagroep E, van der Werf JM, Brinkkemper S, Blom L, van Vliet R (2017) Extending software architecture views with an energy consumption perspective. Computing 99(6):553–573. https://doi.org/10.1007/s00607-016-0502-0
7. Kumar S, Buyya R (2012) Green cloud computing and environmental sustainability. Harnessing Green It Princ Pract:315–339. https://doi.org/10.1002/9781118305393.ch16
8. Oyedeji S, Seffah A, Penzenstadler B (2018) A catalogue supporting software sustainability design. Sustainability 10(7):2296. https://doi.org/10.3390/su10072296

9. Gupta PK, Singh G (2012) Minimizing power consumption by personal computers: a technical survey. Int J Inf Technol Comput Sci 4(10):57–66. https://doi.org/10.5815/ijitcs.2012.10.07
10. Kern E et al (2018) Sustainable software products—towards assessment criteria for resource and energy efficiency. Futur Gener Comput Syst 86(3715):199–210. https://doi.org/10.1016/j.future.2018.02.044
11. Murugesan S, Gangadharan GR (2012) Green IT: an overview. In: Murugesan S, Gangadharan GR (eds) Harnessing green IT: principles and practices. Wiley, pp 1–21
12. Egham (2018) Gartner says worldwide end-user device spending set to increase 7 percent in 2018; global device shipments are forecast to return to growth. Gartner, Press Releases. Accessed 11 Feb 2019. https://www.gartner.com/en/newsroom/press-releases/2018-04-05-gartner-says-worldwide-end-user-device-spending-set-to-increase-7-percent-in-2018-global-device-shipments-are-forecast-to-return-to-growth
13. Penzenstadler B, Femmer H (2013) A generic model for sustainability with process- and product-specific instances. In: Proceedings of the 2013 Workshop on Green by Software Engineering, pp 3–7. doi:https://doi.org/10.1145/2451605.2451609
14. Raturi A, Tomlinson B, Richardson D (2015) Green software engineering environments. In: Green in software engineering. Springer International Publishing, pp 31–59
15. Banerjee A, Chong LK, Chattopadhyay S, Roychoudhury A (2014) Detecting energy bugs and hotspots in mobile apps. In: Proceedings of the 22nd ACM SIGSOFT international symposium on foundations of software engineering - FSE, vol 16–21-Nov, pp 588–598, doi: https://doi.org/10.1145/2635868.2635871
16. Allix K, Bissyandé TF, Klein J, Le Traon Y (2016) AndroZoo: collecting millions of Android apps for the research community. In: Proceedings of the 13th international workshop on mining software repositories - MSR, May 2016, pp 468–471, doi: https://doi.org/10.1145/2901739.2903508
17. Anwar H, Pfahl D (2017) Towards greener software engineering using software analytics: a systematic mapping. In: Proceedings of the 43rd Euromicro conference on software engineering and advanced applications -SEAA, Aug 2017, pp 157–166, doi: https://doi.org/10.1109/SEAA.2017.56
18. Martin W, Sarro F, Jia Y, Zhang Y, Harman M (2017) A survey of app store analysis for software engineering. IEEE Trans Softw Eng 43(9):817–847. https://doi.org/10.1109/TSE.2016.2630689
19. Oliveira W, Oliveira R, Castor F (2017) A study on the energy consumption of android app development approaches. In: Proceedings of the IEEE/ACM 14th international conference on mining software repositories - MSR, May 2017, pp 42–52, doi: https://doi.org/10.1109/MSR.2017.66
20. Rawassizadeh R (2010) Mobile application benchmarking based on the resource usage monitoring. Int J Mob Comput Multimed Commun 1(4):64–75. https://doi.org/10.4018/jmcmc.2009072805
21. Viennot N, Garcia E, Nieh J (2014) A measurement study of google play. ACM SIGMETRICS Perform Eval Rev 42(1):221–233. https://doi.org/10.1145/2637364.2592003
22. Wang H et al (2017) An explorative study of the mobile app ecosystem from app developers' perspective. In: Proceedings of the 26th international conference on World Wide Web, pp 163–172, doi:https://doi.org/10.1145/3038912.3052712
23. Wang H et al (2018) Beyond Google play: a large-scale comparative study of Chinese Android App Markets. ArXiv, vol 1810.07780, Sep 2018. http://arxiv.org/abs/1810.07780
24. Ardito L, Procaccianti G, Torchiano M, Migliore G (2013) Profiling power consumption on mobile devices. In: Proceedings of the third international conference on smart grids, green communications and IT Energy-aware Technologies, pp 101–106
25. Azevedo L, Dantas A, Camilo-Junior CG. DroidBugs: an android benchmark for automated program repair. ArXiv, vol abs/1809.0, 2018 [Online]. http://arxiv.org/abs/1809.07353

26. Chung YF, Lin CY, King CT (2011) ANEPROF: energy profiling for android java virtual machine and applications. In: Proceedings of the international conferences on parallel and distributed systems - ICPADS, pp 372–379, doi: https://doi.org/10.1109/ICPADS.2011.28
27. Kansal A, Zhao F (2008) Fine-grained energy profiling for power-aware application design. ACM SIGMETRICS Perform Eval Rev 36(2):26. https://doi.org/10.1145/1453175.1453180
28. Pathak A, Hu YC, Zhang M (2012) Where is the energy spent inside my app? Fine Grained Energy Accounting on Smartphones with Eprof. EuroSys, pp 29–42, Accessed 04 Apr 2018. https://www.cse.iitb.ac.in/~mythili/teaching/cs653_spring2014/references/energy-eprof-tool.pdf
29. Banerjee A, Roychoudhury A (2016) Automated re-factoring of Android apps to enhance energy-efficiency. In: Proceedings of the international workshop on mobile software engineering and system - MOBILESoft, pp 139–150, doi: https://doi.org/10.1145/2897073.2897086
30. Fernandes TS, Cota E, Moreira AF (2014) Performance evaluation of android applications: a case study. In: Proceedings of the Brazilian symposium on computing system engineering, Nov 2014, vol 1998-Jan, pp 79–84, doi: https://doi.org/10.1109/SBESC.2014.17
31. Fowler M, Beck K (1999) Refactoring: improving the design of existing code. Addison-Wesley
32. Hecht G, Rouvoy R, Moha N, Duchien L (2015) Detecting antipatterns in android apps. In: Proceedings of the 2nd ACM international conference on mobile software engineering and systems, MOBILESoft, Sep 2015, pp 148–149, doi: https://doi.org/10.1109/MobileSoft.2015.38
33. Palomba F, Di Nucci D, Panichella A, Zaidman A, De Lucia A (2017) Lightweight detection of Android-specific code smells: the aDoctor project. In: Proceedings of the 24th IEEE international conference software analysis evolution and reengineering - SANER, pp 487–491. doi: https://doi.org/10.1109/SANER.2017.7884659
34. Rasool G, Ali A (2020) Recovering android bad smells from android applications. Arab J Sci Eng 45(4):3289–3315. https://doi.org/10.1007/s13369-020-04365-1
35. Xu B, An L, Thung F, Khomh F, Lo D (2020) Why reinventing the wheels? An empirical study on library reuse and re-implementation. Empir Softw Eng 25(1):755–789. https://doi.org/10.1007/s10664-019-09771-0
36. Wang H, Guo Y (2017) Understanding third-party libraries in mobile app analysis. In: Proceedings of the IEEE/ACM 39th international conference on software engineering companion, pp 515–516, doi: https://doi.org/10.1109/ICSE-C.2017.161
37. Zhan J, Zhou Q, Gu X, Wang Y, Niu Y (2017) Splitting third-party libraries' privileges from android apps. In Lecture Notes in Computer Science (including subseries Lecture Notes in Artificial Intelligence and Lecture Notes in Bioinformatics), vol 10343 LNCS, Springer, pp 80–94
38. Gkortzis A, Feitosa D, Spinellis D (2019) A double-edged sword? Software reuse and potential security vulnerabilities. In: Lecture Notes in Computer Science (including subseries Lecture Notes in Artificial Intelligence and Lecture Notes in Bioinformatics), vol 11602 LNCS, pp 187–203, doi: https://doi.org/10.1007/978-3-030-22888-0_13
39. Ikram M, Vallina-Rodriguez N, Seneviratne S, Kaafar MA, Paxson V (2016) An analysis of the privacy and security risks of android VPN permission-enabled apps. In: Proceedings of the ACM SIGCOMM internet measurement conference - IMC, vol 14–16-Nov, pp 349–364, doi: https://doi.org/10.1145/2987443.2987471
40. Mazuera-Rozo A, Bautista-Mora J, Linares-Vásquez M, Rueda S, Bavota G (2019) The Android OS stack and its vulnerabilities: an empirical study. Empir Softw Eng 24 (4):2056–2101. https://doi.org/10.1007/s10664-019-09689-7
41. Ogawa H, Takimoto E, Mouri K, Saito S (2018) User-side updating of third-party libraries for android applications. In: Proceedings of the sixth international symposium on computing and networking workshops - CANDARW, Nov 2018, pp 452–458, doi: https://doi.org/10.1109/CANDARW.2018.00088
42. Binns R, Zhao J, Van Kleek M, Shadbolt N (2018) Measuring third-party tracker power across web and mobile. ACM Trans Internet Technol 18(4). doi: https://doi.org/10.1145/3176246

43. Fu J, Zhou Y, Liu H, Kang Y, Wang X (2017) Perman: fine-grained permission management for android applications. In: Proceedings of the IEEE 28th international symposium on software reliability engineering - ISSRE, Oct 2017, vol 2017-Oct, pp 250–259, doi: https://doi.org/10.1109/ISSRE.2017.38
44. Gao X, Liu D, Wang H, Sun K (2016) PmDroid: permission supervision for android advertising. In: Proceedings of the IEEE symposium on reliable distributed systems, vol 2016-Jan, pp 120–129, doi: https://doi.org/10.1109/SRDS.2015.41
45. Jin H et al. (2018) Why are they collecting my data?. In: Proceedings of the ACM on interactive, mobile, wearable and ubiquitous Techniques, Dec 2018, vol 2(4), pp 1–27, doi:https://doi.org/10.1145/3287051
46. Wang H, Li Y, Guo Y, Agarwal Y, Hong JI (2017) Understanding the purpose of permission use in mobile apps. ACM Trans Inf Syst 35(4). https://doi.org/10.1145/3086677
47. Chen K, Liu P, Zhang Y (2014) Achieving accuracy and scalability simultaneously in detecting application clones on Android markets. In: Proceedings of the international conference on software engineering, no 1, pp 175–186, doi: https://doi.org/10.1145/2568225.2568286
48. Li L, Bissyandé TF, Wang HY, Klein J (2019) On identifying and explaining similarities in android apps. J Comput Sci Technol 34(2):437–455. https://doi.org/10.1007/s11390-019-1918-8
49. Soh C, Tan HBK, Arnatovich YL, Wang L (2015) Detecting clones in android applications through analyzing user interfaces. In: Proceedings of the IEEE 23rd international conference on program comprehension, May 2015, pp 163–173, doi:https://doi.org/10.1109/ICPC.2015.25
50. Yuan L (2016) Detecting similar components between android applications with obfuscation. In: Proceedings of the 5th international conference on computer science and networking technologies - ICCSNT, Dec 2016, pp 186–190, doi:https://doi.org/10.1109/ICCSNT.2016.8070145
51. Zhang Y, Ren W, Zhu T, Ren Y (2019) SaaS: a situational awareness and analysis system for massive android malware detection. Futur Gener Comput Syst 95:548–559. https://doi.org/10.1016/j.future.2018.12.028
52. Anwar H, Demirer B, Pfahl D, Srirama SN (2020) Should energy consumption influence the choice of Android third-party HTTP libraries?. In: Proceedings of the IEEE/ACM 7th International conference on mobile software engineering and systems, MOBILESoft, pp 87–97. doi: https://doi.org/10.1145/3387905.3392095
53. Fatima I, Anwar H, Pfahl D, Qamar U (2020) Tool support for green android development: a systematic mapping study. In: Proceedings of the 15th international conference on software technologies - ICSOFT, pp 409–417
54. Fontana FA, Mariani E, Mornioli A, Sormani R, Tonello A (2011) An Experience report on using code smells detection tools. In: Proceedings of the IEEE fourth international conference on software testing, verification and validation workshops, Mar 2011, pp 450–457, doi:https://doi.org/10.1109/ICSTW.2011.12
55. Kaur A, Dhiman G (2019) A review on search-based tools and techniques to identify bad code smells in object-oriented systems. Adv Intell Syst Comput 741:909–921. https://doi.org/10.1007/978-981-13-0761-4_86
56. Singh S, Kaur S (2017) A systematic literature review: refactoring for disclosing code smells in object oriented software. Ain Shams Eng J 9(4):2129–2151. https://doi.org/10.1016/J.ASEJ.2017.03.002
57. Li L et al (2017) Static analysis of android apps: a systematic literature review. Inform Softw Technol 88:67–95. https://doi.org/10.1016/j.infsof.2017.04.001
58. Degu A (2019) Android application memory and energy performance: systematic literature review. IOSR J Comp Eng 21(3):20–32
59. Qiu L, Wang Y, Rubin J (2018) Analyzing the analyzers: FlowDroid/IccTA, AmanDroid, and DroidSafe. In: Proceedings of the 27th ACM SIGSOFT international symposium on software testing and analysis - ISSTA, pp 176–186, doi:https://doi.org/10.1145/3213846.3213873

60. Corrodi C, Spring T, Ghafari M, Nierstrasz O (2018) Idea: benchmarking android data leak detection tools. In: Lecture Notes in Computer Science (including subseries Lecture Notes in Artificial Intelligence and Lecture Notes in Bioinformatics), Jun 2018, vol 10953 LNCS, pp 116–123, doi: https://doi.org/10.1007/978-3-319-94496-8_9
61. Ndagi JY, Alhassan JK (2019) Machine learning classification algorithms for adware in android devices: a comparative evaluation and analysis. In: Proceedings of the 15th international conference on electronics, computing, and computation - ICECCO, Dec 2019, pp 1–6, doi: https://doi.org/10.1109/ICECCO48375.2019.9043288
62. Cooper VN, Shahriar H, Haddad HM (2014) A survey of android malware and mitigation techniques. In: Proceedings of the 11th international conference on information technology: new generations, Apr 2014, pp 327–332, doi: https://doi.org/10.1109/ITNG.2014.71
63. Li L, Bissyande TF, Klein J (2019) Rebooting research on detecting repackaged android apps: literature review and benchmark. IEEE Trans Softw Eng:1–1. https://doi.org/10.1109/tse.2019.2901679
64. Roy CK, Cordy JR, Koschke R (2009) Comparison and evaluation of code clone detection techniques and tools: a qualitative approach. Sci Comput Program 74(7):470–495. https://doi.org/10.1016/j.scico.2009.02.007
65. Wang Y, Li Y, Lan T (2017) Capitalizing on the promise of Ad prefetching in real-world mobile systems. In: Proceedings of the IEEE 14th international conference on mobile Ad Hoc and sensor systems - MASS, Oct 2017, pp 162–170, doi:https://doi.org/10.1109/MASS.2017.46
66. Rasmussen K, Wilson A, Hindle A (2014) Green mining: energy consumption of advertisement blocking methods. In: Proceedings of the 3rd international workshop on green and sustainable software - GREENS, pp 38–45, doi:https://doi.org/10.1145/2593743.2593749
67. Shao Y, Wang R, Chen X, Azab AM, Mao ZM (2019) A lightweight framework for fine-grained lifecycle control of android applications. In: Proceedings of the 14th EuroSys conference - EuroSys, pp 1–14, doi:https://doi.org/10.1145/3302424.3303956
68. Petersen K, Feldt R, Mujtaba S, Mattsson M (2008) Systematic mapping studies in software engineering. In: Proceedings of the 12th international conference on evaluation and assessment in software engineering - EASE, pp 68–77
69. Fowler M (2002) Refactoring: improving the design of existing code. In: Extreme programming and agile methods — XP/Agile universe. Springer, Berlin, pp 256–256
70. Pathak A, Charlie Hu Y, Zhang M (2011) Bootstrapping energy debugging on smartphones: a first look at energy bugs in mobile devices. In: Proceedings of the 10th ACM workshop on hot topics in networks (HotNets-X). Association for Computing Machinery, New York, NY, Article 5, 1–6. doi:https://doi.org/10.1145/2070562.2070567
71. Yasumatsu T, Watanabe T, Kanei F, Shioji E, Akiyama M, Mori T (2019) Understanding the responsiveness of mobile app developers to software library updates. In: Proceedings of the 9th ACM conference on data and application security and privacy - CODASPY, pp 13–24, doi: https://doi.org/10.1145/3292006.3300020
72. Alrubaye H, Mkaouer MW, Khokhlov I, Reznik L, Ouni A, Mcgoff J (2020) Learning to recommend third-party library migration opportunities at the API level. Appl Soft Comput 90:106140. https://doi.org/10.1016/j.asoc.2020.106140
73. Nguyen PT, Di Rocco J, Di Ruscio D, Di Penta M (2020) CrossRec: supporting software developers by recommending third-party libraries. J Syst Softw 161:110460. https://doi.org/10.1016/j.jss.2019.110460
74. Ouni A, Kula RG, Kessentini M, Ishio T, German DM, Inoue K (2017) Search-based software library recommendation using multi-objective optimization. Inf Softw Technol 83:55–75. https://doi.org/10.1016/j.infsof.2016.11.007
75. Saied MA, Ouni A, Sahraoui H, Kula RG, Inoue K, Lo D (2018) Improving reusability of software libraries through usage pattern mining. J Syst Softw 145:164–179. https://doi.org/10.1016/j.jss.2018.08.032

Chapter 8
Architecting Green Mobile Cloud Apps

Key Considerations for Implementation and Evaluation of Mobile Cloud Apps

Samuel Jaachimma Chinenyeze and Xiaodong Liu

Abstract With the resource-constrained nature of mobile devices, and the resource-abundant offerings of the cloud, several promising optimization techniques have been proposed by the green computing research community. Prominent techniques and unique methods have been developed to offload resource-/computation-intensive tasks from mobile devices to the cloud. Most of the existing offloading techniques can only be applied to legacy mobile applications as they are motivated by existing systems. Consequently, they are realized with custom runtimes, which incurs overhead on the application. Moreover, existing approaches which can be applied to the software development phase are difficult to implement (based on manual process) and also fall short of overall (mobile to cloud) efficiency in software quality attributes or awareness of full-tier (mobile to cloud) implications.

To address the above issues, this chapter first examines existing approaches to highlight key sources of overhead in the current methods of MCA implementation and evaluation. It then proposes key architectural considerations for implementing and evaluating MCA applications which easily integrate software quality attributes with the green optimization objective of Mobile Cloud Computing—in other words, minimizing overhead. The solution proposed in the chapter builds on the benefits of already existing software engineering concepts, such as Model-Driven Engineering and Aspect-oriented Programming for MCA implementation, and Behavior-Driven Development and full-tier test coverage concepts for MCA evaluation.

S. J. Chinenyeze (✉)
Edinburgh Napier University, Edinburgh, Scotland, UK
e-mail: sjchinenyeze@gmail.com

X. Liu
Driven Software Engineering Research Group, Edinburgh Napier University, Edinburgh, Scotland, UK
e-mail: x.liu@napier.ac.uk

C. Calero et al. (eds.), *Software Sustainability*,
https://doi.org/10.1007/978-3-030-69970-3_8

8.1 Introduction

Mobile cloud computing (MCC) is a well-known technique used to address the challenges—such as limited performance and constrained power capacity—commonly faced by mobile devices by use of cloud computing as a surrogate. Mobile Cloud Applications (MCA) are applications that leverage the MCC technique. What makes MCC particularly popular is the rich resource offering and high availability of the cloud computing infrastructure. And due to the kind of metrics which MCC uses for mobile devices, such as energy efficiency, it is generally known as a "green software" approach. Research studies propose various MCC techniques for addressing the aforementioned mobile challenges. The aim of this chapter is to present the knowledge base and reasoning around the MCA domain, and some identified challenges in the domain, and consequently offer some directions/key considerations for an improved implementation and evaluation process in real-life scenarios—mainly based on renowned software engineering techniques.

The documentation starts by presenting a background to green software, and further develops the thesis on how MCC is a green software technique. MCA is then introduced, from its architectural composition (what it's made of and how it works) to the green metrics used for MCAs to the taxonomy of apps best suited for MCAs. The work then dives into the details of the MCA architecture, exploring the challenges and proposing solutions. This is similar for the MCA evaluation approach. Finally, the chapter concludes with a summary of the outlined solutions.

8.2 Green Software

8.2.1 Definition

Green software [1–3] is a subset of the green computing initiative, which involves reducing the overall impact of IT on the environment. The initial green computing initiative, however, focused more on the computing hardware—investigating optimal processes for the manufacture, usage, and disposal of computing hardware products with minimal impact on the environment.

As the research on green hardware thrived, the outcome presented the need to apply similar optimal processes to software—i.e., the need for green software—since software is the driver of hardware utilization. For example, research shows that IT hardware has a direct impact on the environment, which is inherent in software, as application inefficiencies such as inefficient algorithms and high resource (e.g., CPU) usage, are sources of high energy consumption [3, 4]. Furthermore, as the total electrical energy consumption by computer equipment increases, there is a consequent increase in greenhouse gas emissions. Moreover, each personal computer in use generates approximately a ton of CO_2 every year [5], and personal computers are only useful in terms of the number and kind of software applications they run. In sum, software has indirect environmental implications.

The term "green software" is therefore used to refer to software applications that efficiently monitor, manage, and utilize underlying resources with minimal or controlled impact on the environment [1–3, 6]. Green Software Engineering is a newly coined term and a branch of software engineering that is increasingly gaining interest, aiming toward improving existing software design and implementation approaches to achieve energy-efficient or resource-efficient software. Green computing presents two key roles played by software in sustainability, which are as follows:

- Software as a "green" enabler: where software contributes as a tool to monitor and help reduce the resource and environmental footprint of hardware systems and other industrial processes. This is also referred to as *Greening by Software*.
- Software as a "greening" target: where software itself is improved or optimized so that its execution and lifecycle meet green initiatives—such as energy efficiency. This is also referred to as *Greening in Software* [3, 7, 8].

Green software engineering focuses on optimizing software in such a way that its process or execution is energy efficient—i.e., greening in software. In the rest of the chapter *green software* will be used to mainly refer to greening in software, except where otherwise stated.

8.2.2 Green Software Objectives

Although software may not have direct environmental impact, it contributes indirectly through resources [6, 9]. For example, several works show direct correlation between efficient resource usage (such as CPU and memory) and improved energy usage by software [9]. More specifically, improvement in application energy usage has often been achieved by better monitoring and utilization of system resources. Therefore, as presented in Sect. 8.2.1:

- A key objective of green software is efficient resource usage and efficient energy usage [10, 11]. Energy efficiency and resource efficiency are often considered as congruent in the research.

The focus of green software on energy and resource efficiency was motivated by the success of green computing research for the datacenters domain [12–14], which largely targeted optimization of data centers for low *energy* and optimum hardware *resource* requirements.

Furthermore, software systems are often presented in terms of functional and non-functional requirements [15]. While functional requirements deal with the functionalities, capabilities, services, or behaviors of the system, non-functional requirements (or quality attributes) deal with requirements that support the delivery of system functionalities. Examples of quality attributes are performance, accessibility, security, and development efficiency, to mention but a few [15].

Thus far energy efficiency and resource efficiency (popular target of green optimization) are often considered in context and conjunction with other software quality attributes. Hence the varying themes of green software research: *performance (response or execution time)* and energy efficiency [16–18], optimal *accessibility* and resource efficiency [16], energy-efficient *secure* systems [19], *development efficiency* and energy efficiency [8], etc. The practice of implementing green metrics as a quality attribute in the context of other quality attributes—such as performance, accessibility, and security—is often imbibed by current research as a means to explore trade-offs of green optimization, while keeping *in sight* software quality assurance. This trade-off capability of the software product to meet the current needs of a set (required) functionality—say resource usage—without compromising the ability to meet future needs—say changing workload/performance—is often referred to as Sustainability of Software [2]. This is a core green software objective. Thus:

- Another key objective of green software is to achieve greenness as a quality attribute while considering other software qualities such as performance and availability. This means, for example; achieving energy efficiency (green metrics) with little or no performance compromise (software quality).

8.2.3 Green Software Approaches

Green software optimization targets two main artifacts: the process and code [2, 6]. Artifacts or assets is the term used to describe what is being optimized. We have classified green approaches into conceptual, algorithmic, and augmentation.

Green optimization approaches which target the software process are often abstract and systemic, and sometimes holistic—these are the *conceptual approaches*. On the other hand, there are approaches which are more specific regarding monitoring and improving code execution—these are examples of *algorithmic approaches*. There is also a third kind, which employs strengths of successful computing paradigms (and can be a hybrid of conceptual and algorithmic approaches) to solve a domain problem. For the purpose of this chapter, we have called these *augmentation approaches*.

8.2.3.1 Conceptual Approaches

Conceptual techniques such as architectures or models present a comprehensive plan required for achieving green computing [20], and can span multiple phases of the Software Development Life Cycle (SDLC). An example is the GREENSOFT model which adopts a layered approach to software sustainability, to structure concepts, strategies or guidelines, activities, and processes for (1) green software and (2) its engineering [6]. The aspect of GREENSOFT model which focuses on the engineering of green software (i.e., (2), as given in statement above) adopts a lifecycle approach to investigate optimization concepts for various phases of the SDLC. In practice, however, existing green software conceptual models do not integrate well with the SDLC, and this means that they may be more effective for business planning but less effective in a software implementation context. The conceptual approach is an area that could be better explored for greening in software in various domains of applications, since software models are effective at solving domain problems. In this chapter we will apply a conceptual model approach to software engineering (i.e., Model-Driven Engineering) to solve MCA domain problems.

8.2.3.2 Algorithmic Approaches

Algorithmic approaches are techniques that directly apply to or make changes to the software code. These include (a) refactoring for efficient resource usage, (b) use of energy-aware custom runtimes which manipulate the program's execution or codebase, and (c) green compilers or IDEs.

(a) *Refactoring techniques*

Refactoring techniques aim to make changes to the structural composition of the system in such a way that the new code base or optimized component uses fewer resources to accomplish the same or even more tasks (e.g., Hsu et al. [21]). In green software, optimizing the code base can warrant structural change which leads to a more optimized architecture. For example, the research in Zhong et al. [22], through comparison of two commonly used distributed architectural patterns, shows that the choice of architecture adopted in a software program affects its energy consumption.

(b) *Custom runtimes*

Custom runtimes are additional codebase—often independent of the functional features of the system—implemented in a software application to aid its efficient use of resources or energy (e.g., Morgan and MacEachern [23] and Tayeb et al. [24]). Custom runtimes are often used for executing custom optimization logic which is otherwise foreign to the base runtime of a program. The runtimes may comprise monitors (power monitor, for energy awareness; or resource monitor, for resource awareness) which monitor different environmental states in order to make an optimization decision at runtime.

(c) *Green compilers*

Green compilers are used for generation of optimized codebase for efficient use of resources or energy [21, 25, 26]. Green compilers are targeted toward specific resources such as CPU optimization or GPU optimization, and are therefore often vendor specific as well as resource-type specific.

8.2.3.3 Augmentation Approaches

Augmentation approaches are techniques which employ the strengths of successful computing paradigms—and can be a hybrid of conceptual and algorithmic approaches—to solve a problem in a domain. Below we present two domains, and their example augmentation approaches found in the research.

(a) *Cloud domain. Example augmentation approaches: Virtualization, Load-balancing, Aspect-orientation, Context-awareness*

The domain problem which green IT aims to solve within the cloud is overutilization (runtime bloat) or underutilization (runtime waste) of resources. This directly translates to increasing energy demands in data centers.

Several approaches have been explored to address this domain problem (e.g., Yanggratoke et al. [27], Hsu et al. [21], Yamini [28], Chen and Kazman [29]). A popular software-based approach is load-balancing, which deals with even distribution of workload across interconnected servers to mitigate overutilization or underutilization of resources [21, 30–32].

Some examples are based on external monitoring and load-balancing, in other words power-aware or resource-aware load-balancing, such as shown in Hsu et al. [21]—an example of *greening by software*. However, there are also works that explore the application of *greening in software* for the cloud. Examples of these are, to name but two: (1) a context-aware *Virtualization* technique—a technique which optimizes resource provisioning by employing benefits from the *Virtualization* paradigm and based on the awareness of user-defined rules [31]; and (2) a context-aware *Aspect-Oriented* model—a technique for resource optimization which employs benefits of the *Aspect-oriented* paradigm and performs load-balancing based on awareness of application real-time resource context [32].

(b) *Mobile domain. Example augmentation approach: Mobile Cloud Computing*

The domain challenge with the mobile device is its limited resources and limited power supply—battery based.

To address the mobile domain problem, quick and easy solutions have been proposed in the form of mobile app tools which provide monitor-and-tweak functionalities to extend battery life. An example is the App power monitor which comes with the Android OS and other power modeling tools [33] to monitor and shut down services which are not being used—these are examples of *greening by software*.

However, with the ever-increasing demand for mobile devices, advances have been made in green software for mobile domains to address the domain challenges and improve user experience, at a more efficient and intuitive scale. A popular technique known as computation offloading has been massively adopted in current works. This is a technique that adopts the augmentation approach by the leveraging cloud computing paradigm to solve the mobile domain problem. In other words, this is achieved by using the cloud as a surrogate for the execution of computation- or data-intensive tasks [34, 35]. This computation augmentation approach is known as Mobile Cloud Computing (MCC). Although many works have shown MCC to be efficient and intuitive, it is not an easy solution to implement, due to difficulty in replicating research propositions in real-world scenarios. Furthermore, some MCC solutions are automated with runtimes, which makes it promising to replicate the solution; however, in many cases, at best the automation becomes the main added value to the real-world application, as the key objective (the gains) of green metrics becomes difficult to replicate in a real-world scenario. In this chapter we propose a solution that makes it easy to apply the natural software engineering paradigm in MCC for implementation and evaluation to ensure ease of adoption and replication. Furthermore, given that Android is the most popularly investigated platform for MCA, as shown in the literature (e.g., Cuervo et al. [36], Chun et al. [37], Kwon and Tilevich [38], Zhang et al. [39], Hassan et al. [40], Justino and Buyya [41]), consequently, MCA applications in this chapter are based on the Android platform.

8.3 Mobile Cloud Applications

Mobile Cloud Applications (MCA) are mobile applications that employ the MCC paradigm, i.e., they are optimized by using *the cloud* as a surrogate for execution of resource-intensive tasks. Thus, MCC transforms mobile applications into two tiers—the mobile and cloud tier—MCA.

The mobile tier of MCA is composed of the mobile device, whereas the cloud tier is popularly implemented as clouds or fogs (cloudlets). Fogs or cloudlets are installations of small data centers at designated locations and connected to the larger cloud server via the Internet. Fogs are much closer to the end-user device than the cloud, with the aim of providing mobility at the cloud tier [35].

A number of MCC studies also propose the use of mobile services at the cloud tier, similar to cloud services but provisioned by a collection of mobile devices. In other words, mobile devices are considered as providers of the cloud, making up a peer-to-peer network as in Zachariadis et al. [42], Marinelli [43], and Wichtlhuber et al. [44]. This is also a form of fog computing; however, the focus is on the use of mobile devices for cloud provisioning, rather than cloudlets.

From a greening in software perspective, MCAs are realized through a key MCC technique known as Offloading [34, 35], as mentioned earlier. Thus, the research explores solutions for MCA through the lens of offloading schemes [36–40].

8.3.1 MCA Offloading Schemes

Computation/task offloading is an algorithmic mobile optimization technique that involves the transfer of computation- or resource-intensive tasks of a mobile

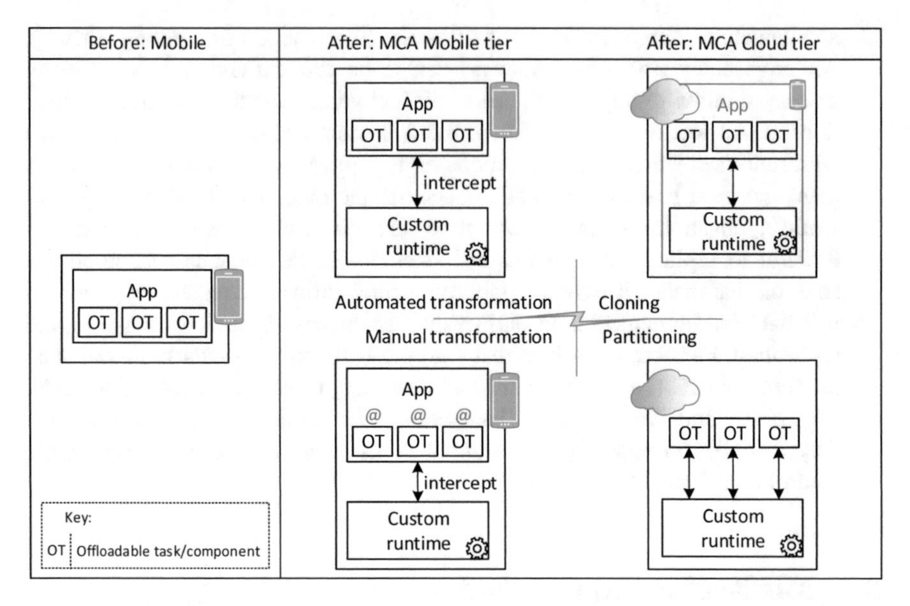

Fig. 8.1 MCA architecture characteristics

application to a remote system (cloud or fog) with higher processing capability for execution [39, 45]. Existing offloading schemes employ both code refactoring techniques and the use of custom runtimes; thus, an algorithmic approach (as presented in Sect. 8.2.3) is used and, more so, an augmentation approach, as offloading is done from a mobile to a cloud computing environment. MCA offloading schemes involve three main activities/features hinged on the offloading task. The activities, in simple terms, are identification, execution, and decision. Figure 8.1 depicts different features of these activities, as will be explained below.

8.3.1.1 *Identification* of Offloadable Task (Manual vs Automated Transformation)

To transform a mobile application into MCA, identification of offloadable tasks is a sine qua non activity. This can be achieved either manually or automatically. Schemes classified by *manual transformation* require source code modification for identification of the offloadable task. In this scenario, *annotations* are used by the developer to identify methods of the code that are resource intensive [36, 38]. The challenge with manual identification of offloadable components is that it is difficult to ascertain which components are actually resource intensive prior to execution/runtime. Moreover, a manually identified task may be tightly coupled to a *resource-dependent code*, making it a difficult and coarse-grained approach to optimization (even if the identified task may actually be resource intensive). *A resource-dependent code* is a code fragment or method which requires a mobile resource to execute and is thus tightly coupled to the resource, e.g., a piece of code that continuously reads GPS.

Schemes classified by *automated transformation* do not require source code modification in the identification of offloadable tasks. The automated transformation approach makes use of *static* and *dynamic analysis* of an application to identify the offloadable tasks [40, 46]. The purpose of the static analysis is to identify *resource-dependent code and filter it out, leaving potentially offloadable units*. Static analysis is achieved by performing a call-graph analysis on the bytecode of the application (whether packaged or not). The purpose of dynamic analysis is to estimate that a statically identified offloadable task yields benefit when executed remotely in the cloud. This estimation is achieved by comparing the local execution time of the offloadable task against its remote execution time. While static analysis does not require execution of the program, dynamic analysis does, and also requires that the offloadable task be set up in the cloud prior to the analysis.

Since automated transformation does not require source code modification (i.e., no need for annotations), the custom runtime stores the method signatures of offloadable tasks and intercepts any methods at runtime which have their signature stored in the repository of the custom runtime [37, 40]. Automated transformation is explored for legacy systems—where source code may not be available for refactoring—but it can also be used for new applications where source code is accessible. However, the manual approach is not only ineffective but also limited to scenarios where the source code is accessible.

8.3.1.2 Remote *Execution* of Offloadable Task (Partitioning vs Cloning)

The feature that achieves the remote execution of offloadable tasks is referred to as the *offloading mechanism* in the literature [38, 40]. This feature describes the structural composition of the cloud tier after the MCA refactoring process. To execute the offloadable tasks remotely, the cloud tier can be set up either as a clone of the mobile device (i.e., cloning) or as independent components executed remotely (i.e., partitioning).

Cloning [36–38] involves the setup of a virtual mobile device in the cloud. The full mobile application is also installed on the virtual device and executes remotely at the same time as the local application. The cloning approach works by state synchronization/checkpointing. In other words, when a check-pointed state (i.e., thread) is reached, a snapshot is created for fault-tolerance and the state of execution is offloaded to the cloud, which continues execution on the virtual device (in the cloud), after which the final state (of remote execution) is synchronized with the local state.

Partitioning [39, 40] involves the setup of identified offloadable tasks as independent components in the cloud. In partitioning, a virtual device is not required. Partitioning works by use of sockets to transmit execution parameters to the cloud. The component in the cloud then listens for socket connections and processes the mobile request using the parameters sent. The response is in turn sent to the mobile tier after execution using socket API.

8.3.1.3 *Decision* Making (Static vs Dynamic Thresholds)

The decision-making feature—*decision maker*—is a feature in offloading schemes which is used to decide when to offload or when not to offload. Decision making can be based on static thresholds, which use fixed values for decisions [38], or dynamic thresholds, which use machine learning algorithms such as multi-layer perceptrons [40]. In either case, these algorithms employ varying environmental factors in offload decision making. Hassan et al. [40] state that the more environmental factors considered in the decision-making process, the greater the depth of accuracy of the decision maker. However, accuracy is traded off for an element of overhead due to much monitoring, as shown in Hassan et al. [40].

8.3.2 MCA Environmental Factors

Offloading schemes makes decisions by monitoring and learning from the environmental factors of MCA. Any number of factors can be used in decision making, ranging from *one*, e.g., data size [38], to *a few*, e.g., network bandwidth and latency [36], to *all factors* [40] depending on the kind of application, as will be explained below. The awareness or monitoring of environmental factors is also referred to as *context awareness* [47–49]. The generally investigated factors impacting MCAs are as follows.

8.3.2.1 Mobile CPU Availability

Mobile CPU availability is measured in percent and is of particular importance for computation-intensive tasks. In other words, the lower the percentage of CPU availability, the greater the chance of mobile energy consumption or performance compromise for a computation-intensive task. Thus, the objective is to only execute a (computation-intensive) task on the mobile device when the percentage of CPU availability is higher or at least above a set threshold. Mobile CPU availability is obtained programmatically by examining the */proc/stat* files in Android to compute the percentage of CPU availability.

8.3.2.2 Mobile Memory Availability

Mobile memory availability is measured in percent and is of particular importance for data-intensive tasks. In other words, the lower the percentage of memory availability, the greater the chance of mobile energy consumption or performance compromise for a data-intensive task. Thus, the objective is to execute a (data-intensive) task on the mobile device only when the percentage of memory availability is higher or at least above a set threshold. A higher-bound mobile CPU and memory availability is useful for determining when to execute a task on a mobile device. Mobile memory availability is obtained programmatically by examining the */proc/meminfo* files in Android to compute the percentage of memory availability.

8.3.2.3 Cloud CPU Availability

Cloud CPU availability is measured in percent and is of particular importance for computation-intensive tasks. In other words, the lower the percentage of CPU availability, the greater the chance of mobile energy consumption or performance compromise for a computation-intensive task. Thus, the objective is to execute a (computation-intensive) task on the cloud when its percentage of CPU availability is higher or at least above a set threshold. The concept is that avoiding offloading to the cloud when the cloud CPU is overworked can curtail mobile performance compromise. Cloud CPU availability is obtained programmatically by examining the */proc/stat* files in a Linux-based server to compute the percentage of CPU availability.

8.3.2.4 Cloud Memory Availability

Cloud memory availability is measured in percent and is of particular importance for data-intensive tasks. In other words, the lower the percentage of memory availability, the greater the chance of mobile energy consumption or performance compromise for a data-intensive task. Thus, the objective is to execute a (data-intensive) task on the cloud when its percentage of memory availability is higher or at least above a set threshold. The idea is that avoiding offloading to the cloud when the cloud memory is overworked can curtail mobile performance compromise. A higher-bound cloud CPU and memory availability is useful for determining when to offload a task to the cloud. Cloud memory availability is obtained programmatically by examining the */proc/meminfo* files in a Linux-based server to compute the percentage of memory availability.

8.3.2.5 Network Bandwidth

Network bandwidth is the average rate of a successful data transfer through a network communication path. It is measured in bits per second and is achieved programmatically by sending packets to and from the server to measure the bandwidth. The objective of monitoring the bandwidth is to offload a task when the bandwidth is higher than a set threshold. The idea is that the higher the bandwidth, the greater the tendency for mobile energy or performance savings.

8.3.2.6 Network Latency

Network latency is the time interval or delay between request and response over a network communication path. It is measured in milliseconds and is similar to bandwidth. It is achieved programmatically by sending packets to and from the server to measure the latency. The objective of monitoring latency is to offload a task when the latency is lower than a set threshold. The concept is that the lower the latency, the greater the tendency for mobile energy or performance savings.

8.3.2.7 Data Size

Data size is the size of the data transmitted over the communication network. It is measured in series of bytes (i.e., B, KB, and MB) and can be achieved programmatically by checking the byte size of the request packet prior to client socket transmission. The objective of monitoring the data size is to offload a task when the data size is lower than a set threshold. The idea is that transmitting larger data packets over the network could result in increased mobile energy usage or performance compromise.

The popularly investigated metrics in the literature for MCA offloading schemes are performance and energy efficiency, and the aforementioned environmental factors are used to optimally attain these green metrics. The following section presents the most frequently explored MCA green metrics, including additional metrics we view as relevant to MCA.

8.3.3 MCA-Associated Green Metrics

Following the green software objective (Sect. 8.2.2), the green metrics are energy and resource efficiency [10]. In the context of MCAs, the core investigated green metric is *mobile energy efficiency*, given that the focus of MCA offloading schemes is the optimization of mobile application. However, MCA is two tiered, i.e., also involving the cloud tier, and thus this research presents *cloud resource efficiency* as another relevant green metric for MCA.

Furthermore, the green software objective also investigates trade-offs based on other software qualities, and the popularly investigated software quality in the literature for MCA offloading schemes is *mobile performance*. In this chapter, *software availability* is also presented (at both mobile and cloud tier) as a relevant software quality for MCAs.

8.3.3.1 Mobile Performance

According to Bass et al. [15], performance is how long it takes an application to respond to an event. The key driver of the advancement in mobile computing is the portability of mobile devices, which is defined by fluidity and ease of operation [45, 50, 51]. From the user perspective, the ease of operation or usability of a mobile application is critically dependent on its performance. Thus performance is a crucial MCA metric, as shown in the literature (e.g., Cuervo et al. [36], Chun et al. [37], Kwon and Tilevich [38], Zhang et al. [39], and Hassan et al. [40]), and is popularly explored in the context of MCAs as a trade-off software quality for mobile energy efficiency.

In MCA, mobile performance is often measured by computing the difference between the time of call (or request) to an offloadable task and the time of result (or response) after execution of the offloadable task. While *call* and *result* refers to a scenario where the offloadable task is executed on a mobile device, *request* and *response* refers to when it is offloaded to the cloud. The *time* is representative of timestamp, measured in *ms*, and often computed programmatically using the Java timestamp utility as in Zhong et al. [22].

8.3.3.2 Mobile Energy

According to Johann et al. [26], energy efficiency is the ratio of useful work done to used energy. In other words, it is the amount of energy incurred for executing a task.

Energy efficiency is derived from three quantities: power, time, and work done [8, 26]—in this way, energy efficiency is used in the comparison of two or more entities where their useful work done is likely to vary, as in the case of Zhong et al. [22] and Johann et al. [26]. However, in a situation of comparison between entities of similar work done or singular evaluation, energy efficiency is congruent with energy usage, which is based on two quantities: power and time. Consequently, the terms energy efficiency and energy usage are used interchangeably in the MCA research. Power Tutor [52] is a popularly adopted tool in the literature [36–40] for mobile power monitoring.

8.3.3.3 Cloud Resource

Achieving cloud resource efficiency in a mobile cloud environment requires care so as not to compromise mobile performance (see the cloud environmental factors in section 8.3.2). Resource efficiency in servers (the cloud) is often achieved through load-balancing [21, 30–32].

Although cloud resource efficiency is not often explored in the research on MCAs, investigating resource efficiency/usage for the cloud can be achieved using the core impacted resource of the cloud, i.e., the CPU and memory resources. Thus, for cloud resource usage of MCAs, percentage of CPU utilization and memory utilization are the key metrics. Percentage of CPU utilization and memory utilization can be measured by examining the */proc/stat* and */proc/meminfo* files in a Linux-based server.

8.3.3.4 Software Availability

According to Bass et al. [15], availability is the probability that a system will be operational when it is needed. In other words, availability is concerned with the consequences of a system failure. Most research does not take software availability into consideration in the implementation of MCA schemes. This category of schemes uses only network exception catch (e.g., Cuervo et al. [36], Chun et al. [37], Zhang et al. [39]). Moreover, several studies which consider availability investigate only at the mobile tier (e.g., Kwon and Tilevich [38]). Availability is achieved at the mobile tier by implementing a time limit for how long the mobile device can wait for the cloud to complete the execution of a request. When the time limit has elapsed, the execution is made on the mobile tier. Availability can also be implemented in a similar manner for the cloud tier.

The mobile tier availability time limit can be obtained by measuring the acceptable network communication time (to and from the cloud) plus the acceptable cloud execution time. The time limit for the cloud tier can be obtained by measuring only the acceptable cloud execution time.

Availability is a software quality relating to performance and therefore realized in *ms*. At the mobile tier, availability and performance are congruent; however, at the cloud tier availability is distinctly specified while performance is not.

8.3.4 Application Taxonomy

The application taxonomy presents the classification for applications in which MCAs have been explored. The MCA taxonomy has been derived by exploring the case studies used in the evaluation of offloading schemes, as shown in Table 8.1. Furthermore, the offloading schemes used to generate the taxonomy span across the characteristic features of MCA architectures (presented in Sect. 8.3.1) and are based on Android apps. The offloading schemes reviewed are: (1) POMAC[1]—categorized under automated transformation schemes, (2) EFDM[2]—categorized under manual transformation schemes and cloning schemes, and (3) DPartner—categorized under partitioning.

As shown in Table 8.1, three taxonomies are identified from the MCA literature:

[1]POMAC: Properly Offloading Mobile Applications to Clouds.

[2]EFDM: Energy-Efficient and Fault-Tolerant Distributed Mobile Execution.

Table 8.1 Application taxonomy

S/N	Case study apps	Taxonomy	Sample offloading schemes
1	Picaso [53]	Data intensive	POMAC [40, 46]
2	MatCalc [54]	Data intensive	POMAC [40, 46]
3	MathDroid [55]	Data intensive	POMAC [40, 46]
4	NQueen [56]	Computation intensive	POMAC [40, 46]
5	Droidslator [57]	Hybrid	POMAC [40, 46], EFDM [38]
6	Mezzofanti [58]	Computation intensive	POMAC [40, 46], EFDM [38]
7	ZXing [59]	Data intensive	POMAC [40, 46], EFDM [38]
8	JJIL [60]	Computation intensive	EFDM [38]
9	OsmAnd [61]	Computation intensive	EFDM [38]
10	Andgoid (Zhang et al. [39])	Hybrid	Dpartner [39]
11	Linpack (Čokulov [62])	Computation intensive	Dpartner [39]
12	XRace [63]	Hybrid	Dpartner [39]

Note that the references are appended to the Apps links to the source code or Google Play app

- Computation-intensive applications.
- Data-intensive applications.
- Hybrid applications.

8.3.4.1 Computation-Intensive Applications

Computation-intensive applications are a class of mobile applications that are highly or significantly dependent on the computing power, i.e., the CPU resource, of the mobile device. The core benefits of offloading schemes are realized with computation-intensive applications, as this class of applications consume the most battery power from the mobile device. Most computation-intensive applications fall within the category of gaming applications and media processing applications—such as face recognition apps and optical character recognition apps—as shown in the literature [36, 40, 46, 64].

8.3.4.2 Data-Intensive Applications

Data-intensive applications are a class of mobile applications that devote most of their processing time to I/O and data manipulation. Due to the focus on manipulation of data, these applications make more use of memory than the processing power. In most scenarios data-intensive applications do not consume significant mobile energy, unless in situations where the data-intensive components also require extensive computation. The literature [38, 46] has emphatically shown that offloading applications that do not consume significant mobile energy (such as data-intensive applications) can result in mobile performance compromise or even slight throttle in energy usage.

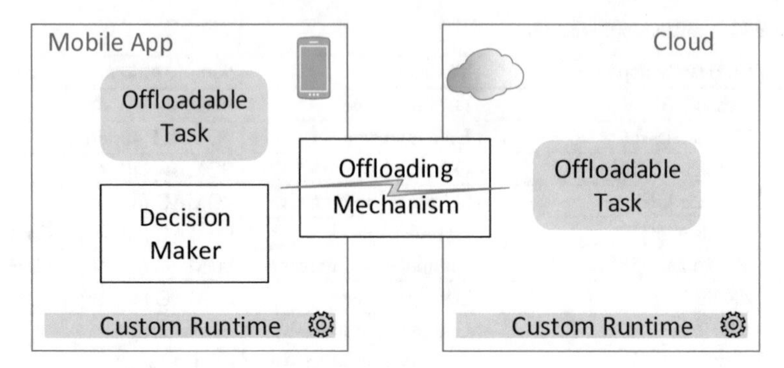

Fig. 8.2 MCA architecture (based on custom runtime)

8.3.4.3 Hybrid Applications

Hybrid applications are applications that are composed of different offloadable tasks with at least one computation-intensive task and at least one data-intensive task. As mentioned earlier, offloading such tasks will likely save energy only if they consume significant mobile energy.

8.4 MCA Optimization Approach

This section identifies the challenges associated with the approaches used for designing MCAs, i.e., the optimization or offloading schemes. Furthermore, key considerations toward an improved solution are presented and also justified.

8.4.1 Gaps in Existing Approaches

This section reviews the main studies that have addressed MCAs, particularly offloading techniques for mobile performance and energy improvement. The challenges in related work are presented in terms of overheads in the components that make up the generic MCA architecture (illustrated in Fig. 8.2). Consequently, the gaps motivating this research are highlighted from the review.

8.4.1.1 Challenges of Offloadable Tasks

The key challenge with offloadable tasks is the *identification phase*. A task is identified for offload if it possesses chances of performance or energy improvement when executed remotely, i.e., its remote execution time is shorter than local time.

Table 8.2 Comparison of offload models

System	Identification of offloadable task		Decision maker		Offloading mechanism	
	Fully automatic	Fine granularity	Thresholds	Parameters	Type	Custom runtime
Kwon et al. [38]	No (annotation)	No	Static	Resource	Cloning	Yes
MAUI [36]	No (annotation)	No	Dynamic	Resource	Cloning	Yes
Hassan et al. [40, 46]	Yes	No	Dynamic	Resource (full)	Partitioning	Partial (machine learning)
Native Offloader [65]	Yes	No	Dynamic	Resource	Partitioning	Yes

N.B. This table is not an exhaustive list of the models explained in the related work, but a list of distinct representative models

A *key constraint* impacting the performance gain of an offloadable task is its dependence on mobile-only resources, such as a sensor or camera.

Zhang et al. [66] adopt a shortest path algorithm to identify an optimal cut which minimizes offloading overhead. However, this does not take into account the aforementioned constraint when identifying an offloadable task. Elicit [46] uses the shortest path approach for identifying offloadable tasks, and taking the constraint into account provides better performance gain. In the literature offloadable tasks can either be identified manually using annotations or automatically through static/dynamic analysis, as shown in Table 8.2. The automatic identification technique is more *development efficient*—as per automated, *and accurate*—as per code analysis [46, 65].

Furthermore, not all offloaded tasks prove to be performance or energy efficient, particularly the data-intensive applications, as shown in the literature [39, 40, 46], thus, raising thus question of the effectiveness of the approach used in *identification* of offloadable tasks. Two concepts can be deduced: either the task in question was wrongly identified as offloadable or there was an overhead during runtime which was unaccounted for during the identification process. The second concept is likely a more valid point, because if decision-making and offloading components add an overhead during runtime which was not considered during identification, then a particular offloading task may never yield performance or energy-efficiency benefits, i.e., in a scenario where the runtime overhead overshadows the offloading gain. Existing work [38–40, 46, 64] to the best of our knowledge does not consider this challenge, as their offloading task identification does not include all MCA components during evaluation. It can be argued that the decision maker would effectively allow such scenarios to execute locally. However, the decision-making process is a required precondition, and thus overhead would have already been made. In other words, the *identification process* offered by existing offload models is not fine grained (as shown in Table 8.2). Thus, an effective identification process must take into account all the MCA components of Fig. 8.2 for fine granularity.

Moreover, with automated transformations, dynamic and static analyses are used independently in the research. Hassan et al. [46] adopted dynamic analysis—i.e., transformation at bytecode level or runtime—for existing applications as there was no access to the source codes of the apps. Thus, the case for exploring the use of both static and dynamic analysis throughout the MCA development cycle has not been well explored.

8.4.1.2 Challenges of the Decision Maker

A good way to understand the decision-making component is as a kind of monitor and comparator, rather than just a set of if else conditions.

As a monitor: Conditions are executed by checking (i.e., monitoring) the *actual* environmental state. In MCA a given environmental state is defined by different factors/parameters: mobile device CPU and memory availability, network bandwidth and latency, cloud CPU and memory availability, and transmitted data size, as used in the literature [39, 40, 46]. Note that the parameters employed by existing current work are resource based (i.e., CPU, memory, and network)—also see Table 8.2. For accuracy in the decision-making process, it is critical that these factors be captured by the decision maker.

Some research such as Hassan et al. [40] takes into account all of the aforementioned factors for decision making, thus providing a more accurate picture of the environmental state. Most works, however, only consider one or a few factors during decision making, for example data size alone [38], or a combination of bandwidth and latency [36]. Whichever combination of factors is used, an overhead is added to the application performance. For example, monitoring network bandwidth and latency requires sending packets to and from the communicating endpoints (e.g., Hassan et al. [40, 46]), which contributes its own overhead. Thus, the more factors considered, the greater the overhead introduced and the greater the accuracy in decision making. This is a trade-off to take into account.

As a comparator (or thresholding mechanism): The actual environmental state, obtained by measuring the aforementioned factors, is compared against a (set of) predetermined value(s). Hassan et al. [40] use a machine learning algorithm, specifically multi-layer perceptron, as a comparator due to the use of multiple environmental factors in the monitoring. Expected environmental factors are obtained as training data, collected for *offload condition*, i.e., when remote execution time is shorter than local time, and *non-offload condition*. Kwon and Tilevich [38] use a single thresholding approach on data size for offload condition. However, this is ineffective, given the varying factors affecting MCA, and the unpredictable nature of the environment. Cuervo et al. [36] use a linear regression model on bandwidth and latency, which also fails to compare other factors.

As noted, the decision maker decides when or when not to offload, based on a satisfied scenario, obtained from monitored environmental factors. The satisfied scenario is the scenario where the remote execution time of an offloadable component is shorter than its local or mobile execution time. Thus, the core factor is the elapsed execution time. However, as this factor cannot be determined explicitly before runtime, the aforementioned factors, with support from dynamic thresholds, e.g., learning models [36, 40], or static thresholds [38], are used to determine the time factor, as shown in the literature. Thus, an effective decision-making process must effectively predict elapsed time for an environmental state with minimal overheads.

8.4.1.3 Challenges of Offloading Mechanism

There are two key categories of the offloading mechanisms used in the literature: cloning and partitioning (see Table 8.2). Many offloading models implement their offloading mechanisms as runtime engines.

As the term implies, cloning involves execution of a virtual device on the cloud. It is based on checkpointing, i.e., adding fault-tolerance by saving snapshots, thus creating more overhead in offloading due to state synchronization. Chun et al. [37] state that cloning can require a data transfer as high as 100 MB.

Partitioning makes use of remote procedure calls. Unlike the cloning approach, which requires the virtual device to run on the cloud, the partitioning approach only requires the offloadable component to execute on the cloud. Thus, it is more efficient—saving energy and time—compared to cloning.

Whether cloning or partitioning, the key challenge is the dependence on custom runtime engine (e.g., Kwon and Tilevich [38] and Justino and Buyya [41]). As such deeply layered frameworks contribute runtime bloat [8, 67, 68], an appropriate mechanism would need to be simplified—and without dependencies on custom runtimes for minimal overheads.

8.4.2 Considerations for Improved Solutions

Following the challenge of offloadable tasks, it is critical that the process for identifying offloadable tasks take into account the cost of decision making and offloading. In other words:

- **Solution 1:** A task should be qualified for offload if and only if the combined overhead of the decision-making component, offloading mechanism, and remote execution is less than local.

Offloading any task which compromises the aforementioned condition will always compromise performance, even if the remote execution time is shorter than that of local.

As mentioned earlier, the core of the decision maker is the ability to predict the elapsed time prior to offload, so as to know if there will be gain or loss given the current environmental state. This prediction is currently obtained from different factors and using learning models (contributing overheads) or inaccurate thresholds. For better decision making:

- **Solution 2:** Adopt the use of an adaptive full-tier time-based decision making (at the mobile and cloud tier).

Adaptive means that the decision maker needs to adapt to various environmental states, and this is achieved by monitoring for environmental factors. Full-tier means that the monitoring is not only on the mobile, but also on the cloud tier. The objective of full-tier is to achieve software qualities at both mobile and cloud tier.

Full-tier time-based decision making means that the comparator or thresholding mechanism is based on a set of predetermined optimal execution times at both the mobile and cloud tier. The aim of using a set of predetermined optimal execution times is to achieve overall target green objectives while eliminating existing challenges of existing counterparts, e.g., the complexity and overhead introduced by machine learning models like MLP [40], and the inaccuracies of static thresholding.

Current works use custom runtimes to handle offloading; this is to dynamically intercept code and perform cloning or partitioning. However, these custom runtimes introduce runtime overhead. Moreover, they increase the size of the application itself. Although these techniques may be useful for existing systems, where there is no access to source code, they are impractical for building MCAs. To address the challenge of overhead caused by custom runtime:

- **Solution 3:** Adopt the use of sockets as offloading mechanism (this is a partitioning technique). Cloning techniques are more expensive and consume a lot of cloud resource as they require virtual mobile devices in the cloud. Partitioning would involve having only a copy of the offloadable code in the cloud.
- **Solution 4:** Adopt Aspects from Aspect-oriented Programming to dynamically intercept the application when it reaches the offloadable task. Aspects are lightweight constructs used to implement cross-cutting concerns [69]. They can be used to implement offloading logic in such a way that the MCA application code is not altered, thus maintaining a clean separation of concerns while eliminating the overhead of custom runtime.

As noted earlier in the chapter, the problem being solved by MCA is a domain problem—a mobile domain problem, an environment of constrained resources. Furthermore, the issues identified with the MCA solutions are also domain problems—involving the mobile and cloud domain, however juxtaposed, but targeting the mobile domain problem. In software engineering, Model-Driven Engineering focuses on exploiting domain models to effectively solve recurring domain problems by providing abstraction through high-level models. Thus, for an effective domain solution for MCA:

- **Solution 5:** Adopt the use of model-driven tools to encapsulate the aforementioned solutions alongside design patterns of the MCA app being built. This will simplify the development process, increase productivity, and enhance repeatability of good solutions [70–73].

8.5 MCA Evaluation Approach

This section presents the gaps and methodology associated with the evaluation techniques applied to MCAs. Consequently, steps and methodologies toward an improved solution are presented and justified.

8.5.1 Gaps in the Existing Approach

In this section, we use examples from the research to highlight the challenges and difficulties of the currently used MCA evaluation approach (i.e., the architecture scenario approach) in the evaluation and comparison of offloading schemes.

A Motivating Example Let us consider a situation in the development of mobile cloud applications. The choice of offloading scheme would be a critical decision, as it is the core functionality which transforms a mobile app to MCA [36, 38–40]. Assuming the development team chooses to use an existing scheme, they will need to evaluate and compare between other offloading schemes or other variations of a scheme. Two offloading schemes have been selected, one based on single thresholding [38]—ST for brevity—and another based on the multi-layer perceptron learning algorithm [40]—known as POMAC. Also selected is an Optical Character Recognition (OCR) Android app called Mezzofanti, which is used in the source literature to validate the schemes [38, 40]. From the source literature the computation-intensive offloadable component is the OCR functionality [38, 40]. The data presented in Table 8.2 is obtained from the source literature using WebPlotDigitizer [74].

To achieve the evaluation of individual schemes and comparison between the schemes—ST and POMAC—the mobile-centric architecture scenarios provided by the source literature are used. The term "mobile-centric" indicates that the approach provides green metrics results for only the mobile tier, i.e., mobile performance and energy usage. Using mobile-centric architecture scenarios which are prevalent in the research [36, 38–40], however, comes with challenges which make it difficult to come to a satisfactory conclusion for both schemes, in terms of evaluation and comparison.

The research challenges identified for mobile-centric architecture scenarios are presented below.

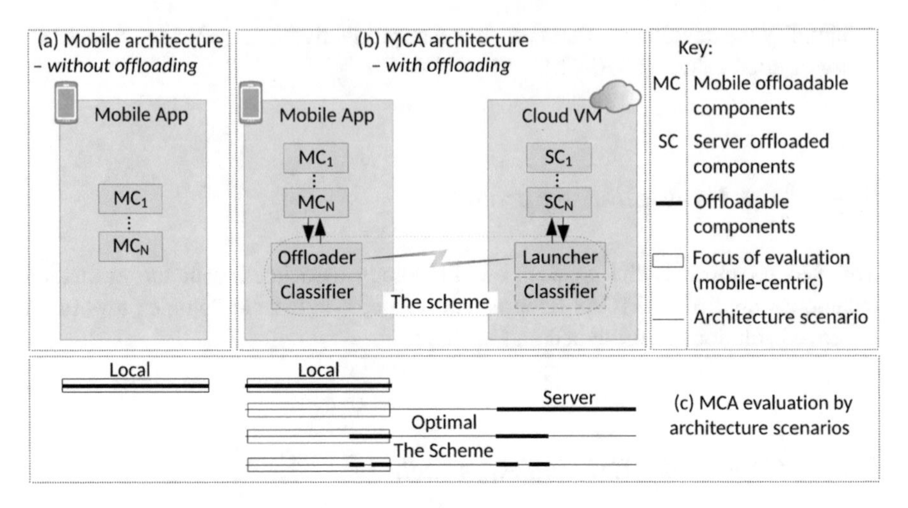

Fig. 8.3 MCA architecture showing mobile-centric architecture scenarios

Table 8.3 MCA evaluation and comparison by architecture scenarios

	ST [38]		POMAC [40]	
Arch.scenarios	Elapsed time (ms)	Used energy (J)	Elapsed time (ms)	Used energy (mJ)
Local	49,331.55	86.59	3930.33	4854.24
Server	27,673.79	63.86	34,873.15	19,839.42
Optimal	17,486.63	44.73	3986.21	4845.10
The scheme	10,347.59	41.33	4242.32	5085.80
Local % diff.	130.65	70.76	−7.64	−4.66
Server % diff.	91.14	42.84	156.62	118.38
Optimal % diff.	51.30	7.90	−6.23	−4.85

Note: *Local % diff.*, *Server % diff.*, and *Optimal % diff.* are the % difference of the scheme in comparison to local, server, and optimal scenarios respectively. A negative value is used to signify loss in energy or performance. Note that the metrics presented, i.e., elapsed time and used energy, are for the mobile tier

8.5.1.1 Inconsistency in Evaluation Results of Scenarios for an Offloading Scheme

To evaluate POMAC, Hassan et al. [40] define four[3] scenarios (illustrated in Fig. 8.3). The efficiency of POMAC is evaluated by comparing the POMAC scheme against other defined architecture scenarios, using % difference. Deducing from Table 8.3, for energy usage we can conclude that POMAC is approximately 5% inefficient compared to both local and optimal scenarios and 118% efficient compared to the server scenario. Although the local and optimal % differences seem to

[3]Four scenarios are defined by Hassan et al. [40] for evaluating POMAC: OnDevice, OnServer, Optimal, and POMAC.

Table 8.4 Variability of architecture scenarios

Summary of architecture scenarios	Scenarios adopted in ST [38]	Scenarios adopted in POMAC [40]
Local	Smartphone only	OnDevice
Server	Offloading w/all objects	OnServer
Optimal	Offloading w/necessary objects (delta)	Optimal
The scheme	Offloading w/threshold check	POMAC

arrive at the same conclusions, there is no clear relationship between the scenarios. This is shown by ST, which has approximately 71%, 43%, and 8% energy improvement based on local, server, and optimal respectively. This challenge makes it difficult to weigh a scheme based on easily verifiable values or conclusions.

8.5.1.2 Variability of Architecture Scenarios (Making It Difficult to Compare Between Offloading Schemes)

Different literature uses varying scenarios to evaluate proposed schemes. For example, Hassan et al. [40] define four[3] scenarios to evaluate POMAC, while Kwon and Tilevich [38] define five[4] scenarios to evaluate ST. Therefore, to establish a basis for comparison, scenarios will have to be matched (as shown in Table 8.4). This process introduces complexity in comparing schemes, especially since scenarios which may be congruent by inference may have slightly different definitions from each other based on the actual literature implementation. This introduces difficulty in communicating varying scenarios between the development teams and also a challenge to comparison.

The summary column of Table 8.4 is used to match the scenarios from the literature [38, 40]. Local is the execution of the application without any offloading. Server is a scenario where all offloadable objects are always executed on the server. Optimal is a scenario where only assessed objects are offloaded. Assessed objects are the objects identified as computation or data intensive. The Scheme is based on extending the previous optimal scenario with decision-making[5] mechanisms for offload. It refers to the proposed offloading schemes in the literature.

[4]Five scenarios are defined by Kwon and Tilevich [38] for evaluating ST: Smartphone only, Offloading w/All objects, Offloading w/Necessary objects, Offloading w/Necessary objects (delta), and Offloading w/Threshold check.

[5]Decision making is the check on the environmental conditions of the communications which influence the offloading. Decision-making mechanisms can be based on single (static) thresholds [38] or predictive learning [40].

8.5.1.3 Coarse Granularity of Evaluation

Different literature uses different levels of experimental rigor. For example, Hassan et al. [40] performed a more rigorous experiment for POMAC evaluation (as the scheme is based on MLP), compared to Kwon and Tilevich [38]'s experiment for ST which is not as rigorous. Comparing ST energy with POMAC (using the optimal scenario as reference) gives an approximately 8% gain in ST and 5% loss in POMAC. The case may be that in adverse environmental conditions the ST scheme fails to save mobile energy. Also, since the analysis is mobile-centric, it fails to provide the overall implications of a scheme's decision regarding whether to offload. This challenge makes it difficult to establish the overall efficiency of a scheme, i.e., the extent to which the scheme is mobile as well as cloud aware.

8.5.2 Methodology for a Solution

To propose a solution for the identified gaps in the existing MCA evaluation approach (i.e., mobile-specific architecture scenarios) we adopt concepts from Behavior-Driven Development and Software Test Coverage (Fine-Grained Testing).

8.5.2.1 Behavior-Driven Development (BDD)

A gap in MCA evaluation identified earlier is the variability of architecture scenarios, which makes it difficult to compare between offloading techniques. Since there is no standard for determining what scenarios to use for justifying the efficiency of an offloading technique, different literature or techniques use different scenarios. To curtail the aforementioned difficulty, an approach can be investigated to capture all the necessary environmental factors surrounding typical MCA scenarios, and perform an evaluation or comparison on the basis of these factors rather than varying scenarios.

For example, for a typical scenario (Table 8.4), whether server, optimal, or scheme, the environmental factors surrounding the efficiency of the application are mobile CPU and memory availability, server CPU and memory availability, and network bandwidth and latency [40, 46]. Rather than evaluate schemes by comparing against different scenarios which are all affected by the aforementioned factors,

- **Solution 1:** Evaluate and compare schemes on the basis of the environmental factors themselves which affect MCA schemes.

Evaluation: The implication of the proposed technique above is that to *evaluate* an offloading scheme S1, a result can be presented thus:

- The performance and energy usage of S1 are x and y respectively, given the aforementioned factors. *This is a simplified (straightforward to interpret) and efficient approach.*

  Rather than:

- The performance and energy usage of S1 is x and y respectively, compared to a scenario A, which is affected by its own uncontrolled factors, and another scenario B, which is also affected by its own uncontrolled factors, and yet another scenario C, which is also affected by its own unique factors. *This is the case when varying scenarios are used to evaluate a scheme.* Apart from introduced complexities, this approach is inefficient.

Comparison: To *compare* a second offloading scheme of interest, say S2 to the previous one, S1, the process would be performed as follows:

- Given that the factors of S1 and S2 are closely related compare S1 to S2.

  Assuming that S1 is more efficient, then the result can be presented as follows:

- S1 is x% and y% more performance and energy efficient than S2 given the factors.

  Rather than:

- Compare S1 to A and S2 to A; then S1 to B and S2 to B; then S1 to C and S2 to C.

  Assuming that S1 is more efficient, then the result can be presented as follows:

- S1 is x% and y% more performance and energy efficient than S2 in A, *and/or.*

  S1 is x% and y% more performance and energy efficient than S2 in B, *and/or*
  S1 is x% and y% more performance and energy efficient than S2 in C.

"And/or" means that in most cases S1 might not be more efficient in all the compared scenarios, and thus it is difficult to establish a concrete result for comparison using varying scenarios.

Note that for the proposed approach, the syntax is "**given** *environmental factors* **then** *assert results*." The above syntax is the core of Behavior-Driven Development (BDD). Thus, to curtail the aforementioned difficulty:

- **Solution 2:** Adopt the use of the BDD technique, which is based on simple clause semantics such as given, when, and then. This will also help to simplify software design decisions and can be automated by tooling.

BDD is a design approach to aid collaboration between non-technical contributors (such as business analysts or users) and software engineers. Consequently, BDD is geared toward a more verifiable and collaborative test process by being able to compare expected behaviors with actual results, following standard simplified scenarios constructed by simple language clauses, such as GIVEN, WHEN, and THEN [75].

8.5.2.2 Full-Tier as the New Fine-Grained Test Coverage for MCA

Also presented as a key challenge to the current multi-scenario approach adopted in the MCA evaluation process is the mobile-centric nature of the evaluation process. Thus, only the impact of an offloading scheme on the mobile device is estimated. However, MCA is composed of mobile and cloud tiers. Therefore, to address the coarse granularity of the current approach, an effective solution must take into consideration the mobile as well as the cloud resource impact of an offloading scheme.

Johann et al. [26] show that a fine-grained approach to energy measurement (using counters) aid reveals specific energy usage in relation to specific points of execution. Furthermore, in software testing, test coverage is a metric used to measure the extent of testing in respect to the code being executed. In MCA the actual execution involves the mobile and the cloud tiers; therefore, to achieve fine-grained testing with acceptable optimum coverage (i.e., a test coverage that reflects both tiers of MCA):

- **Solution 3:** Adopt a full-tier testing to measure across the mobile tier (for mobile metrics, e.g., mobile performance and energy usage) and cloud tier (for cloud metrics, e.g., CPU and memory usage).

With full-tier evaluation one can better understand if a scheme just keeps offloading to server, or if it checks server availability (i.e., robustness), thus ensuring that a scheme is aware of both mobile and cloud resource consumption.

8.6 Summary

This chapter presented the current state of the art in mobile cloud applications development. Consequently, it highlighted the key issues with the domain, in terms of the *optimization approach* and *evaluation approach*.

The chapter presented the gaps in the existing optimization approaches (replicated in Table 8.5, which includes proposed solution) as follows:

- Coarse granularity in identification of offloadable task which leads to unnecessary MCA transformation of mobile applications, which in turn results in performance overhead.
- Multiple parameter-based decision making (with intension of accuracy in environmental prediction), which leads to performance overhead.
- Runtime-dependent and development-inefficient offloading mechanism which incurs performance overhead and implementation complexities.

To address the identified gaps in existing optimization approaches the research proposed the use of a Model-Driven Approach with the following solutions:

Table 8.5 Summary of offload models including proposed solution

System	Identification of offloadable task		Decision maker		Offloading mechanism	
	Fully automated	Fine granularity	Thresholds	Parameters	Type	Custom runtime
Kwon et al. [38]	No (annotation)	No	Static	Resource	Cloning	Yes
MAUI [36]	No (annotation)	No	Dynamic	Resource	Cloning	Yes
Hassan et al. [40, 46]	Yes	No	Dynamic	Resource (full)	Partitioning	Partial (machine learning)
Native Offloader [65]	Yes	No	Dynamic	Resource	Partitioning	Yes
Proposed solution	Yes (static and dynamic analysis)	Yes	Dynamic	Time	Partitioning (aspect-oriented)	No (model-driven)

- Ensure that identified offloadable tasks will most certainly yield benefits, during optimization, prior to final deployment.
- Monitoring and decision making is based on execution times in full-tier as opposed to multiple environmental factors as parameters. Thus, the approach does not seek (or monitor) best path of execution (requiring extensive resource monitoring, thus causing overhead), but adopts a good-fit path with respect to execution time. That is to say that the decision to offload is made based on time, and control of remote execution is achieved based on time (as threshold).
- The scheme is based on Model-Driven Engineering, and thus mitigates the overhead caused by custom runtimes. Furthermore, the proposed scheme is based on partitioning (not cloning), which further eliminates runtime overheads of cloning solutions.

Furthermore, the chapter furthered the case for improving the existing MCA evaluation approach by presenting the inefficiencies in the existing state-of-the-art evaluation approach to MCA, i.e., the scenario-based approach. The current gaps were presented as follows:

- Inconsistency in evaluation results of scenarios for an offloading scheme.
- Variability of architecture scenarios (making it difficult to compare between offloading schemes).
- Coarse granularity of evaluation—focused on mobile implications of an MCA or its offloading scheme.

To address the identified gaps in the scenario-based evaluation approach the research proposed the use of a Behavior-Driven (BDD) approach with the following solutions:

- Use environmental factors as parameters for evaluation rather than varying scenarios. Applying environmental factors in the evaluation process would provide information on the environmental state (i.e., measurements) at a finer granularity, and since this is not within the optimization code, there is no cause for concern about performance overhead. Thus, the evaluation process is best suited for real-time environmental factors, rather than the optimization process, which can cause performance overhead every time offload decisions are made.
- Simplify the evaluation process using simple clauses, as offered by BDD—as exemplified in the BEFTIGRE solution [76].
- Adopt the concept of fine-grained software coverage testing, which takes into account all components of software for higher coverage. Specifically, this means taking into account the mobile tier as well as the cloud tier during MCA evaluation—a full-tier evaluation—as exemplified in the BEFTIGRE solution [76].

References

1. Dick M, Naumann S, Kuhn N (2010) A model and selected instances of green and sustainable software. In: Berleur J, Hercheui MD, Hilty LM (eds) What kind of information society? Governance, virtuality, surveillance, sustainability, resilience. Springer, Berlin, pp 248–259
2. Lami G, Buglione L, Fabbrini F (2013) Derivation of green metrics for software. In: Woronowicz T, Rout T, O'Connor RV, Dorling A (eds) Software process improvement and capability determination. Springer, Berlin, pp 13–24
3. Calero C, Piattini M (2015) Introduction to Green in Software Engineering. In: Green in software engineering. Springer International Publishing, Cham, pp 3–27
4. Rogers D, Homann U (2009) Application patterns for green IT. Archit J Green Comput. https://msdn.microsoft.com/en-us/library/dd393307.aspx. Accessed 18 Jan 2016
5. Murugesan S (2008) Harnessing green IT: principles and practices. IT Prof 10:24–33. https://doi.org/10.1109/MITP.2008.10
6. Naumann S, Dick M, Kern E, Johann T (2011) The GREENSOFT model: a reference model for green and sustainable software and its engineering. Sust Comput Inform Syst 1:294–304. https://doi.org/10.1016/j.suscom.2011.06.004
7. Reimsbach-Kounatze C (2009) Towards green ICT Strategies - assessing policies and programmes on ICT and the environment
8. Capra E, Francalanci C, Slaughter SA (2012) Is software "green"? Application development environments and energy efficiency in open source applications. Inf Softw Technol 54:60–71. https://doi.org/10.1016/j.infsof.2011.07.005
9. Steigerwald B, Agrawal A (2011) Developing Green Software | Intel® Developer Zone. http://software.intel.com/en-us/articles/developing-green-software. Accessed 18 Jan 2016
10. Bozzelli P, Gu Q, Lago P (2013) A systematic literature review on green software metrics. VU University, Amsterdam
11. Taina J, Mäkinen S (2015) Green software quality factors. In: Green in software engineering. Springer International Publishing, Cham, pp 129–154
12. Benini L, De Micheli G (2000) System-level power optimization: techniques and tools. ACM Trans Des Autom Electron Syst 5:115–192. https://doi.org/10.1145/335043.335044
13. Goiri I, Beauchea R, Le K, Nguyen TD, Haque ME, Guitart J, Torres J, Bianchini R (2011) GreenSlot: scheduling energy consumption in green datacenters. In: Proceedings of 2011

International Conference for High Performance Computing, Networking, Storage and Analysis on - SC '11. ACM Press, New York, NY, pp 1–11
14. Vasić N, Bhurat P, Novaković D, Canini M, Shekhar S, Kostić D (2011) Identifying and using energy-critical paths. In: Proceedings of the Seventh COnference on emerging Networking EXperiments and Technologies on - CoNEXT '11. ACM Press, New York, NY, pp 1–12
15. Bass L, Clements P, Kazman R (2003) Software architecture in practice, 2nd edn. Addison-Wesley Professional, Boston
16. You C-W, Chu H (2004) Replicated client-server execution to overcome unpredictability in mobile environment. In: 2004 4th Workshop on applications and services in wireless networks, 2004. ASWN 2004. IEEE, pp 21–29
17. Khan MA, Hankendi C, Coskun AK, Herbordt MC (2011) Software optimization for performance, energy, and thermal distribution: initial case studies. In: 2011 International green computing conference and workshops. IEEE, Orlando, FL, pp 1–6
18. Denti M, Nurminen JK (2013) Performance and energy-efficiency of scala on mobile devices. In: 2013 Seventh International conference on next generation mobile apps, services and technologies. IEEE, pp 50–55
19. Cano M-D, Domenech-Asensi G (2011) A secure energy-efficient m-banking application for mobile devices. J Syst Softw 84:1899–1909. https://doi.org/10.1016/j.jss.2011.06.024
20. Williams J, Curtis L (2008) Green: the new computing coat of arms? IT Prof 10:12–16. https://doi.org/10.1109/MITP.2008.9
21. Hsu C-H, Chen S-C, Lee C-C, Chang H-Y, Lai K-C, Li K-C, Rong C (2011) Energy-aware task consolidation technique for cloud computing. In: 2011 IEEE third international conference on cloud computing technology and science (CloudCom). IEEE, Athens, pp 115–121
22. Zhong B, Feng M, Lung C-H (2010) A green computing based architecture comparison and analysis. In: 2010 IEEE/ACM International conference on green computing and communications & international conference on cyber, physical and social computing. IEEE, pp 386–391
23. Morgan R, MacEachern D (2010) SIGAR - system information gatherer and reporter. https://support.hyperic.com/display/SIGAR/Home. Accessed 18 Jan 2016
24. Tayeb J, Bross K, Bae CS, Li C, Rogers S (2010) Intel energy checker software development kit user guide. https://goo.gl/Yrtbn9. Accessed 1 Jul 2016
25. Naik K, Wei DSL (2001) Software implementation strategies for power-conscious systems. Mob Netw Appl 6:291–305. https://doi.org/10.1023/A:1011487018981
26. Johann T, Dick M, Naumann S, Kern E (2012) How to measure energy-efficiency of software: metrics and measurement results. In: 2012 First international workshop on green and sustainable software (GREENS). IEEE, pp 51–54
27. Yanggratoke R, Wuhib F, Stadler R (2011) Gossip-based resource allocation for green computing in large clouds. In: 2011 7th International conference on network and service management. IEEE, Paris, pp 171–179
28. Yamini R (2012) Power management in cloud computing using green algorithm. In: 2012 International conference on advances in engineering, science and management (ICAESM). IEEE, Nagapattinam, Tamil Nadu, pp 128–133
29. Chen H-M, Kazman R (2012) Architecting ultra-large-scale green information systems. In: 2012 First international workshop on green and sustainable software (GREENS). IEEE, Zurich, pp 69–75
30. Baliga J, Ayre RW, Hinton K, Tucker RS (2011) Green cloud computing: balancing energy in processing, storage, and transport. Proc IEEE 99:149–167. https://doi.org/10.1109/JPROC.2010.2060451
31. Fang D, Liu X, Liu L, Yang H (2013) TARGO: transition and reallocation based green optimization for cloud VMs.In: Proceedings of the 2013 IEEE international conference on green computing and communications and IEEE internet of things and IEEE cyber, physical and social computing, GreenCom-iThings-CPSCom 2013, pp 215–223. doi:https://doi.org/10.1109/GreenCom-iThings-CPSCom.2013.56

32. Chinenyeze SJ, Liu X, Al-dubai A (2014) An aspect oriented model for software energy efficiency in decentralised servers. In: 2nd international conference on ICT for sustainability. Atlantis Press, Stockholm, pp 112–119
33. Zhang L, Tiwana B, Qian Z, Wang Z, Dick RP, Mao ZM, Yang L (2010) Accurate Online Power Estimation and Automatic Battery Behavior Based Power Model Generation for Smartphones. In: Proceedings of the eighth IEEE/ACM/IFIP international conference on Hardware/software codesign and system synthesis - CODES/ISSS '10. ACM Press, New York, NY, pp 105–114
34. Dinh HT, Lee C, Niyato D, Wang P (2013) A survey of mobile cloud computing: architecture, applications, and approaches. Wirel Commun Mob Comput 13:1587–1611. https://doi.org/10.1002/wcm.1203
35. Fernando N, Loke SW, Rahayu W (2013) Mobile cloud computing: a survey. Futur Gener Comput Syst 29:84–106. https://doi.org/10.1016/j.future.2012.05.023
36. Cuervo E, Balasubramanian A, Cho D, Wolman A, Saroiu S, Chandra R, Bahl P (2010) MAUI: making smartphones last longer with code offload. In: Proceedings of the 8th international conference on mobile systems, applications, and services - MobiSys '10. ACM Press, New York, Y, pp 49–62
37. Chun B-G, Ihm S, Maniatis P, Naik M, Patti A (2011) Clonecloud: elastic execution between mobile device and cloud. In: Proceedings of the sixth conference on computer systems - EuroSys'11. ACM Press, New York, NY, pp 301–314
38. Kwon Y-W, Tilevich E (2012) Energy-efficient and fault-tolerant distributed mobile execution. In: 2012 IEEE 32nd international conference on distributed computing systems. IEEE, pp 586–595
39. Zhang Y, Huang G, Liu X, Zhang W, Mei H, Yang S (2012) Refactoring android Java code for on-demand computation offloading. In: Proceedings of the ACM international conference on Object oriented programming systems languages and applications - OOPSLA '12. ACM Press, New York, NY, p 233
40. Hassan MA, Bhattarai K, Wei Q, Chen S (2014) POMAC: properly offloading mobile applications to clouds. In: Proceedings of the 6th USENIX conference on hot topics in cloud computing. USENIX Association, pp 1–6
41. Justino T, Buyya R (2014) Outsourcing resource-intensive tasks from mobile apps to clouds: android and aneka integration. In: 2014 IEEE international conference on cloud computing in emerging markets (CCEM). IEEE, pp 1–8
42. Zachariadis S, Mascolo C, Emmerich W (2004) SATIN: a component model for mobile self organisation. In: Lecture Notes in Computer Science (including subseries Lecture Notes in Artificial Intelligence and Lecture Notes in Bioinformatics). Springer, Berlin, pp 1303–1321
43. Marinelli EE (2009) Hyrax: cloud computing on mobile devices using MapReduce. Carnegie Mellon University, Pittsburgh, PA, p 15213
44. Wichtlhuber M, Rückert J, Stingl D, Schulz M, Hausheer D (2012) Energy-efficient mobile P2P video streaming. In: 2012 IEEE 12th international conference on peer-to-peer computing, P2P 2012. IEEE, pp 63–64
45. Yu P, Ma X, Cao J, Lu J (2013) Application mobility in pervasive computing: A survey. Pervasive Mob Comput 9:2–17. https://doi.org/10.1016/j.pmcj.2012.07.009
46. Hassan MA, Wei Q, Chen S (2015) Elicit: efficiently identify computation-intensive tasks in mobile applications for offloading. In: 2015 IEEE international conference on networking, architecture and storage (NAS). IEEE, pp 12–22
47. Lemlouma T, Layaida N (2004) Context-aware adaptation for mobile devices. In: IEEE International conference on mobile data management, 2004. Proceedings. 2004. IEEE, pp 106–111
48. Huang S, Mangs J (2008) Pervasive computing: migrating applications to mobile devices: a case study. In: 2008 2nd annual IEEE systems conference. IEEE, pp 1–8

49. Miyake S, Bandai M (2013) Energy-efficient mobile P2P communications based on context awareness. In: 2013 IEEE 27th international conference on advanced information networking and applications (AINA). IEEE, pp 918–923
50. Satyanarayanan M (2001) Pervasive computing: vision and challenges. IEEE Pers Commun 8:10–17. https://doi.org/10.1109/98.943998
51. Saha D (2003) Pervasive computing: a paradigm for the 21st century. Computer (Long Beach Calif) 36:25–31. https://doi.org/10.1109/MC.2003.1185214
52. (2013a) PowerTutor. https://github.com/msg555/PowerTutor. Accessed 18 Jan 2016
53. (2013b) Picaso. https://code.google.com/p/picaso-eigenfaces/. Accessed 18 Jan 2016
54. (2012) MatCalc. https://github.com/kc1212/matcalc. Accessed 18 Jan 2016
55. (2013c) MathDroid. https://f-droid.org/repository/browse/?fdid=org.jessies.mathdroid. Accessed 18 Jan 2016
56. (2015) NQueen. https://play.google.com/store/apps/details?id=com.memmiolab.queens. Accessed 18 Jan 2016
57. Droidslator (2010) Droidslator. https://code.google.com/p/droidslator/. Accessed 18 Jan 2016
58. (2009a) Mezzofanti. https://code.google.com/p/mezzofanti/. Accessed 18 Jan 2016
59. (2016a) ZXing. https://github.com/zxing/zxing. Accessed 18 Jan 2016
60. (2009b) JJIL. https://code.google.com/p/jjil/. Accessed 18 Jan 2016
61. (2016b) OSMAnd. https://github.com/osmandapp/Osmand. Accessed 18 Jan 2016
62. Čokulov P (2014) Linpack. https://github.com/pedja1/Linpack. Accessed 18 Jan 2016
63. (2008) XRace. https://code.google.com/p/xrace-sa/. Accessed 18 Jan 2016
64. Gu Y, March V, Lee BS (2012) GMoCA: green mobile cloud applications. In: 2012 First international workshop on green and sustainable software (GREENS). IEEE, Zurich, pp 15–20
65. Lee G, Park H, Heo S, Chang K-A, Lee H, Kim H (2015) Architecture-aware automatic computation offload for native applications. In: Proceedings of the 48th international symposium on microarchitecture - MICRO-48. ACM Press, New York, NY, pp 521–532
66. Zhang W, Wen Y, Wu DO (2013) Energy-efficient scheduling policy for collaborative execution in mobile cloud computing. In: 2013 Proceedings IEEE INFOCOM. IEEE, Turin, pp 190–194
67. Bhattacharya S, Gopinath K, Rajamani K, Gupta M (2011) Software bloat and wasted joules: is modularity a hurdle to green software? Computer (Long Beach Calif) 44:97–101. https://doi.org/10.1109/MC.2011.293
68. Saarinen A, Siekkinen M, Xiao Y, Nurminen JK, Kemppainen M, Labs DT (2012) Can offloading save energy for popular apps? In: Proceedings of the seventh ACM international workshop on mobility in the evolving internet architecture - MobiArch '12. ACM Press, New York, NY, pp 3–10
69. Laddad R (2010) AspectJ in action: enterprise AOP with spring applications, 2nd edn. Manning Publications Co
70. Laguna MA, Gonzalez-Baixauli B (2005) Requirements variability models: meta-model based transformations. In: Proceedings of the 2005 symposia on metainformatics - MIS '05. ACM Press, pp 1–9
71. Cortellessa V, Di Marco A, Inverardi P (2006) Software performance model-driven architecture. In: Proceedings of the 2006 ACM symposium on Applied computing - SAC '06. ACM Press, New York, NY, pp 1218–1223
72. Schmidt DC (2006) Model-driven engineering. Computer (Long Beach Calif) 39:25–31. https://doi.org/10.1109/MC.2006.58

73. Trask B, Paniscotti D, Roman A, Bhanot V (2006) Using model-driven engineering to complement software product line engineering in developing software defined radio components and applications. In: Companion to the 21st ACM SIGPLAN conference on object-oriented programming systems, languages, and applications - OOPSLA '06. ACM Press, New York, NY, pp 846–853
74. Rohatgi A (2016) WebPlotDigitizer - extract data from plots, images, and maps. http://arohatgi.info/WebPlotDigitizer/. Accessed 18 Jan 2016
75. Solís C, Wang X (2011) A study of the characteristics of behaviour driven development. In: Proceedings - 37th EUROMICRO conference on software engineering and advanced applications, SEAA 2011. IEEE, pp 383–387
76. Chinenyeze SJ, Liu X, Al-Dubai A (2017) BEFTIGRE: behaviour-driven full-tier green evaluation of mobile cloud applications. J Softw Evol Process 29:e1848. https://doi.org/10.1002/smr.1848

Chapter 9
Sustainability: Delivering Agility's Promise

Jutta Eckstein and Claudia de O. Melo

Abstract Sustainability is a promise by agile development, as it is part of both the Agile Alliance's and the Scrum Alliance's vision. Thus far, however, not much has been delivered on this promise. This chapter explores the Agile Manifesto and points out how agility could contribute to sustainability in its three dimensions – social, economic, and environmental. Additionally, it provides some sample cases of companies focusing on both sustainability (partially or holistically) and agile development.

9.1 Introduction

The two major agile organizations, the Agile Alliance and the Scrum Alliance, both promise in their vision statements that sustainability is one of their core goals:

- Agile Alliance is a nonprofit organization committed to supporting people who explore and apply Agile values, principles, and practices to make building software solutions more effective, humane, and sustainable [1].
- Scrum Alliance® is a nonprofit organization that is guiding and inspiring individuals, leaders, and organizations with agile practices, principles, and values to help create workplaces that are joyful, prosperous, and sustainable [2].

If we want to support building software solutions to be more effective, humane, and sustainable or to help create workplaces that are joyful, prosperous, and sustainable we have to aim (among other things) for sustainability. However, thus far not much has been done for approaching this aim.

J. Eckstein (✉)
Independent, Braunschweig, Germany
e-mail: jutta@jeckstein.com

C. de O. Melo
International Agency (United Nations), Vienna, Austria
e-mail: research@claudiamelo.org

C. Calero et al. (eds.), *Software Sustainability*,
https://doi.org/10.1007/978-3-030-69970-3_9

In this chapter, we are going to provide a new lens in order to understand the Agile Manifesto under the premise the agile approach wants to fulfill its promise for sustainability, and we will provide various case studies of companies attempting to use agile development to contribute to sustainability.

The chapter is structured as follows: we will at first examine the various definitions of sustainability and explore both how the business and Information and Communication Technologies (ICT) approach and classify sustainability. We will then take a close look at the principles defined by the Agile Manifesto in order to find out how they can support sustainable development [3]. Next, we present various case studies of companies either addressing sustainability partially or holistically by leveraging it with an agile approach. In the conclusion, we will take a critical look at sustainability initiatives before we will provide an outlook on the (hopefully) not so far future.

9.2 Sustainability

There are several definitions for sustainability and nuances across the spectrum of sustainable use, sustainable development, and sustainability [4]. The most famous and frequently adopted definition to frame discussions around sustainability is provided by the Brundtland report [5]:

> Sustainable development is development that meets the needs of the present without compromising the ability of future generations to meet their own needs. It contains within it two key concepts: the concept of "needs," in particular the essential needs of the world's poor, to which overriding priority should be given; and the idea of limitations imposed by the state of technology and social organisation on the environment's ability to meet present and future needs.

The Brundtland report suggests how economic and social development should be defined and calls all countries to action:

> Thus the goals of economic and social development must be defined in terms of sustainability in all countries—developed or developing, market-oriented or centrally planned. Interpretations will vary, but must share certain general features and must flow from a consensus on the basic concept of sustainable development and on a broad strategic framework for achieving it.

Finally, the report describes how a path toward sustainability should look like, bringing the concept of physical sustainability (related to living under the laws of nature and minimizing the impact on the physical environment) and its connection to intra- and intergenerational social equity:

> Development involves a progressive transformation of economy and society. A development path that is sustainable in a physical sense could theoretically be pursued even in a rigid social and political setting. But physical sustainability cannot be secured unless development policies pay attention to such considerations as changes in access to resources and in the distribution of costs and benefits. Even the narrow notion of physical sustainability implies a concern for social equity between generations, a concern that must logically be extended to equity within each generation.

According to these definitions, sustainability is about taking long-term responsibility for your action and reaches further than energy consumption and pollution as it is often casually understood.

Other important concepts that seek to explain sustainability are the three-pillar model and the triple bottom line. The **three-pillar model** depicts sustainability by synthesizing social, economic, and environmental concerns. It is the model most widely used, for example, it is the definition given by Wikipedia and also used on the 2005 World Summit on Social Development [6, 7]. However, as explained in [8], "the conceptual foundations of this model are far from clear and there appears to be no singular source from which it derives."

The **triple bottom line** is an accounting framework that seeks to broaden the notion of a company bottom line by introducing a full cost accounting. A single bottom line is the company's profit (if negative, loss) in an accounting period. A triple bottom line adds social and environmental (ecological) concerns to the accounting. If a corporation has a monetary profit, but it causes thousands of deaths or pollutes a river, and the government ends up spending taxpayer money on health care and river clean-up, the triple bottom line needs to account for these cost-benefit analyses too.

The triple bottom line is also known by the phrase "people, planet, and profit" and was coined by John Elkington in 1994 while at SustainAbility (a British consultancy). A triple bottom line company seeks to gauge a corporation's level of commitment to corporate social responsibility and its impact on the environment over time [9].

Another important framework that also articulates sustainability is the United Nations 2030 Agenda and the 17 Sustainable Development Goals: "The Sustainable Development Goals are a universal call to action to end poverty, protect the planet and improve the lives and prospects of everyone, everywhere" [10].

Also, these sustainable development goals are founded in the three pillar model: social (the aim of ending poverty), environmental (protecting the planet), and economic (improving the lives and prospects of everyone, everywhere). Therefore, throughout this chapter, we will use the three-pillar model as the definition for sustainability, as illustrated in Fig. 9.1.

Even with the definition of the three pillars, it has always to be understood that sustainability is highly interconnected that any elaboration on sustainability requires a holistic perspective because all actions are interdependent [11]:

> All definitions of sustainable development require that we see the world as a system—a system that connects space; and a system that connects time. When you think of the world as a system over space, you grow to understand that air pollution from North America affects air quality in Asia, and that pesticides sprayed in Argentina could harm fish stocks off the coast of Australia. And when you think of the world as a system over time, you start to realise that the decisions our grandparents made about how to farm the land continue to affect agricultural practice today; and the economic policies we endorse today will have an impact on urban poverty when our children are adults.

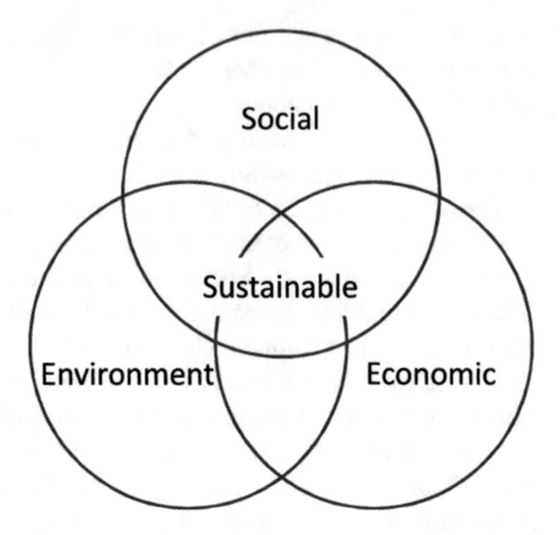

Fig. 9.1 The three dimensions of sustainability [8]

9.2.1 *Business and Sustainability*

According to the dictionary [12], "Business pertains broadly to commercial, financial, and industrial activity, and more narrowly to specific fields or firms engaging in this activity."

However, also the business and its self-conception are changing. In particular, this became visible, by the Business Roundtable, an association of chief executive officers of leading companies in the USA, refining its statement on the purpose of a corporation in 2019. This refined statement made a shift from the focus on satisfying shareholders by making money to a more holistic understanding and aiming for satisfying customers, employees, communities, suppliers, and shareholders equally. This shift comprehends an understanding that financial success is not the sole purpose of business.

For example, the statement says [13]:

> [W]hen it comes to addressing difficult economic, environmental and societal challenges, these companies are starting in their own backyards—partnering with communities to provide the investment and innovative solutions needed to revitalize local economies and improve lives.
>
> These investments and initiatives aren't just about doing good; they're about doing good business and creating a thriving economy with greater opportunity for all.

Although, when talking about sustainability, the Business Roundtable refers only to energy and environment, they show an understanding of sustainability as it is defined by the Brundtland report [5, 14] by pointing out the importance and interdependence of the economic, environmental, and societal challenges.

At the organizational level, there are concrete movements aiming to answer the question of how an organization can be more balanced across the three sustainability pillars. They recognize that capitalism, as it's currently practiced, is starting to run into fundamental structural problems.

In the USA, **B Lab** [15] has worked to create a certification of "social and environmental performance" to evaluate for-profit companies. B Lab certification requires companies to meet social sustainability and environmental performance, and accountability standards, as well as being transparent to the public according to the score they receive on the assessment. Later on, they also developed the concept of "benefit corporations," companies that legally commit themselves to honor moral values, while pursuing the standard capitalist goal of maximizing profits. A few examples of well-known benefit corporations include Method, Kickstarter, Plum Organics, King Arthur Flour, Patagonia, Solberg Manufacturing, Laureate Education, and Altschool.

In Europe, **Economy for the Common Good (ECG)** [16] has similar goals. ECG is an economic model, which makes the Common Good, a good life for everyone on a healthy planet, its primary goal and purpose. The Economy for the Common Good (ECG) was initiated in 2010 by economic reformist and author Christian Felber, together with a group of Austrian pioneer enterprises. ECG is an ethical model for society and the economy with the goal of reorienting the free market economy through a democratic process toward common good values. According to Felber, it seeks to address a capitalist system that "creates a number of serious problems: unemployment, inequality, poverty, exclusion, hunger, environmental degradation and climate change." The solution is an economic system that "places human beings and all living entities at the centre of economic activity."

At the heart of ECG lies the idea that values-driven businesses are mindful of and committed to (1) human dignity, (2) solidarity and social justice, (3) environmental sustainability, and (4) transparency and codetermination. ECG is currently supported by more than 1800 enterprises in 40 countries such as Sparda-Bank Munich, VAUDE, Sonnentor, and taz (German newspaper), about 250 have created a Common Good Balance Sheet. This balance sheet is a scorecard that measures companies based on their preservation of those four fundamental values, considering five key stakeholders: suppliers; owners/equity/financial service providers; employees/coworker employers; customers/other companies; and social environment.

9.2.2 ICT/Technology and Sustainability

The idea of using Information and Communication Technologies (ICT) to address sustainability issues has been investigated in a number of interdisciplinary fields that combine ICT with environmental and/or social sciences. Among these are environmental informatics, computational sustainability, sustainable human-computer interaction, green IT/ICT, or ICT for sustainability [4].

The contributions of these fields are manifold: monitoring the environment; understanding complex systems; data sharing and consensus building; decision support for the management of natural resources; reducing the environmental impact of ICT hardware and software; enabling sustainable patterns of production and consumption; and understanding and using ICT as a transformational technology.

For these different areas, there is a common understanding that ICT can be used to reduce its own footprint and to support sustainable patterns [4]:

- **Sustainability in ICT**: Making ICT goods and services more sustainable over their whole life cycle, mainly by reducing the energy and material flows they invoke.
- **Sustainability by ICT**: Creating, enabling, and encouraging sustainable patterns of production and consumption.

The problem with this differentiation is that often sustainability in and by ICT is intertwined. When analyzing the impact of certain technology, looking for a final positive or negative conclusion, it is not possible to assess it in isolated contexts to make a decision. A positive impact occurs when the effect of a sustainable activity on the social fabric of the community causes well-being of the individuals and families considering the three pillars [17]. Thus, assessing the ICT impact needs to consider that [4]:

> Sustainable development [. . .] is defined on a global level, which implies that any analysis or assessment must ultimately take a macro-level perspective. Isolated actions cannot be considered part of the problem, nor part of the potential solution, unless there is a procedure in place for systematically assessing the macrolevel impacts.

Therefore, the assessment of specific technology impact on sustainable development must consider its nuances and interdependencies on multiple levels that might be on both spaces of "in" and "by." A good example is provided in [18 , p. 284]:

> [H]istory of technology has shown that increased energy efficiency does not automatically contribute to sustainable development. Only with targeted efforts on the part of politics, industry and consumers will it be possible to unleash the true potential of ICT to create a more sustainable society.

9.3 Agile and Sustainability

If agile is really aiming for sustainability, as it is suggested by the vision of both the Agile Alliance and the Scrum Alliance, then we should take a look at the principles of the Agile Manifesto to understand how these principles can contribute to or guide sustainability [3]. Sustainability requires a much broader view to integrate the environmental, economic, and social perspectives.

9.3.1 Our Highest Priority Is to Satisfy the Customer Through Early and Continuous Delivery of Valuable Software

At the core of this very first principle is continuous learning by *focusing constantly on the customer*. Only through continuous delivery, you will be able to keep adjusting the system to the customer's satisfaction. Broadening the perspective, and looking at this principle through a sustainability lens, that is taking also the environmental and social aspect into account, shifts also the meaning of "valuable" software. The value is not only defined by the economic benefit for the customer but also by the social and environmental improvements.

Moreover, by broadening the perspective, we'll find that it will become even harder to come up with a perfect solution right away. For example, measuring the carbon footprint of the system you're building will teach you what aspects you need to adapt, rethink, and redo. Thus, adding two more dimensions (environmental and social) increases the complexity of development, and thus, *continuous learning* is even more important to uncover emergent practices [19].

9.3.2 Welcome Changing Requirements, Even Late in Development: Agile Processes Harness Change for the Customer's Competitive Advantage

This second principle addresses sustainability by ICT, by *focusing on the customer* (like the principle before). At first sight, the "customer's competitive advantage" sounds like it can only aim for economical success. Yet, the advantage can also be built by a reputation of the product as being environmental and social friendly. As a counterexample, the authors have seen websites that require the specification of the first and last name of the user (Fig. 9.2[1]). However, both need to be at least three characters long. If the developers would have been socially aware they would have understood that there are many people (especially in Asia) whose names are only two characters long. This is only a minor example but still shows how broadening the view can make a difference for the customer—by gaining a good or bad reputation.

To provide the customer this kind of competitive advantage, we have to answer:

- How is the product that we're creating helping the world?
- How is it part of a wider sustainable solution?
- How can we ensure the product is inclusive?
- How can the product even improve the environment?
- How can we ensure the product itself is environment friendly?

[1]https://twitter.com/shirleyywu/status/1300628412466298881?s=20

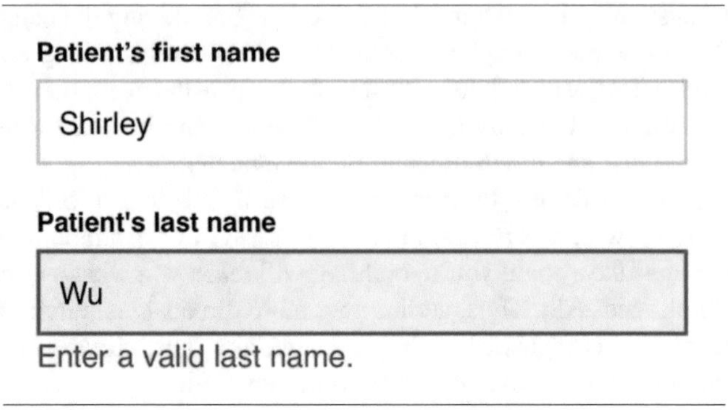

Patient's first name

Shirley

Patient's last name

Wu

Enter a valid last name.

Fig. 9.2 Application requiring two characters for the last name

9.3.3 Deliver Working Software Frequently, from a Couple of Weeks to a Couple of Months, with a Preference to the Shorter Timescale

Similar to the first principle, also this one focuses on *continuous learning*. The difference is that this third principle asks to establish a regular cadence for feedback. The shorter the feedback cycle, the better, because then we can keep the focus on the learning about the changes in the system by the last delivery. This principle is important for tackling sustainability because the knowledge, wisdom, and experience is steadily increasing in all three dimensions (economical, social, and environmental) and this learning has to be reflected in the system.

For example, although we might have designed the system to be highly accessible, only by getting feedback from people who are in need of accessibility will we know how well the system performs. Similarly, we can learn from the real carbon footprint of the system only when we know how it performs in reality.

To continuously learn from the delivery you should regularly ask yourself the following questions [20]:

- Does the system work for people with disabilities?
- Does the system work for people using older devices?
- Does the system provide the best possible performance with the least amount of resources across devices and platforms?

9.3.4 Business People and Developers Must Work Together Daily Throughout the Project

This is the fourth principle of the Agile Manifesto and points out the importance of collaboration between the ones creating the product or service (developers) and the ones who want to offer that product or service to the market. Continuous collaboration enables *self-organization* around the system created and allows to discover and actually meet the market needs.

The collaboration between business people and developers is not market focused only (anymore), but keeps the wider perspective of addressing all three dimensions of sustainability. This includes a reflection on the market we serve, disrupt, and/or create. For example, in the last decades, car manufacturers did create a market for heavy and environmental unfriendly cars (the sport utility vehicles, SUVs) and hesitated on the other hand to create a market for electric cars with the argument that this market does not exist [21].

Thus, this principle is a continuous reminder to consider all aspects of sustainability in our daily work.

9.3.5 Build Projects Around Motivated Individuals. Give Them the Environment and Support They Need, and Trust Them to Get the Job Done

This is the principle number five and directly mentions the environment. First of all, this principle is a call to action for sustainability in ICT. At first sight, the social pillar is emphasizing the importance of supporting the individuals for example, by paying them fair (also despite any difference in race, gender, religion, location, etc.). Providing an environment that helps the individuals to *self-organize* for getting their job done includes the technical tools yet, also the safety of the environment: the individuals are not put at risk by, e.g., a polluted or toxic workspace (think of asbestos, colors, or carpets that evaporate toxic emissions).

Moreover, the environment can also contribute, neutralize, or reduce the carbon footprint. For example, by the energy that is used in the work environment, the availability of natural light, the existence of natural plants, or the commute that is required. Finally, from an economic point of view, this principle also requests that the environment allows every individual in the same way to prosper and grow by offering life-long learning.

9.3.6 The Most Efficient and Effective Method of Conveying Information to and Within a Development Team Is Face-to-Face Conversation

This sixth principle puts direct conversation at the core. Face-to-face conversation provides *transparency* particularly from the social and economic point of view. If the system you are building should support, for example, (individual) growth and development or/and inclusiveness, then you will get qualitatively better feedback from the users if talking to them directly. Direct observation, eye contact, mimic, and gesture provide additional information how well the system is serving the users.

This principle invites as well to build a community with everyone involved in making the product sustainable. The community reflects both on the process and the product. This reflection is an interdependent relationship.

9.3.7 Working Software Is the Primary Measure of Progress

This principle aims for *transparency*. Certainly, it is important to think ideas through however, you will only know if the ideas keep their promises when hitting reality. This includes to regularly examine the system for unintended consequences (also known as the precautionary principle [22]). Therefore, it is also the responsibility of an agile team to prevent and monitor for any unintended consequences in order to address them [23].

Therefore, never underestimate the importance of feedback; for example, on energy consumption on the running system, theory and reality do not always show the same results. From the social perspective, regularly examine the working system guided by the following questions:

- Does the system work in the same way for everyone independent if the gender differs or it is used by people from different races or ethnicity?
- Would a domestic abuser find possibilities to do harm to the system but more so to other people?
- Can the system be used by a government to oppress?

We suggest to take a look at other questions, for example, the ones offered by the Ethical Explorer toolkit [24].

9.3.8 Agile Processes Promote Sustainable Development. The Sponsors, Developers, and Users Should Be Able to Maintain a Constant Pace Indefinitely

This is the eighth of twelve principles and the only one that calls for sustainable development explicitly. It focuses foremost on sustainability in ICT. Thus far, this principle has mainly been understood from the social aspect for example, calls for the '40 hour week' have been argued with the reference to this principle [25].

Using the lens of sustainability, it asks that developers *learn continuously* about all three pillars and take the learning into account: environment, social, and economic. Therefore, having a constant pace in mind, developing the product or service should not lead to burnout of the developers (social), to reduction of resources (environment), or to overspending financially (economic). For example, the features developed should take the diversity and the abilities of the users into account, the energy used for creating the product should be renewable, and the so-called feature-creep should be avoided.

Feature-creep, the inclusion of unnecessary features, and the development of features that do not adhere to the intended design and architecture lead to inefficiency of the overall system. Moreover, the utilized capacity of the CPU in idle mode or the demand for storage are reasons for software getting slower.

Therefore, sustainable development has to be taken into account by the developers and the sponsors to ensure that, for example, the user is not required to invest in a new hardware regularly. Modularity and allowing the user to determine which modules to buy, install, and use take into account that not every user will need every possible innovation (at least not on every device).

9.3.9 Continuous Attention to Technical Excellence and Good Design Enhances Agility

Developments in sustainability are progressing fast—from new understandings of algorithm bias (social aspect), or of where most energy is used and how it could be reduced, i.e., by storing data locally and reducing network traffic (environmental aspect) to better ways for finding out which features are actually used and so—to ensure that money is not wasted by feature-creep.

Thus, this ninth principle is a call for *continuous learning* and for keeping technologically up-to-date and taking into account new learning and development that enable good design: now good in the sense of sustainability. This ensures the principle supports sustainability in ICT.

9.3.10 Simplicity—the Art of Maximizing the Amount of Work Not Done—Is Essential

The tenth principle often requires a second read before it is understood. In general, it addresses the feature-creep mentioned above. So, when developing software, we need to pay attention to what is really needed in the product. Looking through a sustainability lens at this principle, we need to examine, for example, how does this new feature fit in the system without increasing the energy consumption? Using Scrum as an example, sustainability or rather energy consumption needs to be considered during backlog refinement, sprint planning, by the definition of done, as well as monitored through tests.

The German Federal Environment Office created and published in 2020 a label (or certificate) "Blauer Engel" for resources and energy-efficient software stand-alone products, based on a criteria catalogue jointly developed by the universities of

Trier and Zurich [26, 27]. The label focuses on energy-efficiency, conservative resource consumption, and transparent interfaces. The plan is to develop a similar label for cloud-based software. However, already today, the criteria catalogue can guide a team to develop a sustainable system. For example, the following questions should be regularly discussed [27]:

- How much electricity does the hardware consume when the software product is used to execute a standard usage scenario?
- Does the software product use only those hardware capacities required for running the functions demanded by the individual user? Does the software product provide sufficient support when users adapt it to their needs?
- Can the software product (including all programs, data, and documentation including manuals) be purchased, installed, and operated without transporting physical storage media (including paper) or other materials goods (including packaging)?
- To what extent does the software product contribute to the efficient management of the resources it uses during operation?

Thus, this principle requires *transparency* for features that are needed (and used) and those that are not.

9.3.11 The Best Architectures, Requirements, and Designs Emerge from Self-Organizing Teams

The eleventh principle points out the benefits of *self-organizing* teams. This includes that every team member is invited to speak up and make their contribution to the architecture, requirements, and design—independent of other characteristics of that team member (social perspective). Similarly, all team members will get the same fair chance to progress on their career by getting equal support through training, mentoring, or coaching (economic perspective).

9.3.12 At Regular Intervals, the Team Reflects on How to Become More Effective, Then Tunes and Adjusts Its Behavior Accordingly

The twelfth and final principle asks teams to run regular retrospectives. Both the reflection and behavioral adjustments should take (also) sustainability aspects into account: How can the team tune and adjust its behavior to become more effective regarding the three pillars—environmental, social, and economical? Dedicating time regularly for reflecting on sustainability will lead to continuous improvements. This last principle combines the quest for teams to *self-organize* in order to *learn continuously* by making their effectiveness with the *focus on the customer transparent*.

9.3.13 Summary of the Agile Manifesto's Perspective on Sustainability

Implementing sustainable development goals requires approaching wicked problems, i.e., complex, nonlinear, dynamic challenges in situations of insufficient resources, incomplete information, emerging risks and threats, and fast-changing environments [28]. Examining the principles of the Agile Manifesto shows how an agile approach indeed promotes (or can promote) sustainable development. It might be surprising how much guidance the principles can provide although they have been defined originally with the focus on software development only.

Concentrating continuously on inspect and adapt allows sustainable systems to emerge. An agile, cross-functional team integrates different perspectives on the emerging system and has this way the possibility to design solutions for sustainability. The disciplined approach provided by agile development enables the team to permanently learn from their delivery; to measure the outcome according to its environmental, social, and economic impact; and to take necessary actions for adjustments.

However, taking all three dimensions of sustainability into account leads to higher complexity, so an agile approach also comes in handy for addressing this complexity by breaking down the problems and using an inspect and adapt approach for making them simpler.

The role of agility is not to save the world, but to provide a value system based on *transparency, constant customer focus, self-organization, and continuous learning* that leads to sustainable thinking and offers an approach that supports putting this sustainable thinking into action.

9.4 Case Studies: Leveraging Agility for Sustainability

Although the Agile Manifesto [3] originates in 2001 and the Brundtland report [5] in 1987, the combination of agility and sustainability is just in its beginnings. There are some sample companies being conscious of one of the three pillars and even fewer sample companies taking a holistic approach on combining sustainability and agility. In this section, we will provide some sample case studies for both attempts—agility and one of the three pillars as well as company-wide agility and sustainability using a holistic approach.

We want to point out that for the following case studies, as for other examples, there is no such thing as a "perfect" company, neither in terms of sustainability nor in terms of agility. However, these case studies still can serve as examples of possible steps to take in order to leverage agility for sustainability.

9.4.1 Agility and Partial Sustainability

In this section, we explore different examples from companies applying agility on one of the three pillars—social, environmental, and economical—individually. The examples show that being conscious about sustainability can be guided by an agile approach.

9.4.1.1 The Social Pillar

Often, we act as if our responsibility would end by delivering the value to the customer. An agile team typically aims to deliver regularly high value for the customer's advantage. After the (or rather after each) delivery, the team's job is completed (except for maintenance and further development). However, if the team takes full responsibility for their products, then they are also interested in the usage of the product and consider its impact on the world.

For example, some developers working for CHEF (a company providing a configuration management tool with the same name) were also interested in how their customers are using the product and how this is supporting the social good. In this case, these developers learned that one of their customers, the Customs and Border Protection or Immigration and Customs Enforcement (ICE), uses the product at the border between Mexico and the United States of America for running detention centers, ensuring deportation, and implementing the family separation policy. In this case, the developers took the Brundtland definition (sustainability is about taking long-term responsibility for your action) and the social aim of ending poverty by heart and decided that the usage of their product does not confirm with their ethical values and has an unintended social impact.

In the beginning, the developers brought their ethical interest to the attention of CHEF's management. However, the management at first referred to the long-standing contract and to the fact that the product has been used by ICE for many years (and nobody complained). The developers kept trying to convince the management that due to the change in politics also the usage of the product has shifted to the worse. However, the developers could only make a difference once one of the developers decided to delete all the code he contributed to the (Open Source) software. As a consequence, the product was not usable anymore for 2 weeks which created enough pressure for the management to decide on not renewing the contract [29].

It is important to understand that delivering value and satisfying the customer with your product is not an agile team's sole responsibility. An agile team is also responsible for the social impact of the product it is creating. This means for a truly agile team, to stay in touch with the customer for recognizing any differences in the usage of the product. This ongoing connection can be supported by automation such as monitoring, logging, and having tests that observe the usage of the product.

9.4.1.2 The Environmental Pillar

The environmental pillar is mostly connected with the resources consumed. The global e-waste monitor reports that in 2016, 44.7 million metric tons of e-waste were generated—most often because the hardware gets (seemingly) too soon outdated [30]. Additionally, as Nicola Jones reports, by 2030 information technology might exceed 21% of the global energy consumption [31].

Thus, it gets more and more important to consider the energy consumption of the products we are creating. Often it is assumed that hardware is cheap and thus there is no need to pay much attention to performance, because if the software is not performing well enough we request that the hardware is getting faster. This proves Wirth's law from 1995 [32]: "Software is getting slower more rapidly than hardware becomes faster." As elaborated by Gröger and Herterich, one of the reasons for this effect is the feature-creep where features are developed for a product that are unnecessary (not used) and don't fit the intended software architecture [33].

The Mozilla foundation argues for the importance of examining cloud-based software in particular. In their 2018 Internet Health Report, they concluded that data centers have a similar carbon footprint as global air traffic with the latter being 2% of all greenhouse gas emissions [34].

However, while some companies ignore the problem despite the protests of their employees (see Amazon [35]), others address it by shifting toward renewable energy for their data centers (see Google [36]). This means for an agile organization when deciding on a cloud infrastructure that it is essential to also consider the carbon footprint of that data center. Therefore, it is the responsibility of an agile team to bring not only any technical information but also information about energy consumption and the carbon footprint of the infrastructure under question to the attention when the decision is up.

Another example is Mightybytes, a company focusing on developing digital strategies to create the design and user experience for their clients [20]. By doing so, they pay particular attention to how much energy is consumed by the designs they are creating and decide, for example, against including videos with high energy consumption. Additionally, they also take care of the environment the developers are in by ensuring the carpets are not toxic, there is enough space for everyone, natural light, and plants, plus the offices are powered with renewable energy. Finally, other examples of initiatives that explore energy aspects in ICT can be found in [37, Part II].

9.4.1.3 The Economical Pillar

The economic dimension is often understood as the economic balance, e.g., that no nation (or company) grows economically at the cost of another one. Thus, topics like fair trade or paying fairly are often discussed along these lines. While this can be a topic also in (agile) software development, according to our experience this is seldom the case because agile developers are still benefiting from good payments globally (however, this statement is not based on any research). Yet, there is another economic impact for organizations, because as the Cone Communications Corporate Social Responsibility (CSR) study revealed, a company's reputation regarding their sustainability efforts will have an effect on both their market share and their search for talent [38]. For example, as reported in this study:

> Nearly nine-in-10 Americans (89%) would switch brands to one that is associated with a good cause, given similar price and quality, compared with 66 percent in 1993. And whenever possible, a majority (79%) continue to seek out products that are socially or environmentally responsible.

An economic impact can also be made by organizations and teams through sharing learning. One example is Munich Re, one of the world's leading reinsurers, which got concerned about climate change already in the 1970s. At that time, they began collecting and publishing research data about climate change. Protecting the research data for a competitive advantage was never considered by Munich Re because they realized that transparency allows them to learn from others and to improve the data. Transparency, they decided will increase both the general societal awareness of climate change and their own resilience [39]. This insight provided a great foundation for Munich Re's further effort in combining also the other two pillars (environmental and social) with a general agile approach [40, 41].

Transparency is also key for all the lessons learned in the near future on how to make the software we are creating more sustainable—only if we make those learning transparent right away, we can make a huge difference for everyone, everywhere.

9.4.2 Company-Wide Agility and Holistic Sustainability

Implementing agility company-wide comes with a responsibility. Professionals as well as companies who claim to be agile are expected to also "take actions based on the best interests of society, public safety, and the environment" [2]. In the same way, are corporate Agile Alliance members expected "to help make the software industry humane, productive, and sustainable" [1]. This means agile companies are expected to have a systemic view and understand the impact of the own actions and products created. This means an organization implementing company-wide agility has to have a wider perspective than one that is aiming at business agility only, as defined by the Business Agility Institute [42]:

Business agility is the capacity and willingness of an organisation to adapt to, create, and leverage change for their customer's benefit!

Thus, business agility focuses on the customer only whereas agile organizations aim for humanity and sustainability while having the society and the environment in mind. In this section, we will explore companies that made quite some progress in implementing company-wide agility in that sense by having a holistic perspective on all three pillars, thus acting with social, environmental, and economical outcomes in mind.

9.4.2.1 Patagonia

Patagonia, Inc. is an American clothing company that markets and sells outdoor clothing since 1973. The company has become recognized as a leading industry innovator through its environmental and social initiatives, and the brand is now considered synonymous with conscious business and high-quality outdoor wear. In 2019, Patagonia received the 2019 Champions of the Earth award from the United Nations [43], being recognized as an organization that has sustainability at the very core of its successful business model.

Patagonia's mission statement is "We're in business to save our home planet." They implement it by accomplishing a number of initiatives that inspires all levels of the organization, as donating profits from their Black Friday sales (millions of dollars) to the environment through grassroots movements [44], or creating their new office space by restoring condemned building using recycled materials. The company states its benefits as: 1% for the Planet; Build the Best Product with No Unnecessary Harm; Conduct Operations Causing No Unnecessary Harm; Sharing Best Practices with Other Companies; Transparency; and Providing a Supportive Work Environment.

Patagonia has been cited as an example for Agile organizations, not only because it has agile teams, but because they embody a north star across the organization that recognizes the abundance of opportunities and resources available, reducing the mindset of competition and scarcity and moving toward cocreating value with and for all of our stakeholders [45].

Other examples of their practices that demonstrate their concern about the three pillars are: encouraging consumers to think twice before making premature replacements, or overconsuming; designing durable textile yarns from recycled fabric; upholding a commitment to 100% organic cotton sourced from over 100 regenerative small farms; sharing its best practices through the Sustainable Apparel Coalition's Higg Index; paying back an "environmental tax" to the earth by founding and supporting *One Percent For The Planet*; and donating its ten million federal tax cut to fund environmental organizations addressing the root causes of climate change.

As an example of social impact, Patagonia invests in improving the supply chain to alleviate poverty. They screen their partners, as factories and more recently farms,

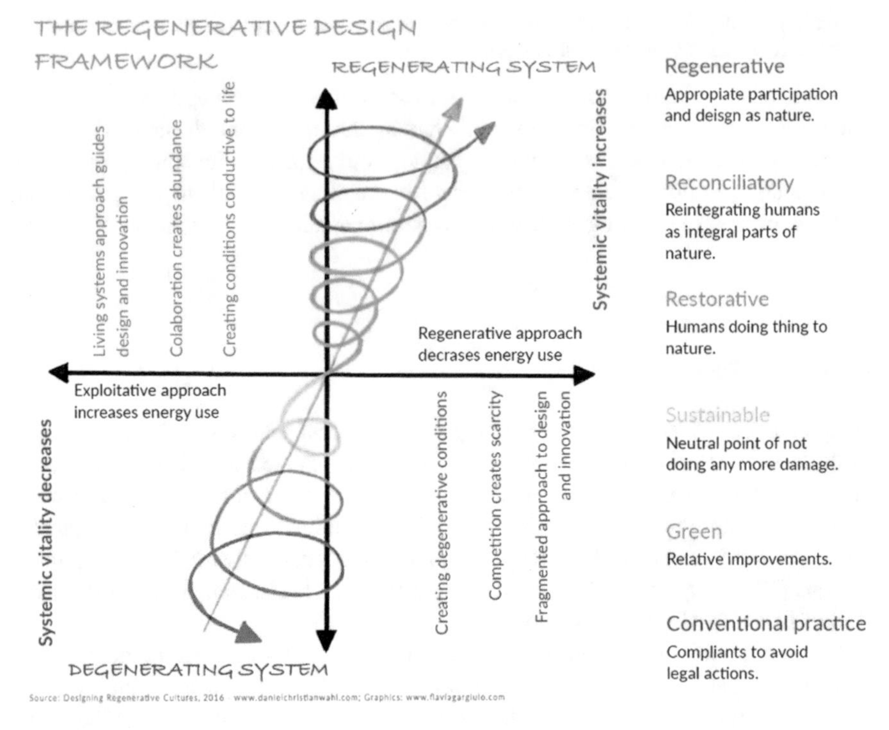

Fig. 9.3 Continuum between conventional, sustainable, and regenerative practices [47]. © D.C. Wahl, 2018; reprinted with permission

using Patagonia staff, selected third-party auditors, and NGO certifiers. They recognize a number of challenges, especially in the farm level [46, p. 30]:

> There can be land management and animal issues, as well as child labour, forced labour, pay irregularities, discrimination, and unsound health and safety conditions. These are often more difficult to resolve because of the complexities that extreme poverty, illiteracy and exploitation bring to this level of the supply chain.
>
> When it comes to land management, we're most concerned with a farm's use of chemicals and the impact its operations have on water, soil, biodiversity and carbon sequestration. For animal welfare, we look at humane treatment and slaughter. And when it comes to labour, we want to see safe and healthy working conditions, personal freedom, fair wages and honest payrolls.

More recently, the company has started to support regenerative agriculture, establishing a goal of sourcing 100% of their cotton and hemp from regenerative farming by 2030. It is important to stress how relevant this initiative is by introducing the meaning of regenerative. While the concept of *sustainable* refers to a neutral point of not harming or damaging, the *regenerative* concept goes beyond, stating that humans are not only doing the right things to nature, but actually are an integral part of it, *learning how to design as nature does* [47]. This means we can reverse the damage we've already done. Figure 9.3 illustrates the continuum between (1) the conventional practices our society adopts in many areas, (2) sustainable practices, and (3) regenerative practices.

Patagonia is a founding member of the Regenerative Organic Alliance,[2] alongside Dr. Bronner's Compassion in World Farming, Demeter, the Fair World Project, and others. They created a certification that showcases whether a product has been made using processes to regenerate the land or not.

Thus, supporting regenerative agriculture is a bold step Patagonia is taking that helps to reverse damage and create abundance. The reason is that agricultural practices are a huge contributor to climate change, accountable for around 25% of global carbon emissions. Regenerative agriculture has the intention to restore highly degraded soil, enhancing the quality of water, vegetation, and land-productivity altogether. It makes it possible not only to increase the amount of soil organic carbon in existing soils, but to build new soil [48]. If more companies follow this example, we will see more ecosystems being restored and communities being benefited.

9.4.2.2 DSM-Niaga

DSM-Niaga is a joint venture of the startup Niaga and the multinational firm Royal DSM. DSM-Niaga's vision is to design for the circularity of everyday products. They started off with carpets and mattresses with the idea to stop these—most often toxic products to go into landfills but instead to decouple the material and use the very same material to go into the next production cycle. To guide this idea, they defined three design principles [49]:

1. Keep it simple: Use the lowest possible diversity of materials.
2. Clean materials only: Only use materials that have been tested for their impact on our health and the environment.
3. Use reversible connections: Connect different materials only in ways that allow them to be disconnected after use.

For ensuring clean materials only and also proving it, they developed, for example, a digital passport for every carpet based on blockchain technology. With the focus on the customer, this passport makes the complete value chain transparent. DSM-Niaga is also sharing their learning and pushing the industry to design for circularity:

> Moving forward, we will continue to focus our efforts and push boundaries to drive transparency and accountability across value chains. Indeed, with designers, producers and recyclers all needing to know what's in a product in order to recycle it, it's only a matter of time before digital product passports are in demand everywhere.

DSM-Niaga is focusing on both sustainability and company-wide agility. The firm is actually a teal organization, that is a company defined by self-organization where, for example, employees are guided by the organization's purpose and not by orders

[2] https://regenorganic.org/

[50]. According to Rhea Ong Yiu, an Agile Coach at DSM-Niaga, the mother company (Royal DSM) is constantly learning from DSM-Niaga. Under the leadership of Feike Sijbesma, CEO and Chairman of the Managing Board, Royal DSM sold its entire petrochemical business. This has also been recognized. As a consequence, Feike Sijbesma has been appointed as Global Climate Leader for the World Bank Group, Co-Chair of the Carbon Pricing Leadership Coalition, and Co-Chair of the Impact Committee of the World Economic Forum where he is one of the originators of the "Stakeholder Principles in the COVID Era." The latter states among other things [51]:

> We must continue our sustainability efforts unabated, to bring our world closer to achieving shared goals, including the Paris climate agreement and the United Nations Sustainable Development Agenda.

Worth mentioning that the business strategy of Royal DSM (and as such as well of DSM-Niaga) is based on the United Nations Sustainable Development Goals. Following are the sustainable development goals that are in particular focus for DSM-Niaga [52]:

- Good health and well-being (Goal 3): Ensuring healthy lives and promoting well-being for all at all ages.
- Responsible consumption and production (Goal 12): Ensuring sustainable consumption and production patterns.
- Climate action (Goal 13): Taking urgent action to combat climate change and its impact.

9.4.2.3 Sparda-Bank Munich

Sparda-Bank is the largest cooperative bank in Bavaria, with more than 300,000 members. The bank maintains on its website a comprehensive description of how they implement all Economy for the Common Good (ECG) values, as well as their Common Good Balance Sheets and the certificate. Sparda was part of the first companies that agreed on ECG goals, back in 2010, being the first—and so far only—bank that operates according to the principles of the common good economy.

At the same time, the company keeps looking for innovation and agility [53]. So the organization needs and it is open to technological solutions. Due to its own values, it would have to carefully examine the need for and impact of the tools that it adopts. In fact, Sparda does have agile coaches and digitized solutions for their clients. They have also regularly had formats such as "World-Café" or smaller events with a "marketplace character" that are carried out in order to obtain the opinion of as many participants as possible and to initiate a dialogue. The design thinking method is adopted to support their product and projects.

The company claims on its website [54] that they are climate neutral and provides an annual CO_2 balance. They reduce greenhouse gas emissions to the extent to what is technically and economically possible, or otherwise by purchasing climate certificates (or permits) in accordance with the Kyoto Protocol, which [55]:

> [O]perationalizes the United Nations Framework Convention on Climate Change by committing industrialized countries and economies in transition to limit and reduce greenhouse gases (GHG) emissions in accordance with agreed individual targets [...]
>
> One important element of the Kyoto Protocol was the establishment of flexible market mechanisms, which are based on the trade of emissions permits.

Sparda-Bank has other initiatives that cover different aspects of the common good matrix. For instance, planting a tree for every new member or agreements to provide green electricity with special tariffs for their clients. When financing an electric or hybrid car or an e-bike, Sparda-Bank Munich customers receive a reduced interest rate. They also state having no customer relationships with or investments in companies whose core business is in the armaments sector, as well as many other restrictions published on their website [56].

There are other examples related to suppliers and employees: they buy dishes and towels from works for the blind and disabled; they work to reduce the difference between the lowest salary and the highest salary (CEO) (which is currently published as 1:13.7 ratio); and finally, they don't pay commissions or establish individual goals related to salary, only goals at a team level.

9.5 Conclusion

This chapter examined that agility can contribute to sustainability. As we have seen, the Agile Manifesto in general and the principles in particular can provide guidance to sustainability [3]. We have presented some achievements of the companies combining agility and sustainability. We explored examples of companies focusing on one of the dimensions only, as well as companies taking all three pillars into account. Certainly, if an agile company takes sustainability seriously, then it has to take a holistic view and look at all three dimensions at once—at people (social), planet (environmental), and profit (economic).

We have seen that most of these sample companies in the case studies are concentrating on using agile development and sustainability, but without a focus on leveraging the one with the other. Especially companies following a holistic approach implement sustainability by ICT. Using an agile approach for implementing sustainability in ICT seems to be a relatively new field.

9.5.1 Criticism

Sustainability became a trend and a symbol for progressive individuals, movements, and organizations, which sometimes leads to the so-called *green-washing*. It happens when it seems to be important to have a reputation of being sustainable, but not everyone who claims to live up to it really does it. Often this is supported by advertising for ones own sustainable reputation as [20] exemplifies:

> One hosting provider even claims in its marketing materials that it plants a tree for every new account, which is wonderful, but doesn't move us closer to an Internet powered by renewable energy.

Sustainability and its deeper implications are still not well-understood by society, despite more diffused, in particular because of the UN 2030 Agenda for Sustainable Development. Initiatives in many sectors, from business to NGOs and universities, can easily distort or simplify it, intentionally or not. Taking the example of companies, we illustrated agile organizations that aim at balancing the three pillars in Sect. 9.4.2, considering BCorps and ECG certified companies. It is important to be aware of the criticism (or limitations) around these models.

The main critique to BCorps and similar movements is that these models still rely on capitalism as the core mechanism—and worldview—for our economy, "ignoring the possibility that capitalism itself, as it is largely practised today, might be at least one cause of the problems we are seeking to solve" [57]. This analysis is also supported by the Nobel Prize-winning economist Joseph Stiglitz [58]:

> Like the dieter who would rather do anything to lose weight than actually eat less, this business elite would save the world through social-impact investing, entrepreneurship, sustainable capitalism, philanthro-capitalism, artificial intelligence, market-driven solutions. They would fund a million of these buzzwordy programs rather than fundamentally question the rules of the game—or even alter their own behavior to reduce the harm of the existing distorted, inefficient and unfair rules.

High expectations are on digitalization for sustainability by ICT. One example is that it is the replacement for paper, but all digital products (thus, also the paper replacements) consume energy. Moreover, as Lorenz M. Hilty points out in [33]:

> So far, neither in the case of air travel nor in the case of lifespan of household goods has the hope been fulfilled that due to digitalization the material and energy intensity of our activities would be reduced. It rather became evident in the digital age that providers turned the principle of intangible value creation through software into its opposite by stimulating or even forcing material consumption through software.

One reason is Wirth's law (software is getting slower more rapidly than hardware becomes faster) that an update in software often requires the exchange of hardware [32]. Another reason is the *rebound effect*: the effect that environmental friendliness is a selling point that leads to overall higher consumption than before. And a third reason (related to the rebound effect) is that the environmental friendly product is often used in addition—and not instead—to the environmental unfriendly product. An example for the latter are many car-sharing offerings that are not used for substituting private cars but, instead, for substituting traveling by (local) public transport [33]. Thus, the rebound effect is the main controversy for all achievements of digitalization regarding sustainability.

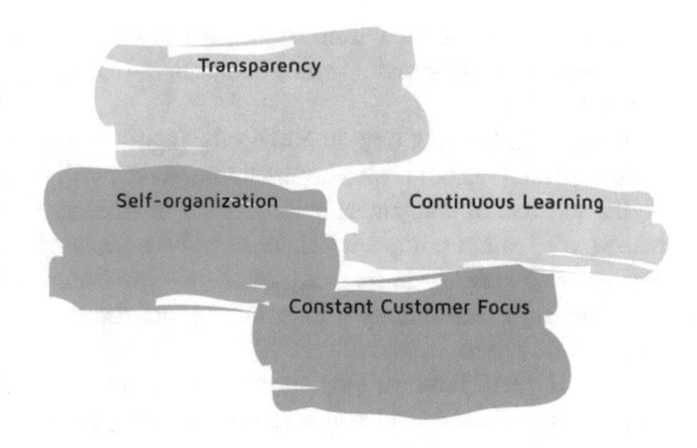

Fig. 9.4 Four values guiding agile companies [59]. © Jutta Eckstein, 2020; reprinted with permission

9.5.2 Outlook

Although agile development promises sustainability for a long time, it has not been addressed sincerely thus far. However, there are some promising developments and also concrete ideas for delivering on agility's promise. Most importantly, we have to increase the awareness of the impact of agility so that at least agile teams can make conscious decisions on effecting the social, environmental, or economical dimensions of sustainability.

An agile team can, for example, consider the energy consumption in their definition of done as well as via respective tests and monitoring. Individual teams and companies might discover ways to improve their own and their ecosystem's sustainability. Yet, only if this learning is shared and the effort improving sustainability is a collaborative one, we can really make a difference. As pointed out by Eckstein & Buck, a company claiming to be agile also has to aim for sustainability and as such needs to live up to the following values [59, p. 196], as illustrated in Fig. 9.4:

- Self-organization: An agile company should understand itself as a part of an ecosystem, belonging to itself, other companies, and the whole society.
- Transparency: An agile company makes its learning and doing transparent for the greater benefit of all.
- Constant customer focus: An agile company understands all aspects of its ecosystem—be it social, environmental, or economic—as its customer.
- Continuous learning: An agile company learns continuously from and with its ecosystem to make the whole world a better place.

Fundamental for agile companies that are sustainability-aware is the need for a connected perspective [59], p. 198: "This connected perspective incorporates the surrounding environment (economic, ecologic, societal, and social) in which companies operate." Thus, companies have to fulfill their role as active members of the society. One way for doing so is by joining networks that focus on improving the economic, social, and environmental aspects of the society. Sample networks are transparency international, global compact, fair labor association, or the climate group [60–63].

Sustainability is not only important for (agile) companies because it's part of the agile's vision [1, 2]. It is also important because it will be the key factor that decides on the survival of companies both in terms of finding talent and clients. This is the reason why some companies have already a sustainability officer in place who ensures sustainability in its many domains—environmental, economic, and social. It is the agile community's task to support the people in this role in making sustainability real.

With digitalization gaining more momentum and with the fact that the core competency of agile is in software development, more needs to be investigated in how agile development can make an important—positive—contribution for achieving higher sustainability. Because, as stated by the Karlskrona Manifesto [64]:

> Software in particular plays a central role in sustainability. It can push us towards growing consumption of resources, growing inequality in society, and lack of individual self- worth. But it can also create communities and enable thriving of individual freedom, democratic processes, and resource conservation.

Acknowledgments We want to thank the members of the Supporting Agile Adoption initiative of the Agile Alliance for their contributions and inspirations to our work.

References

1. Agile Alliance (2020) Agile alliance vision. https://www.agilealliance.org/the-alliance/
2. Scrum Alliance (2020) Scrum Alliance Vision. https://www.scrumalliance.org/
3. Beck K, Beedle M, van Bennekum A, Cockburn A, Cunningham W, Fowler M, Grenning J, Highsmith J, Hunt A, Jeffries R, Kern J, Marick B, Martin RC, Mellor S, Schwaber K, Sutherland J, Thomas D (2001) Manifesto for agile software development. http://www.agilemanifesto.org/
4. Hilty LM, Aebischer B (2015) ICT for sustainability: an emerging research field. In: Hilty LM, Aebischer B (eds) ICT innovations for sustainability. Springer International Publishing, pp 3–36
5. WCED (1987) Our common future. Oxford University Press
6. Wikipedia (2004) Wikipedia sustainability. https://en.wikipedia.org/wiki/Sustainability
7. World Summit (2005) World summit on social development. https://en.wikipedia.org/wiki/2005_World_Summit
8. Purvis B, Mao Y, Robinson D (2019) Three pillars of sustainability: in search of conceptual origins. Sustain Sci 14(3):681–695
9. Investopedia (2020) Triple bottom line. https://www.investopedia.com/terms/t/triple-bottom-line.asp
10. United Nations General Assembly (2015) Transforming our world: the 2030 agenda for sustainable development. Technical report, United Nations
11. Borowski P, Patuk I (2018) Selected aspects of sustainable development in agriculture. In: Proceedings: 2nd international conference on food and agricultural economics, pp 154–160. ICFAEC

12. The American Heritage Dictionary (2020) Definition of business. https://ahdictionary.com/word/search.html?q=business
13. Business Roundtable (2020) Our commitment to our employees and communities. https://opportunity.businessroundtable.org/
14. Business Roundtable (2020) Embracing sustainability challenge. https://www.businessroundtable.org/policy-perspectives/energy-environment/sustainability
15. Benefit Corporation: General questions. https://benefitcorp.net/faq (2020)
16. Economy for the Common Good (2020) What is ECG. https://www.ecogood.org/what-is-ecg/
17. Sousa TC, Melo CO (2019) Encyclopedia of the UN sustainable development goals: industry, innovation and infrastructure (SDG9), Chapter Sustainable infrastructure, industrial ecology and eco-innovation: positive impact on society, pp 1–10. Springer International Publishing
18. Hilty LM, Aebischer B, Andersson G, Lohmann W (eds) (2013) Proceedings of the first international conference on information and communication technologies for sustainability. ETH Zurich, University of Zurich and Empa, Swiss Federal Laboratories for Materials Science and Technology
19. Kurtz CF, Snowden DJ (2003) The new dynamics of strategy: Sense-making in a complex and complicated world. IBM Syst J 42(3):462–483
20. Frick T (2016) Designing for sustainability: a guide to building greener digital products and services. O'Reilly Media
21. Mortsiefer H (2017) The market doesn't want electric cars. https://www.tagesspiegel.de/wirtschaft/auto-der-zukunft-der-markt-will-elektroautos-nicht/19634320.html
22. O'Riordan T, Cameron J (2013) Interpreting the precautionary principle. Taylor & Francis
23. Tenner E (1997) Why things bite back: technology and the revenge effect. Fourth Estate
24. Omidyar Network (2020) Ethical explorer. https://ethicalexplorer.org
25. Beck K (2000) Extreme programming explained. Embrace change. Addison Wesley, Reading, MA
26. Blauer Engel (2020) Ressourcen- und energieeffiziente softwareprodukte. https://www.blauer-engel.de/de/get/productcategory/171/ressourcen-und-energieeffiziente-softwareprodukte
27. Kern E, Hilty LM, Guldner A, Maksimov YV, Filler A, Gröger J, Naumann S (2018) Sustainable software products - towards assessment criteria for resource and energy efficiency. Future Gener Comput Syst 86:199–210. https://doi.org/10.1016/j.future.2018.02.044
28. Melo C d O (2019) Another purpose for agility: sustainability. In: Meirelles P, Nelson MA, Rocha C (eds) Agile methods: 10th Brazilian workshop, WBMA 2019. Springer International Publishing, pp 3–7
29. Chappellet-Lanier T (2019) After protest, open source software company Chef will let ICE contract expire. https://www.fedscoop.com/protest-open-source-software-company-chef-will-let-ice-contract-expire
30. ITU (2017) E-waste monitor. https://www.itu.int/en/ITU-D/Climate-Change/Pages/Global-E-waste-Monitor-2017.aspx
31. Jones N (2018) How to stop data centres from gobbling up the world's electricity. Nature 561:163–166. https://doi.org/10.1038/d41586-018-06610-y
32. Wirth N (1995) A plea for lean software. Computer 28(2):64–68. https://doi.org/10.1109/2.348001
33. Gröger J, Herterich M (2019) Obsolete by software. how to keep digital hardware longer alive. In: Höfner VF (ed) What connects bits and trees: making digitization sustainable. oekom, Munich, pp 58–60
34. Internet Health Report (2018): The Internet uses more electricity than ... https://internethealthreport.org/2018/the-internet-uses-more-electricity-than
35. Matsakis L (2019) Amazon employees will walk out over the company's climate change inaction. https://www.wired.com/story/amazon-walkout-climate-change
36. Google (2018) Moving toward 24x7 carbon-free energy at google data centers: progress and insights. Technical report, Google

37. Hilty L, Aebischer B (2015) ICT innovations for sustainability. Adv Intell Syst Comput 310. doi:https://doi.org/10.1007/978-3-319-09228-7
38. Cone Communications (2017) Cone communications CSR study. https://www.conecomm.com/news-blog/2017/5/15/americans-willing-to-buy-or-boycott-companies-based-on-corporate-values-according
39. Wikipedia (2020) Munich re. https://en.wikipedia.org/wiki/Munich_Re
40. Munich RE (2019) Corporate responsibility report. https://www.munichre.com/content/dam/munichre/global/content-pieces/documents/cr-report-2019.pdf/_jcr_content/renditions/original./cr-report-2019.pdf
41. Jacobson I. Munich re transforms application development with lean and agile practices. https://www.ivarjacobson.com/sites/default/files/field_iji_file/article/munich_re_case_study.pdf
42. Business Agility Institute (2020). https://businessagility.institute/
43. UN Environment (2019) Champions of the earth 2019. https://www.unenvironment.org/championsofearth/laureates?title=&field_award_year_value=2019&field_award_category_target_id=All
44. CNN Money (2016) Patagonia's black Friday sales hit $10 million – and will donate it all. https://money.cnn.com/2016/11/29/technology/patagonia-black-friday-donation-10-million/index.html
45. McKinsey (2019) The five trademarks of agile organisations. https://www.mckinsey.com/business-functions/organization/our-insights/the-five-trademarks-of-agile-organizations
46. Patagonia (2016) Environmental + social initiatives. https://issuu.com/thecleanestline/docs/patagonia-enviro-2016-europe-eng?e=1043061/44692562
47. Wahl DC (2018) Why sustainability is no longer enough, yet still very important on the road to regeneration. https://medium.com/age-of-awareness/sustainability-is-no-longer-enough-yet-still-very-important-on-the-road-to-regeneration-57f5a37e05a
48. Rhodes CJ (2017) The imperative for regenerative agriculture. Sci Prog 100(1):80–129
49. DSM-Niaga (2020) Design out waste. Design for circularity. https://www.dsm-niaga.com/design.html
50. Laloux F (2014) Reinventing organizations: a guide to creating organizations inspired by the next stage in human consciousness. Nelson Parker
51. World Economic Forum (2020) Stakeholder principles in the COVID Era. http://www3.weforum.org/docs/WEF_Stakeholder_Principles_COVID_Era.pdf
52. DSM-Niaga (2020) Design for circularity with Niaga. https://www.dsm.com/corporate/solutions/resources-circularity/design-for-circularity-with-niaga.html
53. Marquard S (2018) What the Sparda Bank learns from Silicon Valley. https://www.stuttgarter-nachrichten.de/inhalt.digitalisierung-in-der-bankenwelt-wie-die-sparda-bank-selbst-kunden-in-malaysia-hilft.0a532acf-fc61-40a0-9c70-34db83e0f7b8.html
54. Sparda-Bank (2020) We are pioneers in climate protection. https://www.sparda-m.de/genossenschaftsbank-umwelt-und-klimaschutz
55. UNFCC (2020) What is the Kyoto protocol? https://unfccc.int/kyoto_protocol
56. Sparda-Bank (2020) Transparency in own investments. https://www.sparda-m.de/gemeinwohl-oekonomie-eigenanlagen/#innernav
57. Gilbert JC (2018) Are B corps an elite charade for changing the world? https://www.forbes.com/sites/jaycoengilbert/2018/08/30/are-b-corps-an-elite-charade-for-changing-the-world-part-1/#67395ea97151
58. Stiglitz JE (2018) Review of the book "winners take all: The elite charade of changing the world". https://www.nytimes.com/2018/08/20/books/review/ winners-take-all-anand-giridharadas.html
59. Eckstein J, Buck J (2020) Company-wide agility with beyond budgeting, open space & sociocracy: survive & thrive on disruption. Jutta Eckstein, Braunschweig
60. Transparency International. https://www.transparency.org

61. United Nations Global Compact. https://www.unglobalcompact.org
62. Fair Labor Association. https://www.fairlabor.org
63. The Climate Group. https://www.theclimategroup.org
64. Becker C, Chitchyan R, Duboc L, Easterbrook S, Mahaux M, Penzenstadler B, Rodríguez-Navas G, Salinesi C, Seyff N, Venters CC, Calero C, Koçak SA, Betz S (2014) The karlskrona manifesto for sustainability design. CoRR

Chapter 10
Governance and Management of Green IT

J. David Patón-Romero, Maria Teresa Baldassarre, Moisés Rodríguez, and Mario Piattini

Abstract Sustainability has become a main pillar for the development of our civilization. It is increasingly evident that achieving sustainable development is not only necessary to have a future, but also helps us create greater value by being more effective and efficient. This has led to more and more organizations implementing sustainable practices across different fields. One of these fields with the greatest impact and which is evolving the most is Information Technology (IT). Through what is known as Green IT, organizations are implementing measures to reduce the environmental impact of their IT, as well as using their IT to be more sustainable in other areas. However, organizations are conducting these Green IT implementations at their own discretion, due to the lack of guidelines, standards, or frameworks in this regard. With the objective of helping organizations, this chapter presents a framework that guides the way in which organizations should properly govern and manage Green IT. To this end, we have developed the *Governance and Management Framework for Green IT* (GMGIT), validating and refining it through different case studies at an international level, obtaining several versions through an iterative and incremental cycle. The results

J. D. Patón-Romero (✉)
University of Castilla-La Mancha (UCLM), Ciudad Real, Spain

University of Bari "Aldo Moro" (UniBa), Bari, Italy

AQCLab, Ciudad Real, Spain
e-mail: JoseDavid.Paton@gmail.com

M. T. Baldassarre
Department of Informatics, University of Bari "Aldo Moro" (UniBa), Bari, Italy
e-mail: mariateresa.baldassarre@uniba.it

M. Rodríguez
AQCLab, Ciudad Real, Spain
e-mail: mrodriguez@aqclab.es

M. Piattini
Alarcos Research Group, Institute of Technologies and Information Systems, University of Castilla-La Mancha (UCLM), Ciudad Real, Spain
e-mail: Mario.Piattini@uclm.es

C. Calero et al. (eds.), *Software Sustainability*,
https://doi.org/10.1007/978-3-030-69970-3_10

of this development show that the GMGIT is a very useful framework for organizations to implement, evaluate, and improve the governance and management of Green IT.

10.1 Introduction

Sustainability [1] has become the biggest challenge and main duty of our time. From all fields of knowledge, experts and professionals join forces to find the best sustainable solutions and practices. Information technology (IT) is an area that is exponentially expanding [2], and, therefore, with the increasing pollution it entails, can be considered an enemy of the environment. Thus, the so-called Green IT practices are researched and developed. This field of Green IT has received multiple definitions since its inception [3, 4], but the one that best fits is the following (adapted from [5]):

> Green IT is the study and practice of design, build and use of hardware, software and information technologies with a positive impact on the environment.

From this definition we can see that Green IT is a much broader field than is normally imagined. On the one hand, we find different areas of application and development, such as hardware, software, and IT as a whole. And, on the other hand, we have the idea that sustainable practices should not only be applied to these elements but also used as sustainability mechanisms. The latter responds to the idea proposed by Erdélyi [6] to differentiate between two great perspectives of Green IT:

- ***Green by IT***: through which it is intended to provide the necessary tools to perform diverse kind of tasks in different areas in a sustainable manner for the environment (i.e., IT understood as a capacitator or enabler [7]).
- ***Green in IT***: through which it is intended to reduce the negative impact that IT has on the environment, due to its energy consumption and the emissions it produces (i.e., IT understood as a producer).

This idea of Green IT increasingly attracts organizations of all kinds (not only dedicated exclusively to IT) because of the great benefits and advantages that it entails [8–11]. The number of organizations that implement some type of sustainable practice in and/or by IT is increasing [4, 12]. However, they put the cart before the horse. They begin to implement Green IT practices in a disorganized manner, without clear objectives and without any control. Unlike other business areas, organizations do not establish adequate governance and management bases to implement, maintain, and improve these practices.

For this reason, we have developed the "Governance and Management Framework for Green IT" [13, 14], which establishes the characteristics and elements of governance and management that organizations should consider when implementing, assessing, and improving Green IT.

In the next sections, the "Governance and Management Framework for Green IT" is presented, as well as the necessary characteristics (based on the experience obtained) to assess/audit and implement improvement plans in organizations using this framework.

10.2 "Governance and Management Framework for Green IT" (GMGIT)

The "Governance and Management Framework for Green IT" (GMGIT, from now on) [13, 14] is a framework based on COBIT 2019 [15, 16], one of the most widespread and used frameworks for the governance and management of IT. From COBIT 2019 we have only taken as a basis the structure of components that it defines, and, for each of these components, we have defined and established the characteristics and elements applicable to Green IT. The following subsections show the most relevant characteristics of the GMGIT, such as the structure of the framework and the components of Green IT defined.

10.2.1 Framework Structure

The GMGIT is divided into four main sections that address all the characteristics for the implementation, assessment, and improvement of Green IT in organizations:

- **Section I.** This first section includes the conceptual basis needed to understand the context of the framework. Thus, it offers an overview of what Green IT is, as well as what COBIT 2019 is and how the architecture of this framework can be adapted to the specific needs of Green IT.
- **Section II.** This section is the main part of the framework, through which the elements to implement a correct governance and management of Green IT are defined and established. To this end, for each of the seven generic components established by COBIT 2019 (cf. Sect. 10.2.2), the applicable Green IT characteristics and elements have been detailed. It is also important to highlight that in this section the ISO 14000 family of standards [17] is also applied to IT, through the specific characteristics of Green IT defined in the "processes" component.
- **Section III.** This third section proposes a Green IT audit framework, which includes the steps that an auditor must follow, as well as the aspects that must be considered to audit Green IT. In the same way, a total of 648 Green IT audit questions (based on the activities defined in each of the practices of the "processes" component) are also included to help auditors conduct their work.
- **Section IV.** The fourth and final section includes a maturity model for Green IT based on the ISO/IEC 33000 family of standards [18]. Through this model, each of the processes defined in Sect. II is organized by maturity levels, so that an organization can conduct the implementation, assessment, and/or improvement of Green IT in a progressive and systematic manner. In Sect. 10.3, we can see how the application of this maturity model for Green IT has been performed through different audits conducted using the GMGIT. Likewise, Section 10.4 also shows the application of the said model through improvement plans.

10.2.2 Governance and Management Components of Green IT

The following subsections show an overview of the components of Green IT defined in the GMGIT, highlighting the characteristics that organizations should consider to implement a proper governance and management of this area.

10.2.2.1 Principles, Policies, and Procedures

The principles, policies, and procedures represent the guidelines established by an organization to govern and manage all its members and stakeholders toward a desired direction and behavior in a specific area.

On the one hand, the principles of Green IT serve to communicate the rules established by the board of directors and the executive management, giving support to governance objectives and organizational values. In this regard, in the GMGIT we have identified nine principles organized into three groups:

- **Give support to the business.**
  - Give quality and value to the stakeholders.
  - Comply with relevant legal requirements and regulations.
  - Provide convenient and precise information on the functioning of Green IT.
  - Evaluate current and future IT capabilities.
  - Promote ongoing improvement in Green IT.
- **Reduce the environmental impact.**
  - Adopt a strategy that is based on the efficient use of IT resources.
  - Develop the systems in a sustainable way.
- **Foster responsible behavior toward and in favor of the environment.**
  - Act professionally and ethically.
  - Foster a positive culture of Green IT. Table 10.1 shows an example of how the characteristics of each of these groups and principles have been defined in the GMGIT.

On the other hand, the policies of Green IT provide a more detailed guide as to how to put the principles of Green IT into practice, as well as indicating how these will influence decision-making. In this regard, we have identified the following six policies that should be considered, adapted, and implemented in Green IT.

- Policy of Green IT.
- Policy of acquisition, development, and maintenance of IT systems.
- Policy of resource management.
- Policy of compliance.
- Policy of conduct.
- Policy of asset management.

Table 10.1 Description of the principles related to "Reduce the environmental impact"

Principle	Objective	Description
Reduce the environmental impact		
Adopt a strategy that is based on the efficient use of IT resources	Ensure that resources are managed in a way that is consistent and effective	There needs to be a strategy established that assures the effective and efficient use of IT resources in terms of sustainability; it will design, implement, and manage the mechanisms needed for this strategy and the goals associated with it to be put into action
Develop the systems in a sustainable way	Build systems of high quality which are economically viable and that are committed to the environment, providing value to stakeholders	Systems that meet quality and sustainability standards need to be designed, built, and put in place. These should make it possible for the goals set by the organization to be fulfilled

Table 10.2 Definition of the main characteristics of the "Policy of Green IT"

Scope/goals	Stakeholders/people with responsibility
Policy of Green IT	
• Definition and vision of Green IT for the organization, including the goals and appropriate metrics. • Strategic plans for Green IT. • Explanation of the alignment of the policy of Green IT with the other high-level policies. • Identification and development of specific aspects of Green IT (management of consumption, compliance with legal obligations, etc.) • Management of the budget and costs of the life cycle of Green IT. • Responsibilities associated with Green IT.	The policy of Green IT is addressed to all the employees and stakeholders in the organization Those responsible for the development, maintenance, and updating of the policy of Green IT are the members of the Sustainability Steering Committee (SSC) and the Chief Sustainability Officer (CSO)

Of course, not all relevant policies are written, nor must all of them necessarily be applied in any one particular organization. Each organization should consider its own specific context, alongside other external factors, and make appropriate modifications in its specific policies of Green IT.

Table 10.2 shows as an example the definition and description of the "Policy of Green IT" that has been performed in the GMGIT.

10.2.2.2 Organizational Structures

The organizational structures are the key elements in the decision-making of an organization when it comes to decisive aspects and areas. Regarding Green IT, in the

Table 10.3 CSO role details

CSO: mandate, operation principles, and scope	
Area	Description
Mandate	The CSO is entirely responsible for the program of Green IT in the organization
Operation principles	The obligations and principles of the CSO are: • Possesses an exact knowledge of the strategic vision of the organization. • Has the ability to translate the objectives and goals of the organization into the requirements of Green IT. • Is an effective communicator and receiver. • Acts as a link between the executive management and the program of Green IT. • Depending on different factors in the organization (such as its organizational structure), reports on matters connected with Green IT to the CEO, to the members of the SSC, or to other executives in the management of the organization. • Builds effective relationships with the board of directors, executive management, and stakeholders. • Communicates and coordinates with the stakeholders, so that their needs in Green IT are met.
Span of control	The CSO is responsible for: • Designing, implementing, and managing a plan and strategy of Green IT. • Developing, maintaining, and updating the principles, policies, and procedures related to Green IT. • Monitoring and managing the correct performance of Green IT.
Authority level	The CSO has responsibility for the approval of decisions that have to do with the correct development, implementation, and management of Green IT in the organization, and has the capacity to do so
Delegation rights	The CSO should delegate specific tasks to the personnel in the organization who are in charge of aspects such as the deployment of a system of energy control, monitoring at all times that this task is carried out correctly
Escalation path	The CSO should report the key problems associated with Green IT to the SSC

GMGIT we have identified two main roles that must be established to be in charge of the Green IT functions:

- **Chief Sustainability Officer (CSO).** Its role is being the main representative of Green IT in the organization. The CSO has the responsibility of controlling and supervising the management of Green IT.
- **Sustainability Steering Committee (SSC).** This committee should be responsible for verifying that Green IT performs correctly, as well as that the policies, plan, strategy, and other governance aspects in this regard are applied and followed effectively and efficiently. It can be formed by different roles in the organization, such as the CSO, CIO, CTO, business owners, other representatives of Green IT, etc.

Table 10.3 shows the example of the role of the CSO, in which the different characteristics to be established are detailed, as identified in the GMGIT.

10.2.2.3 People, Skills, and Competencies

This component identifies the skills and competencies that the people who are in charge of/responsible for Green IT should have. To implement Green IT effectively and efficiently, the following levels or areas of skills and competencies must be considered and covered:

- Governance of Green IT.
- Strategy of Green IT.
- Architecture of Green IT.
- Operations of Green IT.
- Evaluation, tests, and compliance of Green IT.

Table 10.4 shows by way of example the skills and competencies that should be considered and covered in the area of "Strategy of Green IT," as included in the GMGIT.

10.2.2.4 Culture, Ethics, and Behavior

The culture, ethics, and behavior are the patterns of conducts, beliefs, suppositions, attitudes, and ways of conducting activities/practices correctly in the quest of achieving the success of, in this case, Green IT.

Table 10.4 Description of the skills and competencies to be covered in the area of "Strategy of Green IT"

Strategy of Green IT: description, experience, knowledge, and skills	
Area	Description
Description	The roles in this area should define and implement the vision, mission, and objectives of Green IT, always maintaining the alignment with the strategy and organizational culture
Experience	• Experience in Green IT and management in areas of the organization. • Definition, implementation, and management of the strategy and governance of Green IT. • Alignment of the strategy, principles, and best practices of Green IT with the rest of the organization.
Knowledge	• Standards, regulations, frameworks, guidelines, and best practices of Green IT. • Legal and regulatory requirements of Green IT. • State of the art, services, and emerging disciplines of Green IT.
Skills	• Capacity of defining management aspects (management of resources, management of processes, management of performance, etc.) of Green IT that are applicable in the organization. • Leadership (conflict resolution, excellent communication skills, etc.) • Business orientation. • High-level strategic thinking.

Table 10.5 Definition of "Behavior 1" and "Behavior 2"

Organizational ethics	Individual ethics
Behavior 1. Green IT is put into practice in day-to-day operations	
Green IT is included as a key area in the establishment and achievement of the organizational objectives	The best practices of Green IT are followed, since the individuals are committed both to Green IT and to the success of the organization
Behavior 2. The importance of the policies and principles of Green IT is respected	
The board of directors and the executive management support the policies and principles of Green IT, approving them, checking them, and communicating them to the rest of the organization at regular intervals	The policies and principles of Green IT are known and understood, and the guidelines that they establish are followed

Behavior is the key pillar, because it determines the culture of the organization itself and the approach to Green IT. For this area of Green IT, in the GMGIT we have defined eight behaviors that organizations should include in their culture:

- **Behavior 1.** Green IT is put into practice in day-to-day operations.
- **Behavior 2.** The importance of the policies and principles of Green IT is respected.
- **Behavior 3.** The members and stakeholders are provided with enough detailed guidelines on Green IT, and compliance with these is encouraged.
- **Behavior 4.** The members and stakeholders of the organization are responsible for the proper use of Green IT.
- **Behavior 5.** The members and stakeholders of the organization identify and communicate new Green IT needs.
- **Behavior 6.** The members and stakeholders of the organization are receptive when identifying and managing new Green IT challenges.
- **Behavior 7.** The organization is committed to, and aligned with, Green IT.
- **Behavior 8.** The organization acknowledges the value brought to it by Green IT.

Table 10.5 shows as an example the definition that has been conducted in the GMGIT about "Behavior 1" and "Behavior 2."

10.2.2.5 Information

This component identifies how the information coming from the different systems and processes of the organization may be used to govern and manage Green IT. In this regard, the GMGIT identifies the following groups or types of information that an organization must have in relation to Green IT, in order to conduct an appropriate decision-making in the implementation, operation, and maintenance of this area:

- Policies and principles of Green IT.
- Plan and strategy of Green IT.
- Requirements of Green IT.

Table 10.6 Main characteristics of the "Scorecard of Green IT"

Scorecard of Green IT: goals, life cycle, and best practices	
Area	Description
Goals	The scorecard of Green IT has the task of providing the information needed for appropriate decisions to be taken and for correct management of Green IT in the organization to be carried out. To that end, it should contain all the events and relevant information (at a level that is appropriate for decisions to be taken) about the function of Green IT
Life cycle	The scorecard of Green IT should be updated regularly, since this information is needed if those responsible for Green IT are to manage this area and direct it so that it works correctly
Best practices	The CSO is responsible for gathering the relevant information on Green IT, expressing it in the scorecard of Green IT. For that to happen, the information that should be considered is the following: • Effectiveness and efficiency in the function of Green IT. • Progress in the scope of the objectives of Green IT and their relationship with the organizational objectives. • Costs of Green IT. • Action needed to improve Green IT.

- Budget of Green IT.
- Awareness material of Green IT.
- Review reports of Green IT.
- Scorecard of Green IT.

Table 10.6 shows as an example the main characteristics of the information that should be considered in the "Scorecard of Green IT," as established in the GMGIT.

10.2.2.6 Services, Infrastructure, and Applications

The services, infrastructure, and applications are translated into a set of service capacities in order to provide a proper functioning of Green IT in the organization. Regarding Green IT, the following activities or services that should be provided for Green IT to perform correctly have been identified in the GMGIT:

- Provide an architecture of Green IT that is appropriate to the needs and capabilities of the organization.
- Provide awareness of, and training in, Green IT.
- Provide evaluations and tests of Green IT.

Table 10.7 shows by way of example the service capacities included in the GMGIT that should be considered in "Awareness of, and training in, Green IT."

Table 10.7 Description of the service capacities for "Awareness of, and training in, Green IT"

Awareness of, and training in, Green IT: service capabilities	
Service capability	Description
Establish a system of communication and distribution of relevant information on Green IT	Provide a system of communication and distribution of relevant information on Green IT for the different members of the organization and the stakeholders, one that will allow them to fulfill their responsibilities correctly while also raising awareness of the importance of Green IT both within the organization and outside it
Manage the program of awareness and training and keep it up to date	Establish a program to raise awareness and to train in Green IT; this should be one that will help the members of the organization and the stakeholders to understand and become familiar with the importance of Green IT and its relevant features; that will mean that these parties carry out their responsibilities properly, leading to a correct function of Green IT in the organization

10.2.2.7 Processes

The processes are the main component or the core of the GMGIT, since through them the necessary aspects to conduct the implementation and assessment of the rest of the components, among other characteristics and key elements, are established.

In order to perform the definition of the Green IT processes, in the GMGIT we have not invented any new process, but of the 40 processes defined by COBIT 2019 [16], we have chosen and taken as a basis a total of 38 processes that we consider affect or are affected by Green IT. We have excluded two processes, "APO13. Manage security" and "DSS05. Manage security services," since they are processes that are very focused in the field of security, and among their practices and activities, they have no direct relationship with sustainability.

Likewise, it is important to highlight that COBIT 2019 identifies objectives, but they represent and describe the corresponding processes, so we decide in the GMGIT to maintain processes as the main focus.

Thus, in the GMGIT we have adapted each of the 38 selected processes to Green IT, defining and developing the following characteristics:

- **Goals and metrics.** The Green IT goals that the organization should achieve in the context of the process in question and the metrics that can be used to verify whether these goals are achieved. Table 10.8 shows an example of this characteristic in the "EDM03. Ensure risk optimization" process.
- **RACI matrix.** Regarding the position of each of the specific roles of Green IT (defined in the "Organizational structures" component) and other relevant roles in relation to the specific practices of the process in question. Table 10.9 includes as an example the RACI matrix of the "DSS03. Manage problems" process.
- **Practices, inputs and outputs, and activities.** The Green IT practices specific to the process, identifying the inputs and outputs of each practice, as well as the

Table 10.8 Goals and metrics that are specific to Green IT of the "EDM03. Ensure risk optimization" process

EDM03: Goals and metrics of the process that are specific to Green IT	
Goals of the process that are specific to Green IT	Related metrics
1. The risks derived from Green IT are identified, communicated, and managed effectively and efficiently.	• Number of risks of Green IT identified and managed. • Percentage of risks of Green IT that are mitigated effectively.
2. The strategy and management of risks of Green IT are aligned with the strategy and overall risk management of the organization.	• Level of alignment between the risks of Green IT and the business risks. • Percentage of risks of Green IT related to the business risks.

specific Green IT activities (differentiated between activities specific to *Green by IT* and activities specific to *Green in IT*), which will come to define the actions to be assessed to verify if a Green IT implementation complies with the process in question or not. Table 10.10 shows through an example the "APO02.02" practice of the "APO02. Manage strategy" process with all these elements.

- **Related guidance.** Identifying which specific standard and which reference within the standard are directly related to the practices and activities defined in the process in question. In this regard, only the ISO 14000 family of standards [17], related to environmental management, has been applied [19]. Thus, complying with the practices and activities of a process, the practices of these references of the standard would also be fulfilled and vice versa. Table 10.11 illustrates as an example the related guidance regarding "BAI01. Manage programs."

10.2.3 *Evolution of the GMGIT*

The GMGIT is a framework that has emerged following a long and thorough study. Indeed, a process of several years and different versions has been necessary to refine and achieve a solid, coherent, and adapted framework to the current context of organizations and Green IT.

In Fig. 10.1 we can see, as a summary, an overview of the evolution of the GMGIT through its three versions.

From the first version, we started making a proof of concept with only 15 processes, which resulted in a total of 122 audit questions. After the two validations we conducted for this version, we also made a proof of concept developing a maturity model for the framework based on the ISO/IEC 15504 standard [20].

From the lessons learned obtained at the validations of the first version, the second version of the framework emerged. In this second version, we included 20 new processes and established the difference between *Green by IT* and *Green in IT*, giving rise to a total of 600 audit questions. We also updated the maturity

Table 10.9 RACI matrix that is specific to Green IT of the "DSS03. Manage problems" process

DSS03: RACI matrix of the process that is specific to Green IT										
Key management practice	Board of Directors	Chief Executive Officer (CEO)	Chief Financial Officer (CFO)	Chief Information Officer (CIO)	Chief Technology Officer (CTO)	Business owners	Sustainability Steering Committee (SSC)	Chief Sustainability Officer (CSO)	Auditing	Compliance with laws and regulations
DSS03.01 Identify and classify problems				C	C	C	A	R	I	I
DSS03.02 Investigate and diagnose problems				R	R		A	R		
DSS03.03 Raise known errors				C	C		A	R		
DSS03.04 Resolve and close problems				C	C	C	A	R	C	C
DSS03.05 Perform pro-active prob-lem management				C	C	C	A	R		

R: Responsible; A: Accountable; C: Consulted; I: Informed

Table 10.10 "APO02.02" practice, with its inputs, outputs, and activities that are specific to Green IT of the "APO02. Manage strategy" process

APO02: Practices, inputs/outputs, and activities of the process that are specific to Green IT				
Management practice	Inputs specific to Green IT		Outputs specific to Green IT	
APO02.02 Assess current capabilities, performance, and maturity of Green IT of the organization Assess the performance of the business, and the capabilities and outsourced services of Green IT, so as to develop an understanding of the enterprise architecture with respect to Green IT. Identify the problems that are being experienced and produce recommendations in the areas that can benefit from these improvements Consider the distinguishing features and options as regards service providers, as well as financial impact, potential costs, and benefits of using outsourced services	APO01.09	Evaluation of compliance of the policies and procedures of Green IT	Capabilities of Green IT	APO02.03 APO04.04 APO08.05 APO09.05 APO11.01 BAI01.01 BAI02.01 BAI04.01 BAI11.01
	APO02.01	Alignment of Green IT with the strategies, objectives, challenges, stakeholders, and organizational context		
Activities specific to Green by IT				
1. Define and establish some basic capabilities of Green by IT, aligned with the organization's own capabilities. 2. Carry out a SWOT (Strengths, Weaknesses, Opportunities, and Threats) analysis of the context, capabilities, and current performance of Green by IT, in an effort to understand its current performance and identify inconsistencies and/or possibilities of alignment with the context and capabilities of the organization.				
Activities specific to Green in IT				
1. Define and establish some basic capabilities of Green in IT, aligned with the organization's own capabilities. 2. Carry out a SWOT (Strengths, Weaknesses, Opportunities, and Threats) analysis of the context, capabilities, and current performance of IT with respect to Green IT, in an effort to understand its current performance and identify inconsistencies and/or possibilities of alignment with the context and capabilities of the organization.				

model to the new ISO/IEC 33000 standard [18] and conducted four validations in organizations at the international level.

And finally, in the third version we included 3 new processes (making a total of 38 processes and 648 audit questions), and we applied the ISO 14000 family of

Table 10.11 Related guidance that is specific to Green IT of the "BAI01. Manage programs" process

BAI01: related guidance that is specific to Green IT	
Related standard	Detailed reference
ISO 14001, ISO 14004	• 6.1. Actions to address risks and opportunities • 7.5. Documented information • 9.1. Monitoring, measurement, analysis, and evaluation
ISO 14005	• 4. Undertaking an environmental-related project to secure management support and commitment to begin the phased implementation of an Environmental Management System (EMS) • 5.5. Documentation • 6.8. Environmental performance evaluation, including monitoring and measurement
ISO 14006	• 5.4. Implementation and operation • 5.5. Checking
ISO 14031	• 4. Environmental performance evaluation
ISO/TS 14033	• 4. Use of quantitative environmental information • 5. Principles for generating and providing quantitative environmental information • 6. Guidelines

standards [17], as well as other changes derived from the adaptation to the new COBIT 2019 [15, 16].

On the other hand, we are currently working on conducting more validations through new case studies [21], as well as applying the framework through improvement plans in the organizations that we have already audited.

10.3 Auditing the Green IT with the GMGIT

When conducting the implementation of the governance and management of Green IT, it is important to know not only the best practices on which to base but also how to assess/audit this area. Auditing the governance and management of Green IT is crucial to identify the problems in this regard, propose solutions, and perform progressive improvements, among others. Therefore, the GMGIT also includes an audit framework of Green IT (which corresponds to Section III of the GMGIT), as well as a maturity model developed specifically for Green IT (Section IV of the GMGIT), whose main characteristics are shown in the following subsections.

10.3.1 Audit Framework of Green IT

The audit framework of Green IT included in the GMGIT is divided into two main parts, through which it is intended to support the work of the auditors.

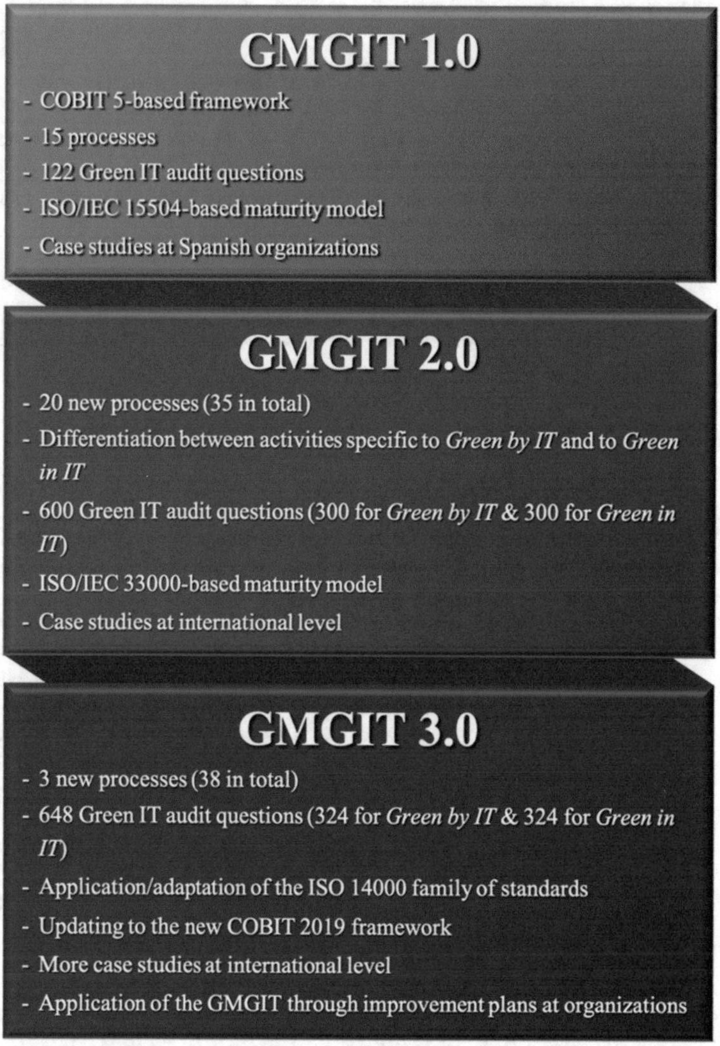

GMGIT 1.0

- COBIT 5-based framework
- 15 processes
- 122 Green IT audit questions
- ISO/IEC 15504-based maturity model
- Case studies at Spanish organizations

GMGIT 2.0

- 20 new processes (35 in total)
- Differentiation between activities specific to *Green by IT* and to *Green in IT*
- 600 Green IT audit questions (300 for *Green by IT* & 300 for *Green in IT*)
- ISO/IEC 33000-based maturity model
- Case studies at international level

GMGIT 3.0

- 3 new processes (38 in total)
- 648 Green IT audit questions (324 for *Green by IT* & 324 for *Green in IT*)
- Application/adaptation of the ISO 14000 family of standards
- Updating to the new COBIT 2019 framework
- More case studies at international level
- Application of the GMGIT through improvement plans at organizations

Fig. 10.1 Evolution of the GMGIT through its three versions

First, the different stages to be followed during a Green IT audit are identified and explained in detail (taking as a reference the stages defined by COBIT) [22].

Then, the second part of the audit framework of Green IT includes a total of 648 audit questions that cover each of the processes defined in Section II of the GMGIT. These audit questions are directly related to the activities to be performed in each of the practices identified in the different processes. That is why they are divided into two large groups (324 questions in each), depending on whether they are activities specific to *Green by IT* or to *Green in IT*.

Table 10.12 Green by IT audit questions for the "APO05. Manage portfolio" process

Process	Questions
APO05. Manage portfolio	Are the investment options, internal and external, necessary to carry out investments in Green by IT identified and analyzed?
	Do the programs of Green by IT have adequate financing to cover the investment needs in this regard?
	Is the performance of investments in Green by IT periodically monitored, evaluated, and optimized?
	Are the funding and adequate resources maintained and updated with respect to the investments in Green by IT and to the services and assets of Green by IT?
	Is the achievement of benefits of the investments conducted in Green by IT evaluated?

Table 10.13 Green in IT *audit questions for the* "MEA03. Manage compliance with external requirements" *process*

Process	Questions
MEA03. Manage compliance with external requirements	Are the new legal, regulatory, and contractual requirements of sustainability that may affect the IT continually identified, implemented, and monitored?
	Are the policies, principles, requirements, objectives, and solutions of IT aligned with the legal, regulatory, and contractual requirements of sustainability that are applicable?
	Is there assured conformance and compliance of the policies, principles, requirements, objectives, and solutions of IT with the legal, regulatory, and contractual requirements of sustainability that are applicable?
	As regards the data related to the fulfillment of the external compliance requirements of sustainability that are applicable to IT: are these obtained and verified?
	Are corrective measures taken to align the IT with the external compliance requirements of sustainability?

As an example, Table 10.12 contains the *Green by IT* audit questions defined for the "APO05. Manage portfolio" process, while Table 10.13 includes the *Green in IT* audit questions defined for the "MEA03. Manage compliance with external requirements" process.

10.3.2 ISO/IEC 33000-Based Maturity Model for Green IT

All of the above, the governance and management components and the audit framework of Green IT, are very important characteristics that organizations should consider and implement in this context. However, based on our experience, it is totally unfeasible to try to implement or assess/audit all these characteristics at once;

it is necessary to perform a systematic and progressive process in this regard. That is why we have also developed and included a maturity model adapted to the characteristics defined in the GMGIT, and for which we have followed the ISO/IEC 33000 family of standards [18].

Thus, to develop this maturity model, we have taken as a basis the five maturity levels (plus level 0) established by the ISO/IEC 33000, and we have adapted them to the specific context of Green IT:

- **Level 0 (Incomplete).** The organization does not consider sustainability and no Green IT practice is defined.
- **Level 1 (Initial).** The organization considers sustainability and carries out Green IT practices in the most critical aspects related to sustainability.
- **Level 2 (Managed).** The Green IT practices are clearly defined, established, and managed throughout the different business areas, contributing to sustainability in and/or by IT.
- **Level 3 (Established).** The organization follows recognized standards and best practices of Green IT (Green IT is correctly managed and governed), as well as identifies and ensures in a continuous manner the compliance with the external requirements.
- **Level 4 (Predictable).** The organization performs the monitoring, evaluation, and measurement of the implemented Green IT practices, through a set of sustainability metrics established for that purpose.
- **Level 5 (Innovating).** The organization is fully committed to sustainability and is oriented toward the continuous improvement of the implemented Green IT practices, by means of, for example, detailed performance reports, exhaustive use of sustainability metrics, and management of the innovation process in sustainability.

On the other hand, we have also organized the 38 processes defined in Sect. II of the GMGIT at each of these maturity levels, depending on which processes are more basic and should be implemented and evaluated first and which processes are more complex and belong to more advanced levels of implementation.

Likewise, we have established the necessary mechanisms to evaluate the capability of each of the processes. To do this, we have adopted the five capability levels established by the ISO/IEC 33000, as well as the process attributes and process attribute results of each of these levels, since they are fully compatible and adaptable to the GMGIT. Furthermore, we have identified and established the relationship between the capability levels and the maturity levels, i.e., the capability levels that each of the processes must meet to reach a certain maturity level.

All this can be seen in Fig. 10.2, where the organization of the different processes at the maturity levels is shown, as well as their correspondence with the capability levels. This organization has been performed following the example of application of the software development life cycle processes at the maturity and capability levels of the ISO/IEC 33000 standard, conducted by the "Software Engineering Maturity Model MMIS 2.0."

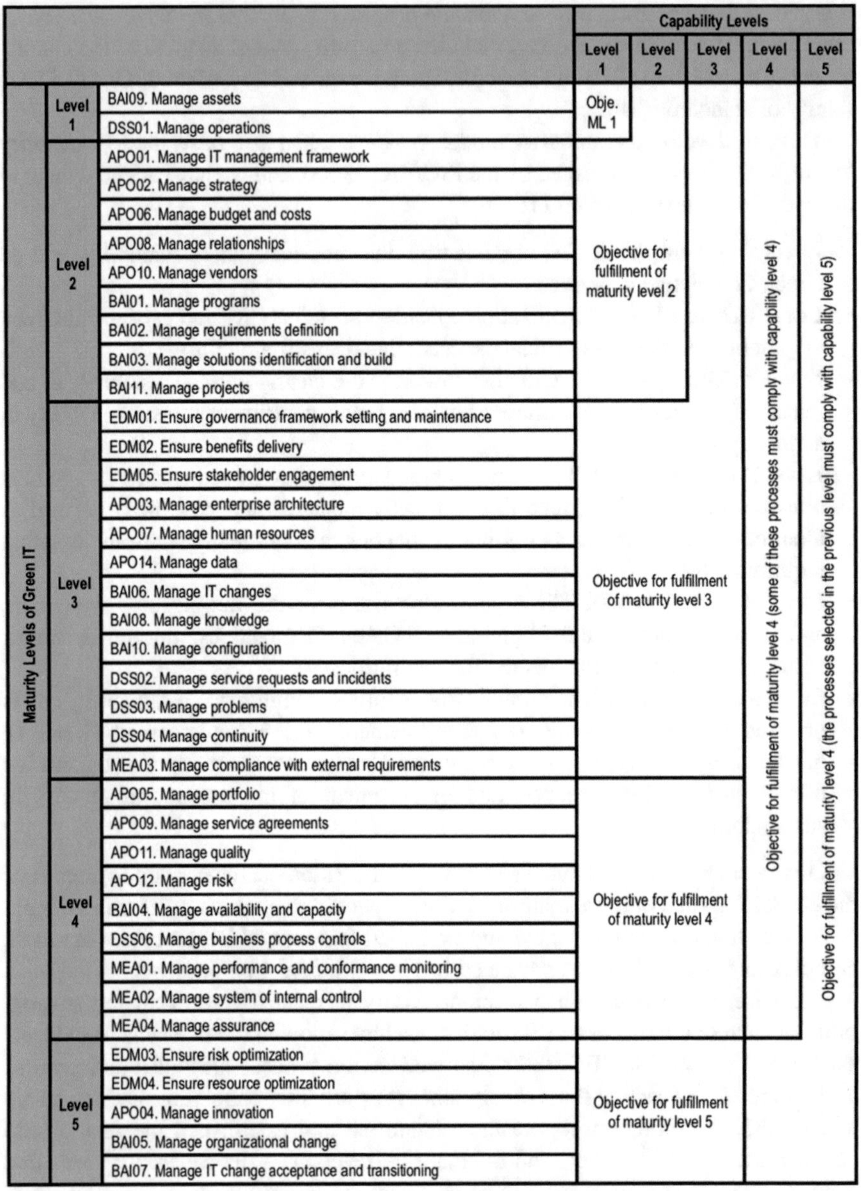

Maturity Levels of Green IT		Capability Levels: Level 1	Level 2	Level 3	Level 4	Level 5
Level 1	BAI09. Manage assets	Obje. ML 1			Objective for fulfillment of maturity level 4 (some of these processes must comply with capability level 4)	Objective for fulfillment of maturity level 4 (the processes selected in the previous level must comply with capability level 5)
	DSS01. Manage operations					
Level 2	APO01. Manage IT management framework	Objective for fulfillment of maturity level 2				
	APO02. Manage strategy					
	APO06. Manage budget and costs					
	APO08. Manage relationships					
	APO10. Manage vendors					
	BAI01. Manage programs					
	BAI02. Manage requirements definition					
	BAI03. Manage solutions identification and build					
	BAI11. Manage projects					
Level 3	EDM01. Ensure governance framework setting and maintenance	Objective for fulfillment of maturity level 3				
	EDM02. Ensure benefits delivery					
	EDM05. Ensure stakeholder engagement					
	APO03. Manage enterprise architecture					
	APO07. Manage human resources					
	APO14. Manage data					
	BAI06. Manage IT changes					
	BAI08. Manage knowledge					
	BAI10. Manage configuration					
	DSS02. Manage service requests and incidents					
	DSS03. Manage problems					
	DSS04. Manage continuity					
	MEA03. Manage compliance with external requirements					
Level 4	APO05. Manage portfolio	Objective for fulfillment of maturity level 4				
	APO09. Manage service agreements					
	APO11. Manage quality					
	APO12. Manage risk					
	BAI04. Manage availability and capacity					
	DSS06. Manage business process controls					
	MEA01. Manage performance and conformance monitoring					
	MEA02. Manage system of internal control					
	MEA04. Manage assurance					
Level 5	EDM03. Ensure risk optimization	Objective for fulfillment of maturity level 5				
	EDM04. Ensure resource optimization					
	APO04. Manage innovation					
	BAI05. Manage organizational change					
	BAI07. Manage IT change acceptance and transitioning					

Fig. 10.2 Organization of the processes at the different maturity and capability levels of Green IT, defined in the ISO/IEC 33000-based maturity model developed for the GMGIT

And, finally, we have defined each of the 38 processes as the ISO/IEC 33000 is established, determining its id, name, description, purpose, results, base practices, and work products. Table 10.14 includes the example of how this definition was performed in the "DSS01. Manage operations" process.

Table 10.14 Description of the "DSS01. Manage operations" process in the ISO/IEC 33000-based maturity model developed for the GMGIT

DSS01: process attributes		
Attribute	Description	
ID	DSS01	
Name	Manage operations	
Description	Coordinate and execute the activities and operational procedures needed for the delivery of IT services, both internal and outsourced, including the execution of predefined standard operating procedures and the required monitoring activities	
Purpose	Deliver the results of the operational IT service as planned	
Results	As a result of the successful implementation of "Manage operations": The operations of Green IT are carried out following the policies, principles, strategy, and goals of Green IT. The standards, regulations, and best practices of Green IT have been identified and implemented and are being complied with	
Base practices	DSS01.BP1: Perform operational procedures. Maintain and execute operational procedures and tasks of Green IT reliably and consistently [Result: 1] DSS01.BP2: Manage outsourced services. Manage the operation of outsourced services so as to maintain their reliability and their consistency with Green IT [Result: 1] DSS01.BP3: Monitor IT infrastructure. Monitor the IT infrastructure and events related to it, in an effort to ensure the alignment of all of them with Green IT. Store enough chronological information in the operations logs of the organization to allow the reconstruction, review, and examination of the time sequences of the operations, as well as of the activities associated with the support to those operations [Result: 2] DSS01.BP4: Manage the environment. Maintain measures for protection against environmental factors. Install specialized equipment and devices to monitor and control the environment from a Green IT perspective [Result: 2] DSS01.BP5: Manage facilities. Manage the facilities according to the laws, regulations, guidelines, and other requirements related to Green IT [Result: 2]	
Work products	Inputs	Outputs
	Services of the architecture of Green IT [Result: 1]	Operational procedures of Green IT [Result: 1]
	Policies of Green IT [Result: 1]	Reports on the compliance of Green IT by third parties [Result: 1]
	Policies of the management of the environment [Result: 2]	Reports on the performance of the infrastructure of the IT, from the point of view of Green IT [Result: 2]
	Policies of the management of the facilities [Result: 2]	Alignment of Green IT with the management of the environment [Result: 2]
		Alignment of Green IT with the management of the facilities [Result: 2]

10.3.3 *Audits Performed During the Development of the GMGIT*

The development of a framework such as the GMGIT must be accompanied by an empirical validation to corroborate its coherence, adequacy, and applicability in the real context. For this reason, we have followed the case study methodology [21, 23], through which we have performed different validations based on audits in six different organizations.

The audits performed have been conducted through the different versions of the GMGIT, in order to gradually refine, expand, and validate the characteristics of the framework. Thus, the first version of the GMGIT was validated auditing two organizations in Spain [13, 24], while the second version of the GMGIT was validated auditing four organizations at the international level (Spain, Italy, Mexico, and Colombia) [14, 25]. Likewise, we are currently validating the third version of the GMGIT through improvement plans in the organizations previously audited. Among these organizations, it is worth mentioning the Colombian organization, through which we have already conducted the first phases of the plan in which we have obtained promising results [26].

Regarding the results and main findings we have obtained through all these validations, three points of view are worth highlighting: the GMGIT, organizations, and Green IT.

First, regarding the GMGIT, we have succeeded in strengthening its validity and applicability by confirming that the developed versions have evolved satisfactorily, maintaining the consistency and coherence, and expanding and improving the characteristics of the framework, such as the scope, components, processes, practices, etc. All the validations and results obtained demonstrate that the GMGIT is an applicable framework for organizations, becoming the first guide to define, implement, assess/audit, and improve the governance and management of Green IT in organizations.

On the other hand, from the point of view of the organizations, we realized that they are disoriented in Green IT. This is mainly due to the fact that they do not have guidelines to help them implement sustainable practices in this area, and therefore, they demand standards or frameworks such as the GMGIT to conduct such implementations.

And, finally, regarding sustainability and Green IT, we continue to corroborate and highlight the growing importance they have in our society and in organizations, and their indispensability. It is not something optional, it is a duty we have for our planet and ourselves.

10.4 Using the GMGIT for Green IT Improvement

When applying the components of a framework such as the GMGIT, it is very important to follow a methodology that allows for a progressive and systematic implementation, based on continuous improvement. For this, the ISO/IEC TR 33014 standard [27] provides the characteristics and phases to perform the improvement of the processes following a continuous improvement plan.

All these characteristics defined by the ISO/IEC TR 33014 standard are fully applicable to the GMGIT. It should not be forgotten that the GMGIT is a framework that not only establishes the necessary components to properly govern and manage the Green IT in organizations, but also includes an audit framework, as well as a maturity model that organizes the Green IT processes at different levels depending on their need and complexity of implementation. That is why applying the ISO/IEC TR 33014 standard to the GMGIT is very simple following, mainly, the maturity model established for the progressive assessment and implementation of the Green IT processes.

Thus, taking the GMGIT and the ISO/IEC TR 33014 standard, we have conducted different improvement plans in the organizations that we have audited during the development of the GMGIT. Of these organizations it is important to highlight the Colombian organization in which we have already applied the first phases of the improvement plan, and we obtained very good results in the progress of the implementation of the different Green IT processes [26].

To establish and conduct the improvement plan in the Colombian organization, the first thing we did was to analyze the results obtained from the *Green in IT* audit that we performed [28] (covering in this way the strategic and tactical levels of the ISO/IEC TR 33014 standard [27]). Based on these results, we identified that the said organization was partially at maturity level 1 of *Green in IT*, since, although it fully complied with the practices of the "BAI09. Manage assets" process, it partially complied with the practices of the other process at maturity level 1, the "DSS01. Manage operations" process (cf. Sect. 10.3.2). During the audit we also performed the assessment of the processes of maturity level 2, and we verified that the organization did not comply with any of the established practices in all of these processes.

Once these results have been obtained and analyzed, we continue with the operational level of the ISO/IEC TR 33014 standard [27], establishing with the organization the objective of conducting a first improvement plan to comply with the defined practices in the processes of the first two maturity levels of *Green in IT*.

Going into detail in the improvement plan, we decided to divide it into two cycles, in order to perform the improvement in a more staggered and affordable way. In each of these two cycles, the processes of maturity levels 1 and 2 of *Green in IT* were addressed, as shown below:

Table 10.15 Improvement actions to be performed in the "DSS01. Manage operations" process for the Colombian organization

DSS01. Manage operations—improvement actions		
Problem	Solution	Work products
It is not monitored whether the IT infrastructure and all those sustainable elements/aspects of it are properly adapted to Green IT	• Ensure that the sustainable aspects of the IT infrastructure, such as the operations and the use of sustainable solutions in IT, the optimal use of IT resources, etc., are monitored, making sure that they are properly suited to Green IT.	Outputs: • Reports on the performance of the infrastructure of IT, from the point of view of Green IT.
It does not consider meeting the requirements of Green in IT in the management of the environment and in the management of the facilities	• Ensure that the management of the environment considers the requirements of Green in IT and that it meets those requirements. • Ensure that the management of the facilities considers the requirements of Green in IT and that it meets those requirements.	Inputs: • Policies of the management of the environment. • Policies of the management of the facilities. • Outputs: • Alignment of Green IT with the management of the environment. • Alignment of Green IT with the management of the facilities.

- **First cycle:**
  - DSS01. Manage operations.
  - APO01. Manage IT management framework.
  - APO02. Manage strategy.
  - APO06. Manage budget and costs.
- **Second cycle:**
  - APO08. Manage relationships.
  - APO10. Manage vendors.
  - BAI01. Manage programs.
  - BAI02. Manage requirements definition.
  - BAI03. Manage solutions identification and build. Likewise, as a calendar to conduct the implementation of the improvements of each cycle, a flexible implementation period of several months was agreed upon, after which we performed an assessment to verify the implementation of the practices at the target processes.

Finally, within this improvement plan, we included the improvement actions to be conducted by the organization in question, based on the problems identified in the audit performed. An example of these improvement actions that were identified to comply with the practices in which there was a problem in the process "DSS01. Manage operations" is shown in Table 10.15.

10.5 Conclusions

We are in a period full of vertiginous changes—necessary and natural changes in search of an increasingly evolved and advanced civilization. But changes, in most cases, have involved an involution regarding life and the preservation of our environment, without which absolutely nothing would be possible.

This has led society, organizations, and governments around the world to wake up and come together to defend a sustainable development [29, 30]. However, these efforts are insufficient and are not giving the expected results. We continue to see how countries and organizations keep on polluting the environment above the agreed limits, how the global temperature continues to rise, and how the number of ecosystems in which life is no longer compatible because of the damage caused by humans increases. The initiatives and agreements reached by the governments will remain inefficient, while organizations, the main sources of these problems, do not know how to conduct the practices and actions in this regard.

Thus, we must contribute from all areas of knowledge to guide organizations to achieve the agreed sustainable goals and, if possible, go further. It is for this reason that we have developed the "Governance and Management Framework for Green IT" (GMGIT) [13, 14].

The GMGIT advocates to help organizations to establish the governance and management bases that will support the implementation, assessment, and improvement of sustainable practices in and by IT (well-known as Green IT). To this end, the GMGIT establishes and provides a set of components and best practices that organizations should consider in this regard, as well as the necessary mechanisms to assess and improve in a progressive and systematic manner the implementation of these components and best practices of Green IT.

Through the application of the GMGIT to validate and refine it in real environments, we have been able to verify that organizations are disoriented in relation to Green IT. They are starting to work in this area, but, like all things at the beginning, they are sometimes confused and do not know precisely what to do. They begin by implementing isolated best practices of Green IT, but without a control and management that allows them to advance properly. Now, thanks to the GMGIT, the organizations in which it is being applied have a much clearer and more organized vision, properly governing and managing the Green IT practices implemented and obtaining satisfactory results in this regard.

However, we are at the beginning of a long road. The GMGIT is just a small grain of sand within Green IT and sustainability, and it must continue to evolve and contribute to the advancement of these areas. And, like the GMGIT, other frameworks that guide organizations to establish, develop, and improve the bases of different business areas must emerge.

We must all stay together doing the right thing for our planet, for the environment, and for the future of these and of the humankind.

Acknowledgments This work is the result of a PhD cotutelle agreement between the University of Castilla-La Mancha and the University of Bari "Aldo Moro." It is also part of the Industrial PhD DI-17-09612, funded by the Spanish Ministry of Science, Innovation and Universities; of the ECD project (PTQ-16-08504), funded by the "Torres Quevedo" Program of the Spanish Ministry of Economy, Industry and Competitiveness; of the SOS project (SBPLY/17/180501/000364), funded by the Ministry of Education, Culture and Sports of the JCCM (Regional Government of Castilla-La Mancha) and the ERDF (European Regional Development Fund); of the "Digital Service Ecosystem" project (PON03PE_00136_1), funded by the Italian Ministry of University and Research; and of the "Auriga2020" project (T5LXK18), funded by the Apulia Region.

References

1. Brundtland G, Khalid M, Agnelli S, Al-Athel S, Chidzero B, Fadika L et al (1987) Our common future ("Brundtland report"). Oxford University Press, Oxford
2. Schwab K (2017) The fourth industrial revolution. The Crown Publishing Group, Danvers
3. Calero C, Piattini M (2017) Puzzling out software sustainability. Sustain Comput Inform Syst 16:117–124. https://doi.org/10.1016/j.suscom.2017.10.011
4. Deng Q, Ji S (2015) Organizational green IT adoption: concept and evidence. Sustainability 7 (12):16737–16755. https://doi.org/10.3390/su71215843
5. Calero C, Piattini M (eds) (2015) Green in software engineering. Springer International Publishing AG, Cham
6. Erdélyi K (2013) Special factors of development of green software supporting eco sustainability. In: Proceeding of the IEEE 11th international symposium on intelligent systems and informatics (SISY 2013), pp 337–340
7. Unhelkar B (2011) Green IT strategies and applications: using environmental intelligence. CRC Press, Boca Raton, FL
8. Brodkin J (2008) Economy driving green IT initiatives. Network World 25(49):16
9. Epstein MJ, Buhovac AR (2014) Making sustainability work: best practices in managing and measuring corporate social, environmental, and economic impacts, 2nd edn. Berrett-Koehler Publishers, San Francisco
10. Hertel M, Wiesent J (2013) Investments in information systems: a contribution towards sustainability. Inform Syst Front 15(5):815–829. https://doi.org/10.1007/s10796-013-9417-x
11. Wimmer W, Lee KM, Quella F, Polak J (2010) ECODESIGN. The competitive advantage. Springer, Dordrecht. https://doi.org/10.1007/978-90-481-9127-7
12. Simmonds DM, Bhattacherjee A (2014) Green IT adoption and sustainable value creation. In: Proceeding of the 20th Americas conference on information systems (AMCIS 2014), pp 2550–2565
13. Patón-Romero JD, Baldassarre MT, Piattini M, García Rodríguez de Guzmán I (2017) A governance and management framework for green IT. Sustainability 9(10):1761. https://doi.org/10.3390/su9101761
14. Patón-Romero JD, Baldassarre MT, Rodríguez M, Piattini M (2019) A revised framework for the governance and management of green IT. J Univ Comp Sci 25(13):1736–1760. https://doi.org/10.3217/jucs-025-13-1736
15. ISACA (2018) COBIT 2019 framework: introduction and methodology. ISACA, Rolling Meadows
16. ISACA (2018) COBIT 2019 framework: governance and management objectives. ISACA, Rolling Meadows
17. ISO (2015) ISO 14000 (Environmental management systems). International Organization for Standardization, Geneva

18. ISO (2015) ISO/IEC 33000 (Information technology — process assessment). International Organization for Standardization, Geneva
19. Patón-Romero JD, Baldassarre MT, Rodríguez M, Piattini M (2019) Application of ISO 14000 to information technology governance and management. Comp Stand Inter 65:180–202. https://doi.org/10.1016/j.csi.2019.03.007
20. ISO (2003) ISO/IEC 15504 (Information technology — process assessment). International Organization for Standardization, Geneva
21. Yin RK (2017) Case study research and applications: design and methods. Sage, Los Angeles
22. ISACA (2013) COBIT 5 for assurance. ISACA, Rolling Meadows
23. Runeson P, Höst M, Rainer A, Regnell B (2012) Case study research in software engineering: guidelines and examples. Wiley, Hoboken
24. Patón-Romero JD, Baldassarre MT, Rodríguez M, Piattini M (2018) Green IT governance and management based on ISO/IEC 15504. Comput Stand Interfaces 60:26–36. https://doi.org/10.1016/j.csi.2018.04.005
25. Patón-Romero JD, Baldassarre MT, Rodríguez M, Runeson P, Höst M, Piattini M (2020) Governance and management of green IT: a multi-case study. Inform Softw Technol. https://doi.org/10.1016/j.infsof.2020.106414
26. Patón-Romero JD, Baldassarre MT, Rodríguez M, Pérez-Canencio JG, Ojeda-Solarte ML, Rey-Piedrahita A, Piattini M (2020) Application of ISO-IEC TR 33014 to the improvement of green IT processes. Comp Stand Inter (Under Peer Review)
27. ISO (2013) ISO/IEC TR 33014 (Information technology — process assessment — guide for process improvement). International Organization for Standardization, Geneva
28. Patón-Romero JD, Baldassarre MT, Rodríguez M, Pérez-Canencio JG, Ojeda-Solarte ML, Rey-Piedrahita A, Piattini M (2019) Application of ISO/IEC 33000 to green IT: A case study. IEEE Access 7:116380–116389. https://doi.org/10.1109/access.2019.2936451
29. European Commission (2017) Report from the commission to the European parliament, the council, the European economic and social committee and the committee of the regions on the implementation of the circular economy action plan. European Commission, Brussels
30. United Nations (2015) Transforming our world: the 2030 agenda for sustainable development. In: Seventieth session of the United Nations general assembly, resolution A/RES/70/1

Chapter 11
Sustainable Software Engineering: Curriculum Development Based on ACM/IEEE Guidelines

Alok Mishra and Deepti Mishra

Abstract Climate change risk and environmental degradation are the most critical issues of our society. Our technology-influenced daily lifestyle involves many types of software and apps which are used by society at large, and their use is increasing more than ever before. Sustainability is a significant topic for future professionals and more so for software engineers due to its impact on society. It is crucial to motivate and raise concern among students and faculty members regarding sustainability by including it in the Software Engineering (SE) curriculum. This chapter discusses how sustainability can be included in various courses of the SE curriculum by considering ACM/IEEE curriculum guidelines for the SE curriculum, literature review, and various viewpoints so that SE students can attain knowledge on sustainable software engineering. It also includes an assessment of key competences in sustainability for proposed units in the SE curriculum.

11.1 Introduction

Software has become an integral part of our everyday life, gradually impacting human beings and society. The current industrial growth and the increasing adoption of ICT threaten the future of sustainability and cause environmental issues [1, 2]. Sustainability is becoming a crucial concern in information technology and software for our future. Sustainability management is one of the upcoming movements of the twenty-first century, but, until now, it is not getting as much attention from software engineering as it should. Furthermore, ICT has a major role in sustainable

A. Mishra
Molde University College, Molde, Norway

Department of Software Engineering, Atilim University, Ankara, Turkey
e-mail: alok.mishra@himolde.no; alok.mishra@atilim.edu.tr

D. Mishra (✉)
Department of Computer Science, Norwegian University of Science and Technology, Gjøvik, Norway
e-mail: deepti.mishra@ntnu.no

C. Calero et al. (eds.), *Software Sustainability*,
https://doi.org/10.1007/978-3-030-69970-3_11

development, specifically in software and green computing [3]. It is important that environmental concerns are addressed in the development, implementation, and operation of software. In this respect, the contribution of ICTs for energy and environmental sustainability has attracted attention of both researchers and professionals [1] as software contributes significantly to every aspect of our lives.

Dick et al. [4] proposed the first definitions for sustainable software and sustainable software engineering (SSE) in 2010, which have become the foundation for later explanations. According to Ray et al. [5], the term "sustainable" applies to both the longer life and greener aspects of software. Brooks et al. [6] suggest three dimensions to sustainability: environmental, economic, and social. These are interrelated and these should be selected in a way to attain optimum arrangement and alliance. Sustainable software is defined as the software whose direct and indirect negative influence on economy, society, human beings, and environment that result from development, implementation, and usage of the software are minimum [7]. Green software generates the minimum amount of e-waste during its operation and development [8].

Sustainable software engineering is developing software through a sustainable software engineering process which satisfies the purpose of sustainability by reducing the environmental impact of software implementation and operations to human beings and society at large. Sustainable software engineering (SSE) is based on the foundation of designing and developing software by taking into consideration various dimensions of sustainability which are economic, environmental, individual, social, and technical [9, 10]. A number of recent studies were performed to find out how sustainability is identified and included in software engineering process towards sustainable software development [11], which can reduce its environmental impact on society.

In many research studies, sustainability and energy efficiency are observed as crucial expertise for future software engineers [12–15]. However, a recent survey by Manotas et al. [16] of 3860 software professionals from Google, ABB, IBM, and Microsoft revealed that present higher educational programs do not prepare professionals to undertake sustainability, although they are inclined to learn about it [16]. Further, they noticed extensive significance of greenability and sustainability. Another study on teaching sustainability in software engineering also supports that sustainability is not included in the software engineering (SE) courses and that the present focal point is on energy efficiency issues [12, 14, 15]. Scientists have recently recognized that issues related to sustainable software engineering should be part of the discipline that has a significant role in the future of human beings.

Software engineers presently perform many tasks that may ensure sustainability, for instance, Agarwal et al. [3] consider the capabilities and benefits of green software and suggest more efficient algorithms will take less time to execute and that this will lead to sustainability. However, sustainability is considered as an additional feature in many software projects as software engineers are tied by time-to-market pressure and are often less inclined to administer sustainable methods and techniques [17]. For now, apart from cost, factors such as environment, social, and human sustainability are required to be considered in any planning, implementation, and running initiative related to software systems [18]. Organizations are now beginning to understand that not only cost efficiency, but also long-term and continued prosperity can be gained from sustainability. Therefore, apart from factors like cost, time, and quality, sustainability has become one of the significant objectives in developing, configuring, operating, and working software systems. Therefore, there is a need to support the transition to sustainability and incorporate it into software systems and other underlying business processes [19].

Green and Sustainable Software Engineering (GSSE) is the art of developing green and sustainable processes [20]. The objective of sustainable software process is to reduce the environmental impact of software solutions and their deployment on human beings, society, economy, and environment [21]. Presently, the effect of IT on sustainable advancement—in particular, of software—is an emerging issue due to the global concern on climate change. The education sector has to play a significant role in ensuring future software engineers understand sustainability dimensions and integrate them into the SE curriculum. Gibson et al. [22] supported that the educational sector has an important part to play in ensuring software professionals understand sustainability issues in software development. Therefore, there is a need to integrate sustainability in the software engineering discipline curricula. Gibson et al. [22] further observed that it is mentioned just once even in ACM/IEEE guidelines and twice in SWEBOK with respect to the software economics area. They argued in the current scenario there is need for sustainable software engineering education guidelines and components in such curricula for future software engineers. Therefore, this chapter advances effort in this direction. It extends our previous work [23] by following ACM/IEEE guidelines for SE curriculum along with authors' long academic experience to first list and categorize relevant SE courses offered in SE programs. Later sustainability-related units are introduced in existing SE courses followed by an assessment of these units with respect to key competences in sustainability.

The rest of the chapter is organized as follows: In Sect. 11.2 related work to describe the relationship between software quality and sustainability along with initiatives to include sustainability in SE higher education programs are presented. Section 11.3 introduces curricula development on sustainable software engineering. Section 11.4 includes points of discussion, concluding with a brief viewpoint for future direction.

11.2 Related Work

11.2.1 Software Quality and Sustainability

Sustainability is usually referred to as a nonfunctional requirement in software systems [19]. Nonfunctional requirements are also known as quality requirements. Although organizations have recognized that sustainability can be incorporated in quality issues, for instance, maintainability, usability, and agility, they could not do so due to time and budget constraints in software management [24].

Amri and Bellamine Ben Saoud [25] proposed Generic Sustainable Software Star Model (GS3M) to examine sustainable software and noticed some studies consider sustainability as a part of quality, while others observe quality and sustainability as different concepts and use quality attributes to support sustainability. Calero et al. [26] and Calero [27] applied the hypothesis that sustainability is a factor of the software quality, thus, unified it as a quality characteristic with three other subcharacteristics: energy consumption, resource optimization, and perdurability. Calero [26] also noticed that operationalization in this way includes introducing some modifications in the ISO quality standard ISO/IEC 25010 to support sustainability as a quality component. Albertao et al. [28] and Kern et al. [29] also identified quality attributes to define sustainability. Interestingly, Albertao et al. [28] formulated software project sustainability characteristics into development-related features (modifiability, reusability, portability, and supportability), usage-related attributes (performance, dependability, usability, and accessibility), and process-related attributes (predictability, efficiency, and project's footprint). Kern et al. [29] endorsed a quality model for sustainable software which constructs sustainability criteria into three categories: common quality criteria which are well-known and standardized issues (such as efficiency, reusability, modifiability, and usability); directly related benchmark (such as energy efficiency, framework entropy, functional types, hardware obsolescence, adaptability, feasibility, accessibility, usability, and organization's sustainability); and indirectly related yardstick that demonstrate the effects of software on other products and services and cover the effects of use as well as systemic effects, such as the fit for aim, elegance, and reflectivity.

11.2.2 Sustainability in SE Curricula

Sustainable software engineering is getting limelight among professionals and researchers [13, 30]. However, researchers have noted that sustainability is underrepresented in the curricula [11], hence the need to include the concept of sustainability in the university curriculum of computer science, software engineering, and information systems. Mann et al. [31] presented a framework for educators to design sustainability-centered education while Sammalisto and Lindhqvist [32] observed on the integration of sustainability in higher education based on different sustainability dimensions like environmental, economic, social, and technical. Gibson et al. [22] studied the significance of requirements engineering in ensuring sustainability in software development in the UK. Groher and Weinreich [33] studied how sustainability is perceived by software professionals in projects and found that professionals mainly linked it to maintainability and extensibility of software. Renzel et al. [34] contributed a detailed strategy for projects in sustainable software engineering.

Chitchyan et al. [35] reviewed sustainability related with Software Product Line Engineering (SPLE) and suggested the focus be on technical and social sustainability issues along with social sustainability related to organizations. Lutz et al. [36] also specified characteristics of sustainability in SPLE. Mohankumar and Anand Kumar [37] proposed a green-based model for sustainable software engineering. Recently Penzenstadler et al. [38] proposed a blueprint for a course on software engineering for sustainability.

11.2.3 Key Competencies in Sustainability

The major challenges in the incorporation of sustainability in the university education are in the field of teaching [39–42]. Therefore identifying key competences in sustainability may be the first step towards sustainability inclusion in higher education [43]. Wiek et al. [44] defines competence as a functionally linked complex of knowledge, skill, and attitude that enables successful task performance and problem solving. Competence is a quality developed through practice and not an end state [45].

In 2002, the Organisation for Economic Co-operation and Development (OECD) identified key competencies needed for an individual to lead an overall successful and responsible life and for contemporary society to face present and future challenges [46]. The OECD key competencies are divided into three categories: subject and methodological, social, and personal. The OECD study on key competencies and comprehensive educational objectives reveals sustainability's significance for the future [47]. Subsequently, multiple studies have introduced key competencies for education for sustainable development in the formal education sector to help assess the learning outcomes of pupils, and an overview is provided in the Table 11.1.

This study will assess proposed units integrated in SE courses based on the approach by Giangrande et al. [52] and Wiek et al. [44]. Wiek et al. [44] proposed a framework of key competencies in sustainability by categorizing competencies into clusters, which was found to be useful by Giangrande et al. [52] who further extended the framework.

11.3 Sustainable Software Engineering Curricula Outline

Sustainability knowledge should be integrated in a curriculum by linking the concept of sustainability to a particular field of study [32] rather than offering separate courses on sustainability. Considering the suggestion, Fig. 11.1 presents an approach to integrate sustainability education in SE curriculum. First, the ACM/IEEE guidelines for SE curriculum development has been followed to include an initial set of courses SE students should take in order to later practice their profession successfully. ACM/IEEE guidelines 2014 for SE curriculum development consists of a set of SE competencies that every SE graduate must possess and provides guidance to academic institutions and accreditation agencies about the knowledge and skills

Table 11.1 Key competencies for education for sustainable development

	Key competencies
De Haan [48]	Foresighted thinking Interdisciplinary work Cosmopolitan perception, transcultural understanding and cooperation Participatory skills Planning and implementation skills Empathy, compassion and solidarity Self-motivation and motivating others Distanced reflection on individual and cultural models
Barth [49]	Self-motivation Capacity for empathy, compassion and solidarity Reflection on individual Motivating others Participatory skills Foresighted thinking Interdisciplinary work Cosmopolitan perception, transcultural understanding and cooperation Reflection on cultural models Planning and implementation
Sleurs [50]	Values and ethics Emotions Systems thinking Knowledge Action
Roorda [51]	Responsibility Emotional intelligence Systems orientation Future orientation Personal involvement Action skills
Wiek [44]	Interpersonal Anticipatory Systemic working Normative Strategic
Giangrande [52]	Intrapersonal Interpersonal Future thinking Systems thinking Disciplinary and interdisciplinary Normative and cultural Strategic

fundamental to software engineering education [53]. Subsequently, additional courses have been included to reflect current advancements in SE education. Furthermore, the final set of courses are organized in different categories based on the structure of academic programs in major universities: fundamental courses of sustainability, core SE courses, technical electives, nontechnical electives, project-based courses, and industrial practice. Finally, the information gained from literature

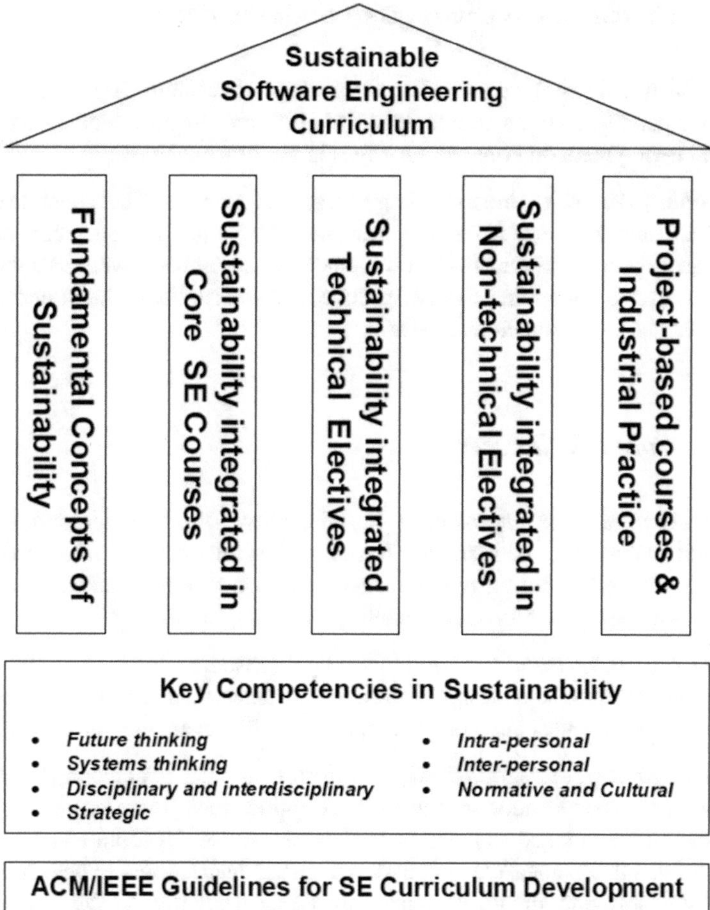

Fig. 11.1 An approach to sustainability inclusion in the Software Engineering Curricula

review along with the authors' long academic experience in SE discipline facilitated the inclusion of sustainability competence in the form of flexible units within existing courses in the SE curricula.

The program should include the following units in the existing courses of the SE curricula so that students can get sufficient exposure to different components of sustainability issues in the software development life cycle.

11.3.1 Fundamental Concepts of Sustainability

Sustainability Theory Understanding the concept of sustainability in software and its various parts so as to be able to apply it in different stages of software development and deployment and operations stages in the organization.

Sustainability Analysis Understanding of rigorous analyses of sustainability issues in software development, from cost estimation to project management, software maintenance, and evolution. It should include, in general, software systems to be developed from a comprehensive perspective to sustainability and long-term consequences on the environment and society.

11.3.2 Core SE Courses

Software Requirements Engineering Sustainability inclusion in requirements elicitation and analysis process is crucial. Therefore, it is important to understand how to include sustainability during requirements elicitation process. Stakeholder modelling, goal modelling, and system modelling can assist in this part.

Software Architecture and Design How to apply sustainability in different kinds of software architecture and design issues, for instance database, human computer interaction, and modules interconnection, and in software architecture development.

Human-Computer Interaction Design Human-Computer Interaction (HCI) is part of many information technology and software applications. Therefore, sustainability issues should be included as a component in this course. Nyström and Mustaquim [54] suggested that persuasive system design can influence users to behave and live more sustainably and should be related to the sustainability of the environment. Sustainable HCI should address WCED's (World Commission on Environment and Development) sustainability view "... that it meets the needs of the present without compromising the ability of future generations to meet their own needs" [55]. Sustainable system design principles can be included in HCI, software system design, and industrial software development project curriculum.

Software Modelling and Analysis It is important that system modelling for complex software systems should be done from a sustainability perspective by using available tools. Software modelling and analysis (using UML diagrams) can assist in understanding how to incorporate sustainability into stakeholders' requirement scenarios. It should also include trade-offs and conflict resolution in the requirements of different stakeholders from a sustainability view.

Software Process Software process improvement should include sustainable software engineering processes along with Agile and DevOps approaches. It should also, knowledge of applicable tools, methods, and technologies to facilitate the sustainable software engineering processes. Energy and resource utilization are the main components that impact sustainability. Therefore, these should be determined from the initiation of the process. Eco-design of digital services to ensure reducing environmental impacts to develop digital services that are more sustainable, consume less resources and energy, and produce less waste. Further knowledge of relevant tools, methods, and technologies should be introduced to facilitate the sustainable software development process.

Software Verification and Validation An optimized approach in ensuring sustainability in software engineering is the software verification and validation process, including different types of testing and operation of the software product. Specification systems and automated verification tools can be helpful in this regard.

Software Quality Assurance Sustainability issue should be part of software process improvement and quality assurance process. Configuration management tools and software inspection tools can be complementary in this regard. Knowledge of standards of eco-design (ISO 14006, ISO 14062) should be imparted.

Software Project Management It includes the planning and controlling phases of sustainability activities along with sustainability policies to ensure an efficient process. Appropriate project management tools and agile methods management tools can facilitate in ensuring sustainability practices in software project management. Eco-design of digital services towards ensuring reduced environmental impacts to develop digital services that are more sustainable, consume less resources and energy, and produce less waste.

Software Construction and Evolution Software evolution is a continuous process. Refactoring tools, automated testing tools, and configuration management tools along with project management tools aid in ensuring sustainability in software construction and evolution.

Software Security Security and safety during the development of complex software systems is crucial. So security and safety are now an integral part of the set of nonfunctional requirements which lead to software quality. ISO/IEC 25010:2011 included safety as an explicit characteristic in software while ISO/IEC 9126-1:2001 ensures security in software. Safety and security are called out and treated specifically because they are significant characteristics. Penzenstadler et al. [56] supported that the same is true for sustainability, specifically the dimension of environmental sustainability, and there is a need to find suitable means to analyze, support, verify, and validate sustainability requirements in software engineering.

11.3.3 Technical Elective Courses

Internet of Things (IoT) IoT has the ability to combat climate change towards green environment. It could impact sustainability in different areas, such as water use and energy efficiency. According to the World Economic Forum, IoT could be a game changer for sustainability [57]. IoT helps in applying waste management strategies and in circular economy. IoT deployments can help in addressing many of the Sustainable Development Goals (SDGs) of the UN. IoT technology can provide tangible benefits to sustainability [57]. Many IoT initiatives may help accomplish sustainability in the future [58]. Therefore students should be made aware of such IoT applications that can be applied to achieve sustainability by including relevant case studies, white paper, discussion, seminar, etc. in the curriculum.

Cloud Computing Cloud computing provides more efficient use of computing power and is advantageous for environmental sustainability. Application of cloud computing ensures social, business, and environmental sustainability. It can include discussions, case studies, seminars, projects, company visits, etc.

Web and Mobile Systems Sustainability and page speed are correlated. When your website runs more efficiently it consumes less processing power thus less energy and leaves a lower carbon footprint [59]. Also, a sustainable design is more efficient and accessible. Sustainable mobile apps and their users may contribute to achieving environmental goals, and mobile devices are enablers of sustainable actions due to their huge potential for scalability [60]. Mobile applications that have even a little effect on resource efficiency or the reduction of greenhouse gas emissions could result in a greater impact as these are used every day [60]. This can include concepts on social software interface with sustainability issues, green software development and usage practices, and promotion of technologies, development frameworks, and tools which facilitate sustainability in web and mobile systems development. These could be included as real-life projects, cases studies, and seminars and lectures from industry practitioners.

Sustainable Data Center Green data centers or sustainable data centers help in reducing carbon footprint, design and deployment of data store, and applications to operate in energy-efficient ways. Therefore, the course should include real-life case studies, seminars, and discussions on how sustainability can be incorporated in this regard.

Tools for Software Sustainability Tools must be introduced to assist different stages of software development (requirements, design, testing, configuration management, etc.) towards ensuring sustainability. This can be a part of a sustainable or green software engineering laboratory program.

11.3.4 Nontechnical Elective Courses

Global Professional Practice/Social Responsibility Students should be aware how carbon footprint, CO_2 emissions, global warming is a matter of concern. Therefore global professional practice should include environmental issues arising from software engineering products and their use. These can be included as case studies, seminars, and group discussions to analyze environmental degradation cases and to explore mitigation plans, global environmental challenges, sustainable software, energy management, and green computing standards in the context of software applications.

11.3.5 Project-Based Courses and Industrial Practice/ Internships

Project-based courses Most universities have final-year projects or thesis for students to explore real-world challenges. Universities sometimes also require their students to do industrial internships of 1–2 months so that students can gain experience in professional projects in real-world settings. Sustainability can be included as a learning outcome for such courses. Projects involving sustainability in software engineering during summer internships or such mini projects should be part of the course.

Table 11.2 presents how the integration of these units into current curriculum will help SE professionals to acquire key competencies in sustainability.

11.4 Discussion

Sustainable software engineering is an emerging paradigm and significant for society in terms of the environment. Sammalisto and Lindhqvist [32] argued that a proper feedback system is required between educators and university administrators to show the value and significance of the integration of sustainability. Torre et al. [11] observed a top 10 universities curriculum analysis that none of the engineering courses explicitly addresses sustainable software engineering or the status of green sustainable software engineering. The vast majority of the survey respondents (97%) expressed there is need for more courses related to sustainability.

The present industrial production and increasing use of ICT may endanger prospective sustainability and lead to environmental problems [1, 2]. In a recent study, Salam and Khan [61] classified 20 success elements towards the evolution of green and sustainable software. Out of these, green software design and efficient coding was found to be the most significant factor (71%) followed by power-saving software methods (70%). Mahmoud and Ahmad [20] proposed green model for sustainable software engineering. Naumann et al. [7] proposed sustainable software engineering process and quality models and suggested nine successive stages: Requirements, Design, Unit Testing, Implementation, System Testing, Green Analysis, Usage, Maintenance, and Disposal. Lami et al. [62] found that sustainability-related processes are missing in ISO/IEC 12207 and proposed three processes:

Table 11.2 Assessment of key competencies in sustainability with respect to proposed units in SE curriculum

Units	Key competencies in sustainability [44, 52]						
	Intrapersonal	Interpersonal	Future thinking	Systems thinking	Disciplinary and interdisciplinary	Normative and cultural	Strategic
Sustainability Theory	X		X	X	X		X
Sustainability Analysis	X	X	X	X	X		X
Software Requirements Engineering	X	X	X	X	X		X
Software Architecture and Design		X	X	X	X		X
Human-computer interaction design		X	X	X	X	X	X
Software Modelling and Analysis		X	X	X	X		X
Software Process		X	X	X	X		X
Software Verification and Validation		X	X	X			X
Software Quality Assurance		X	X	X			X
Software Project Management		X	X	X			X
Software Construction and Evolution		X	X	X	X		X
Software Security		X	X	X			X
Internet of Things (IoT)			X	X	X		X
Cloud Computing			X	X	X		X
Web and Mobile Systems		X	X	X	X		X
Sustainable Data Center			X	X			X
Tools for Software Sustainability		X	X	X	X	X	X
Global Professional Practice/Social Responsibility	X	X	X	X	X	X	X
Project-Based Courses		X	X	X	X	X	X

Sustainability Management Processes, Sustainability Engineering Process, and Sustainability Qualification Process.

The purpose of SSE is to curtail the energy footprint of computers as well as minimize other environmental impacts related to software systems. Software is now a pervasive part of the society as even mobile phone and social media users are in billions. It is the responsibility of software engineering educators to prepare SE professionals by equipping them with skills to meet the expectations of the software industry [63]. Therefore, it is significant to include sustainability in courses for future software engineers so that it can be achieved while developing, deploying, and maintaining all kinds of software in the future. Professional practices should be part of the SE curriculum [64], which can include a sustainability component. Moreover, sustainability has the potential to attract more students to the SE discipline due to its indispensable significance for the future [65]. Programs that address environmental sustainability in information technology are sometimes also referred to as green information technology. Green IT refers to information technology and system initiatives and programs that address environmental sustainability [66] and manage energy consumption as well as waste associated with the use of hardware and software, which tend to have a direct and positive impact on sustainability [67].

The proposed curriculum development can be easily customized and introduced as part of an undergraduate- or graduate-level software engineering curriculum. Since only a limited number of undergraduate and graduate programs on sustainability have been introduced in the last decade in certain institutions, the curricula proposed here can be a useful contribution to the body of knowledge for software engineering educators. As requirement specifications are the base input for software architecture and design, they have an impact on sustainability. With increasing global concern regarding climate change, the time has come to include "Sustainability" as a nonfunctional requirement towards quality software for future generations.

11.5 Conclusion and Outlook

Due to climate changes in the last decade and proliferation of information technology, software, and apps in daily life, there is a crucial need to develop and deploy green software. Therefore, there is a need to train future software engineers in such a manner that they will be able to include sustainability in each stage of the software development life cycle. Here, important units of sustainability inclusion in software engineering curricula have been described according to the recent ACM/IEEE curriculum guidelines for SE curriculum along with literature review on sustainable software engineering approaches, concepts, and tools. Software engineering undergraduate and graduate programs should include at least one foundation course on sustainability in their curriculum. This chapter also included appraisal of key competencies in sustainability for proposed units in SE curriculum.

This work can be extended by a survey and interviewing software engineering professionals to know in a more detailed manner how the SSE course can be developed and improved in the future into a more practice-oriented approach so that future software engineers will be able to produce eco-friendly and sustainable software.

References

1. Cai S, Chen X, Bose I (2013) Exploring the role of IT for environmental sustainability in China: an empirical analysis. Int J Prod Econ s146(2):491–500
2. Sissa G (2010) Green software. UPGRADE: Eur J Inf Prof 11:53–63
3. Agarwal S, Nath A, Chowdhury D (2012) Sustainable approaches and good practices in green software engineering. Int J Res Rev Comput Sci 3(1):1425
4. Dick M, Naumann S (2010) Enhancing software engineering processes towards sustainable software product design. In: EnviroInfo. pp 706–715
5. Ray S (2013) Green software engineering process: moving towards sustainable software product design. J Glob Res Comput Sci 4(1):25–29
6. Brooks S, Wang X, Sarker S (2012) Unpacking green IS: a review of the existing literature and directions for the future. In: Green business process management. Springer, pp 15–37
7. Naumann S, Dick M, Kern E, Johann T (2011) The greensoft model: a reference model for green and sustainable software and its engineering. Sustain Comput Inf Syst 1(4):294–304
8. Erdelyi K (2013) Special factors of development of green software supporting eco sustainability. In: 2013 IEEE 11th International Symposium on Intelligent Systems and Informatics (SISY). IEEE, pp 337–340
9. Becker C et al (2015) Requirements: The key to sustainability. IEEE Softw 33(1):56–65
10. Penzenstadler B (2013) Towards a definition of sustainability in and for software engineering. In: Proceedings of the 28th Annual ACM Symposium on Applied Computing. pp 1183–1185
11. Torre D, Procaccianti G, Fucci D, Lutovac S, Scanniello G (2017) On the presence of green and sustainable software engineering in higher education curricula. In: 2017 IEEE/ACM 1st International Workshop on Software Engineering Curricula for Millennials (SECM). IEEE, pp 54–60
12. Lago P, Damian D (2015) Software engineering in society at ICSE. STC Sustain Computing Newsl 4(1)
13. Lago P, Kazman R, Meyer N, Morisio M, Müller HA, Paulisch F (2013) Exploring initial challenges for green software engineering: summary of the first GREENS workshop, at ICSE 2012. ACM SIGSOFT Softw Eng Notes 38(1):31–33
14. Pang C, Hindle A, Adams B, Hassan AE (2015) What do programmers know about software energy consumption? IEEE Softw 33(3):83–89
15. Penzenstadler B, Fleischmann A (2011) Teach sustainability in software engineering? In: 2011 24th IEEE-CS Conference on Software Engineering Education and Training (CSEE&T). IEEE, pp 454–458
16. Manotas I et al (2016) An empirical study of practitioners' perspectives on green software engineering. In: 2016 IEEE/ACM 38th International Conference on Software Engineering (ICSE). IEEE, pp 237–248
17. Durdik Z, Klatt B, Koziolek H, Krogmann K, Stammel J, Weiss R (2012) Sustainability guidelines for long-living software systems. In: 2012 28th IEEE International Conference on Software Maintenance (ICSM). IEEE, pp 517–526
18. Raisian K, Yahaya J, Deraman A (2016) Current challenges and conceptual model of green and sustainable software engineering. J Theor Appl Inf Technol 94:428–443
19. Betz S, Caporale T (2014) Sustainable software system engineering. In: 2014 IEEE Fourth International Conference on Big Data and Cloud Computing. IEEE, pp 612–619

20. Mahmoud SS, Ahmad I (2013) A green model for sustainable software engineering. Int J Softw Eng Applic 7(4):55–74
21. Amsel N, Ibrahim Z, Malik A, Tomlinson B (2011) Toward sustainable software engineering: NIER track. In: 2011 33rd international conference on software engineering (ICSE). IEEE, pp 976–979
22. Gibson ML et al (2017) Mind the chasm: a UK fisheye lens view of sustainable software engineering
23. Mishra A, Mishra D (2020) Sustainable software engineering education curricula development. Int J Inf Technol Secur 12(2):47–56
24. Chitchyan R et al (2016) Sustainability design in requirements engineering: state of practice. In: Proceedings of the 38th International Conference on Software Engineering Companion. pp 533–542
25. Amri R, Saoud NBB (2014) Towards a generic sustainable software model. In: 2014 Fourth International Conference on Advances in Computing and Communications. IEEE, pp 231–234
26. Calero C, Bertoa MF, Moraga MÁ (2013) Sustainability and quality: icing on the cake. In: RE4SuSy@ RE. Citeseer
27. Calero C (2013) Sustainability as a software quality factor. In: Proceedings of the IBM Conference Day
28. Albertao F, Xiao J, Tian C, Lu Y, Zhang KQ, Liu C (2010) Measuring the sustainability performance of software projects. In: 2010 IEEE 7th International Conference on E-Business Engineering. IEEE, pp 369–373
29. Kern E, Dick M, Naumann S, Guldner A, Johann T (2013) Green software and green software engineering–definitions, measurements, and quality aspects. In: First International Conference on Information and Communication Technologies for Sustainability (ICT4S2013), 2013b ETH Zurich. pp 87–91
30. Naumann S, Kern E, Dick M, Johann T (2015) Sustainable software engineering: process and quality models, life cycle, and social aspects. In: ICT innovations for sustainability. Springer, pp 191–205
31. Mann S, Muller L, Davis J, Roda C, Young A (2010) Computing and sustainability: evaluating resources for educators. ACM SIGCSE Bull 41(4):144–155
32. Sammalisto K, Lindhqvist T (2008) Integration of sustainability in higher education: a study with international perspectives. Innov High Educ 32(4):221–233
33. Groher I, Weinreich R (2017) An interview study on sustainability concerns in software development projects. In: 2017 43rd Euromicro Conference on Software Engineering and Advanced Applications (SEAA). IEEE, pp 350–358
34. Renzel D, Koren I, Klamma R, Jarke M (2017) Preparing research projects for sustainable software engineering in society. In: 2017 IEEE/ACM 39th International Conference on Software Engineering: Software Engineering in Society Track (ICSE-SEIS). IEEE, pp 23–32
35. Chitchyan R, Groher I, Noppen J (2017) Uncovering sustainability concerns in software product lines. J Softw Evol Process 29(2):e1853
36. Lutz R, Weiss D, Krishnan S, Yang J (2010) Software product line engineering for long-lived, sustainable systems. In: International Conference on Software Product Lines. Springer, pp 430–434
37. Mohankumar M, Kumar MA (2016) Green based software development life cycle model for software engineering. Indian J Sci Technol 9(32):1–8
38. Penzenstadler B et al (2018) Blueprint and evaluation instruments for a course on software engineering for sustainability. arXiv preprint arXiv:1802.02517
39. Buckler C, Creech H (2014) Shaping the future we want: UN Decade of Education for Sustainable Development; final report. UNESCO
40. Lazzarini B, Perez-Foguet A, Boni A (2018) Key characteristics of academics promoting Sustainable Human Development within engineering studies. J Clean Prod 188:237–252
41. Mulder KF, Segalàs J, Ferrer-Balas D (2012) How to educate engineers for/in sustainable development. Int J Sustain Higher Educ

42. Wals AE (2014) Sustainability in higher education in the context of the UN DESD: a review of learning and institutionalization processes. J Clean Prod 62:8–15
43. Mishra D, Mishra A (2020) Sustainability inclusion in informatics curriculum development. Sustainability 12(4):5769
44. Wiek A, Withycombe L, Redman CL (2011) Key competencies in sustainability: a reference framework for academic program development. Sustain Sci 6(2):203–218
45. Vare P et al (2019) Devising a competence-based training program for educators of sustainable development: lessons learned. Sustainability 11(7):1890
46. Rychen DS, Salganik LH (2002) Definition and Selection of Competencies (DESECO): theoretical and conceptual foundations. Strategy paper. Swiss Federal Statistical Office, Neuchatel, Switzerland
47. De Haan G (2010) The development of ESD-related competencies in supportive institutional frameworks. Int Rev Educ 56(2–3):315–328
48. De Haan G (2006) The BLK '21'programme in Germany: a 'Gestaltungskompetenz'-based model for Education for Sustainable Development. Environ Educ Res 12(1):19–32
49. Barth M, Godemann J, Rieckmann M, Stoltenberg U (2007) Developing key competencies for sustainable development in higher education. Int J Sustain Higher Educ
50. Sleurs W (2008) Competencies for ESD (Education for Sustainable Development) teachers. A framework to integrate ESD in the curriculum of teacher training institutes. CSCT Project (Comenius 2.1 project 118277-CP-1-2004-BE-Comenius-C2.1), Brussels, Belgium
51. Roorda N (2010) Sailing on the winds of change: the Odyssey to sustainability of the universities of applied sciences in the Netherlands. Doctoral dissertation, Maastricht University
52. Giangrande N et al (2019) A competency framework to assess and activate education for sustainable development: addressing the UN sustainable development goals 4.7 challenge. Sustainability 11(10):2832
53. Ardis M, Budgen D, Hislop GW, Offutt J, Sebern M, Visser W (2015) SE 2014: Curriculum guidelines for undergraduate degree programs in software engineering. Computer 11:106–109
54. Nyström T, Mustaquim MM (2014) Sustainable information system design and the role of sustainable HCI. In: Proceedings of the 18th International Academic MindTrek Conference: Media Business, Management, Content & Services. pp 66–73
55. B Commission (1987) Report of the World Commission on Environment and Development: our common future, vol 10. [Online]. https://sustainabledevelopment.un.org/content/documents/5987our-common-future.pdf
56. Penzenstadler B, Raturi A, Richardson D, Tomlinson B (2014) Safety, security, now sustainability: the nonfunctional requirement for the 21st century. IEEE Softw 31(3):40–47
57. Arias R, Lueth K, Rastogi A (2018) The effect of the Internet of Things on sustainability. In: World Economic Forum. https://www.weforum.org/agenda/2018/01/effect-technology-sustainability-sdgs-internet-thingsiot/. Accessed 14 Mar 2019
58. Lazarevich K (2018) 10 IoT initiatives for a more sustainable future. [Online]. https://www.iotforall.com/10-iot-environment-initiatives-sustainable-future/
59. The building blocks of sustainable web design. https://sustainablewebdesign.org/. Accessed 3 Mar 2020
60. Brauer B, Ebermann C, Hildebrandt B, Remané G, Kolbe LM (2016) Green by app: the contribution of mobile applications to environmental sustainability. In: Pacific Asia Conference On Information Systems (PACIS). Association for Information System
61. Salam M, Khan SU (2016) Developing green and sustainable software: success factors for vendors. In: 2016 7th IEEE International Conference on Software Engineering and Service Science (ICSESS). IEEE, pp 1059–1062
62. Lami G, Fabbrini F, Fusani M (2012) Software sustainability from a process-centric perspective. In: European Conference on Software Process Improvement. Springer, pp 97–108
63. Mishra A, Ercil Cagiltay N, Kilic O (2007) Software engineering education: some important dimensions. Eur J Eng Educ 32(3):349–361

64. Mishra A, Mishra D (2012) Industry oriented advanced software engineering education curriculum. Croat J Educ 14(3):595–624
65. Özkan B, Mishra A (2015) A curriculum on sustainable information communication technology. Problemy Ekorozwoju–Prob Sustain Dev 10(2):95–101
66. Mishra A, Akman I (2014) Green information technology (GIT) and gender diversity. Environ Eng Manag J 13(12)
67. Mishra A, Yazici A, Mishra D (2012) Green information technology/information system education: curriculum views. TTEMTechnics Technol Educ Manag 7(3):679–686

Chapter 12
The Impact of Human Factors on Software Sustainability

Asif Imran and Tevfik Kosar

Abstract Software engineering is a constantly evolving subject area that faces new challenges every day as it tries to automate newer business processes. One of the key challenges to the success of a software solution is attaining sustainability. The inability of numerous software to sustain for the desired time length is caused by limited consideration given to sustainability during the stages of software development. This chapter presents a detailed and inclusive study covering human factor-related challenges of and approaches to software sustainability. Sustainability can be achieved by conducting specific activities at the human, environmental, and economic level. Human factors include critical social activities such as leadership and communication. This chapter groups the existing research efforts based on the above aspects. Next, how those aspects affect software sustainability is studied via a survey of software practitioners. Based on the findings, it was observed that human sustainability aspects are important, and that taking one into consideration and ignoring the other factors will threaten the sustainability of software products. Despite the noteworthy advantages of making a software sustainable, the research community has presented only a limited number of approaches that contribute to improving the human factors to achieve sustainability. To the best of our knowledge, these representations require further research. In this regard, an organized, structured, and detailed study is required on existing human factor-related sustainability approaches which will serve as a one-stop-service for researchers and software engineers who are willing to learn about these.

12.1 Introduction

Software sustainability is an important area of software engineering research today. The goal of software sustainability engineering is to ensure that software continues to achieve its goals despite updates, modifications, and evolution [1]. We consider

A. Imran (✉) · T. Kosar
University at Buffalo, Buffalo, NY, USA
e-mail: asifimra@buffalo.edu; tkosar@buffalo.edu

C. Calero et al. (eds.), *Software Sustainability*,
https://doi.org/10.1007/978-3-030-69970-3_12

the definition of sustainability provided by the Software Sustainability Institute, which states, "software you use today will be available—and continue to be improved and supported in the future" [2]. Other definitions of software sustainability consider the age of software and social aspects. Software sustainability can help us achieve a number of useful goals. Some notable goals of software sustainability are mentioned below:

- ***Operational efficiency***: Sustainability of software used both in industries and by individuals should be a natural part of the overall performance management practice [3]. If the software possess the capability to sustain for a long time, there is no need to train researchers on new type of software [4]. Researchers will become more efficient if they use the same software for a long time, thereby increasing their operational efficiency [4]. Also, an individual using a software for a significant amount of time is likely to stick to that software rather than move to a new one.
- ***Desirable reputation of software product***: To remain competitive, companies need to make innovation their top priority [5]. For example, if the software developed by a company is sustainable from human, environment, and economic perspectives, they can state that their software are long lasting and ensure high-quality output [6]. Hence, consumers will find the software reliable and have more confidence in using it. This, in turn, will provide the company with the capacity to build a desirable reputation.
- ***Reduced cost***: If a software used by an industry or an individual for day-to-day activities is technologically sustainable, then that industry or individual does not need to invest in a new software in the near future, and so their capital expenditure is reduced, unless a new software is procured which offers increased benefits and better fits the business needs [7]. On the other hand, if the software is not sustainable and it needs to be replaced within a short time, the users (both industry and individual) need to spend on procuring a new software, installing it on the computers, arranging for training on the use of the new software, etc. Hence, both the capital and current expenditures will rise due to the lack of sustainability of software [8]. From a business perspective, investing in a software which is sustainable will guarantee cost reduction and profit increase in the long run [8, 9].
- ***Accelerated progress of scientific software***: The influence of digital technology in modern research is manifold, where data and publications are being produced, shared, analyzed, and stored using various types of scientific software [10]. Although research software plays an important role in the field of science, engineering, and other areas, in most cases they are not developed in a sustainable way [11]. The researchers who develop them may be well versed in their own discipline; however, they may not have the required knowledge on the best practices of software maintainability and sustainability which are needed for reproducibility of simulation results [11]. As stated by the US Research Software Sustainability Institute (URSSI), there is a need for a strategic plan that will conduct the necessary activities of training, prototyping, and implementation,

with a goal to create improved and more sustainable software [10]. This software in turn will accelerate the progress of science.

There are different kinds of activities involved in software sustainability engineering. Depending on their complexity and applications, conceptually, sustainability can be divided into three broad levels: human, environmental, and economic. Human sustainability encompasses the development of skills and human capacity to support the functions and sustainable software development of an organization and to promote sustainable software development and usage practices. In this chapter we focus on the areas of concern regarding human sustainability, identify what is being currently done, determine the pillars of human sustainability in software, and conduct a human-factor-based study on the impact on sustainability.

A previous survey paper by Penzenstadler et al. [12] covered several low-level components for understanding software sustainability, focusing mostly on the environmental and economic attributes. However, there is a need to identify the human factors as well to provide a holistic viewpoint for software engineers and researchers. Their extended review on sustainability [13] focused on the sustainable design of software. They stated that secured software design and testing are important to develop sustainable software. They limited their work to specific programming components such as commenting code and following coding standards. However, the authors did not analyze the effect of important aspects like requirement prioritization, code smell detection, change management, etc., which equally play a role in ensuring sustainability. Calero et al. [14] provided a review on software sustainability which is primarily based on environmental friendliness of software. However, human and economic aspects like documentation skills, sustainability manifestos, funding, and leadership skills of the project manager were not considered.

The objective of this chapter is to provide a systematic and comprehensive overview of how stakeholders view human factors to be impacting sustainability. We discuss various types of human activities which aim to make software sustainable. We compare and contrast how these approaches apply from the perspective of software engineers.

Our findings show that software practitioners view certain human factors as critical to sustainability. However, most of the sustainability techniques are rarely applied due to lack of knowledge of the software community [10]. Currently, organizations like the US Research Software Sustainability Institute (URSSI) [10] and The Software Sustainability Institute [2] are investigating to address those issues. Hence extensive research is required to solve the impact of human factors on sustainability.

Based on our findings, we argue that under present circumstances there is still room for improvement in the field of sustainable software development regarding human factors. The open issues emerging from this study will provide input to researchers who are willing to develop improved techniques for addressing human factors to achieve software sustainability. We conclude that to achieve sustenance, human, environmental, and economic factors need to be considered simultaneously.

Considering one and ignoring the others will not provide long-term sustenance of software.

Based on the above information, the major contribution of this chapter can be stated as follows:

- Integrate software practitioner's feedback to identify the negative impact of the smells on architectural debt.

The rest of the chapter proceeds as follows. Section 12.2 illustrates the research questions and identifies the survey questionnaire to collect expert opinions. Section 12.3 describes the implementation of tools for architectural smell detection and recording. It also describes how Spectral clustering is used to group smells. Section 12.4 provides the obtained results and analyzes them. Section 12.5 discusses related work, and Sect. 12.6 concludes the chapter and discusses future research directions.

12.2 Empirical Study Setup

The workflow of the chapter proceeds as follows. Figure 12.1 shows the alignment of the various human factors of this chapter to the core elements of sustainability [15]. As seen in the figure, we identified six human factors, which are related to software sustainability. We took expert opinion via a survey based on the assumption that the impact of the human factors on sustainability for one software can be applied to other, similar software [16]. We studied the feedback of software practitioners to analyze impact since this added intelligence cannot be obtained from software (source code, design documents, blogs, and QA reports).

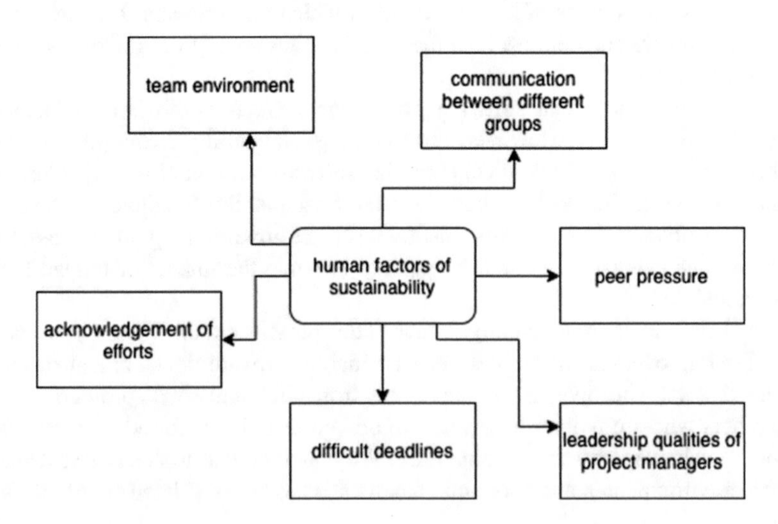

Fig. 12.1 Representation of human factors affecting sustainability

This study focuses on detecting the important human aspects to sustainability. Next, their negative impact is analyzed via feedback from software practitioners. Specifically, the following human factors are identified:

1. Team environment: The environment of the team should be such that all members should own the project and believe in its scope, schedule, and chances of success. If the teammates are not motivated to work towards success of the project, that severely affects the sustainability as well [17].
2. Communication: Lack of effective communication can break the sustainability of a software project. Even if all other aspects of the team are ideal, when communication is lacking, you will have sub-par sustainability. Effective communication can also allow teams to overcome many less-than-ideal circumstances. Here, it must be noted that for sustainability the power of the team members to listen carefully to each other is critically important.
3. Leadership qualities: Leadership includes the skills through which a project manager handles change management, timeline management, cost management, employee turnover, etc. [14].
4. Difficult deadlines: These are stressful for the humans involved in the software project, and if there are more than one, then it can be overwhelming. This leads to software engineers trying to complete the software development rather than focusing on the sustainability manifestos.
5. Peer pressure: Similar to difficult deadlines, peer pressure can create unnecessary stress on software developers which can lead to the developers trying to finish the project early to get into superiors' good books, bypassing the sustainability benchmarks of the software.
6. Acknowledgment of efforts: The appreciation of one's hard work to follow sustainability benchmarks while designing and developing software systems can go a long way to ensure that the critical attributes of sustainability are preserved.

12.2.1 Research Question

As software evolves from initial to matured phases, its sustainability may incur significant interest. Hence, we should be able to identify which human factors have a higher negative impact on sustainability based on expert opinion. This will help to prioritize smells for refactoring. More specifically, the major research question is as follows:

RQ: Which human factors have a greater impact on sustainability according to software practitioners?

For impact determination, we gave the developers an option to determine whether the impact on sustainability is high, low, or no impact. We provided a questionnaire to the developers seeking answers for their selected impact for a human factor. The questionnaire asked about the factors which influenced the developers to assign a specific impact to a smell. These factors included the *impact of mental pressure to meet deadlines [18], the causal effect, and the context dependency of a factor, which resulted in more factors [19], and aspects focusing on team environment, and communication issues [17]*. Hence, the human factor and the presence of specific community contexts were considered for impact on sustainability. These factors are described in greater detail in the following section. The answers to these questions can help team leaders address community issues which threaten sustainability.

12.2.2 Survey Structure

We designed an online survey questionnaire with a minimal number of questions. The goal was to reduce the cognitive complexity and at the same time obtain the required information [20]. The questionnaire aimed to define what is meant by software sustainability and provide the human factors impacting sustainability to the developer and, at the same time, ask them to tag the adverse impact of a factor as high, low, or no impact. Also, we tried to obtain justification behind the tagging, trying to identify the context and any human factor if present. To ensure that the respondents are well aware of what is meant by sustainability, we conducted a training session where we remotely presented the definitions, examples, and scenarios where human factors threatening sustainability may occur and tried to determine the impact caused by them. After the training, we provided them the questionnaire. Then we identified the following questions based on motivation from Palomba et al. [17].

Q1 Are you aware of the identified human factors? If yes, are those handled at your company?
Q2 Will you tag the factor as having high, moderate negative impact, or no impact on sustainability?
Q3 Was your answer to Q2 based on a specific software context, time, or effort?
Q4 Did any friction in your development community affect impact determination? Explain.
Q5 How does the negative affect on the mindset of software engineers due to COVID-19 impact sustainability of the software?

After detection of the issues, we present the factors to the software practitioners together with the questionnaire. The software practitioners who worked the most on the critical parts were asked the questions [17]. This was determined from the Git commits. The respondents included system architect and senior developers in the team who worked with the critical software components and were under stress to meet deadlines. If there were multiple developers who worked equally on a class, we identified the developer who worked solely on that class and no other classes. This is based on the assumption that the developer who worked only on that class is possibly the owner of the class and knows it in depth, and will be the best individual to know whether any kind of human factors threatened the sustainability of the class.

12.3 Survey Exercise

This section describes the implementation of the survey mechanism to determine human factors and their impact on software sustainability. There is a need to group sustainability factors based on human activities. The exercise has been applied to a real-life software on a testing phase and feedback was collected from the developers.

As a result, we need to introduce context awareness. This is done via questionnaire tagging, and we have an additional option where the developer can flag a smell as "having no negative impact" based on the context. We trust the software practitioner's knowledge and expert opinion in this regard. For clustering, we obtained the refactored dataset and conducted the analysis.

For our impact analysis of human factors on software sustainability, we chose OneDataShare [21] since we have full access to the developers of this software to get the required response in the survey. OneDataShare is an open-source software which started in 2016. This software have been studied earlier to analyze impacts in terms of performance [22]. However the effect of human components on the sustainability of the software has not been studied earlier. In this chapter, we aim to fill that gap.

Altogether, we collected data from 10 software practitioners who worked in the OneDataShare project. In total, there were 14 developers, yielding a response rate of 71.43%. OneDataShare is a research software for fast data transfer, which is a flagship project, hence easy access could be obtained to the respondents. Next, the 10 software practitioners who were surveyed worked the most with the classes suffering from smells, and they were the most concerned, hence readily responded.

12.4 Results

In order to reliably state that the smells detected by the tool negatively impact Architectural Technical Debt (ATD), we first need to establish reliance on the tool, followed by taking community feedback to analyze the impact of smells. First, it is highly important to justify that the tool is capable of accurately detecting smells, then to compare the smell detection output of the tool by applying it to the software with a predefined set of smells that are frequently used for experimental purposes in the existing literature. Also, it is equally important to show that the tool can be generalized and used to detect smells in software applications other than OneDataShare. We answer the identified research questions which will generalize our findings.

12.4.1 Answer to RQs

This section describes the results obtained for each of the survey questions to analyze the impact on ATD.

1. *Are you aware of the identified human factors? If yes, are those handled at your company? [17]*

The following awareness issues could be determined by analyzing the responses.

(a) **Lack of awareness:** The surveyed software practitioners were not aware of the impact of certain factors described, such as peer pressure and acknowledgment of efforts, on sustainability. They developed the software using asynchronous event-driven network application framework which enabled quick and easy development and did not consider other factors related to human behavior to be significant catalysts for maintaining sustainability. Besides the lack of knowledge of existing human resources, shorter time to market causes excess pressure from management to deliver the software on-time which is another reason for not paying attention to the human factors.
(b) **Awareness of human factors:** For certain software companies, the respondents were well aware of the remaining factors. They rated that leadership of project managers is the most important aspect of ensuring that the projects sustain. They stated that many of those factors had been addressed from earlier stages of the development life cycle. The respondents stated that regular team meetings, team coffee sessions, sharing of ideas, and sharing a common goal of providing sustainable software services are critical to eliminating the negative impacts of the mentioned human factors in sustainability. During COVID-19, the technical leads were responsible for conducting virtual sessions with team members, learn their thoughts, and eliminate such factors. Even top-level directors tried to communicate regularly with junior employees and interns to motivate them towards building sustainable software.

2. *Will you tag the factor as having high, moderate negative impact, or no impact on sustainability?*

The impact of the identified human factors on sustainability was determined based on the opinions from experts. The response from various respondents regarding the score given to each factor is shown in Fig. 12.2. We summarize the findings as follows:

(a) **Factors with a high negative impact on sustainability**: We see that the software practitioners have flagged the factor called "Leadership" as having a high negative impact on sustainability. "Communication" has been identified as an important factor as well. Many teams who responded were not aware of human factors "peer pressure" and "acknowledgment of efforts." Hence this survey was an eye opener for them to address these issues.
(b) **Factors with a moderate negative impact on ATD**: The survey reported that two types of smells called "Team environment" has a low impact on sustainability according to numerous respondents. The reason was that this factor did not affect a large number of modules of the sustainability software and required relatively lesser effort to solve, given there is good leadership. We received comments from software practitioners, such as, "In some cases like COVID-19, it is the way the engineers are forced to work from home and during such times

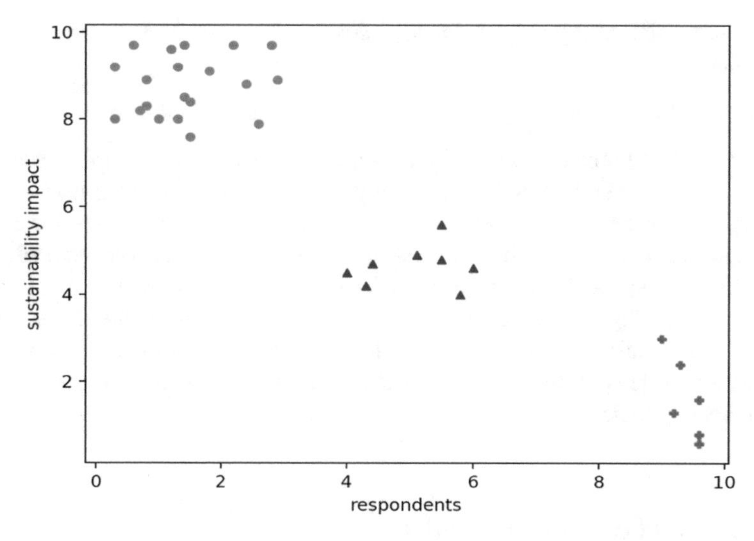

Fig. 12.2 Representation of human factors affecting sustainability

good leadership and continuous communication is more important. Team environment is not important given good leadership since that happens automatically if good leadership is present, and during COVID-19 teams are unable to meet physically for the environment factor is absent. In other cases, we find that effective communication is more important to result in good team environment." The respondents felt that given the short time to market and the COVID-19 situation when this survey was conducted, these factors should be given lower priority compared to the others with a higher negative impact on sustainability.

(c) **Factors with no negative impact:** None of the identified factors were found to have negligible impact on sustainability. In fact, the respondents said that other human factors like gender-based discrimination and racial discrimination should be included in future studies as other important human factors which affect sustainability.

3. *Was your answer to Q2 based on a specific software context, time, or effort?*

During the survey, the following answers were obtained:

(a) **Time and effort**: In line with the survey the responses were motivated by the time and effort required to solve them. Also, lack of awareness was another aspect in this regard. For "difficult deadlines," the deciding factor was time as well.

(b) **Software context**: For the "Leadership qualities" factor, the experts primarily based their answer on software context, stating that for large software, such factors can have tremendous negative impact on sustainability. The reason being that mostly this factor occurs when many employees work on a big project and all of them are not clearly aware of its sustainability goals.

4. *Did any friction in your development community affect impact determination? Explain.*

   For this question, all responses were the same, as discussed below:

(a) For all the human factors which were presented to the respondents, they responded saying they did not have any communication issues or team friction. Capable leadership and good communication may be a reason for such a response. At the same time, the teams which were interviewed worked under the same roof, and they were not affected by any communication gap which happens during a collaboration between remote teams. Also, the hierarchy was flat, which ensured swift decisions. However, studying the impacts of community smells [17] on sustainability with respect to this type of software can be an interesting topic.

12.4.2 Implications of Results

The analyses provided in this chapter can be applied to software researchers and practitioners.

- **Usefulness to researchers:** Researchers can use this information to identify which human factors occur specifically and cause a greater adverse impact on sustainability. This will provide them a useful lead on which type of human factors to focus their research on based on the type of software. Addressing human factors of sustainability opens a new doorway of research in this area as it can reduce the challenges to make a software sustainable.
- **Usefulness to software practitioners:** Firstly, managers in software companies can apply the results on their projects, collect the data, and analyze results to improve their sustainability practices. Also, by keeping the results of the chapter in mind, attention can be paid to specific factors so as to incur minimum impact on sustainability.

By surveying the impact of human factors on sustainability, this chapter allows engineers and researchers to refactor only a subset of factors at a time.

12.5 Related Work

A model depicting the collective efforts required to detect architectural smells is proposed based on the study of related literature [23]. The process of prioritizing architectural smells based on the impact of those on ATD has been studied by Martini et al. [24]. The authors analyzed how users perceive the smells to affect architectural debt. Although the model considers important aspects of architectural smells, it does not study how those smells affect sustainability. Also, the human factors which influence those smells were not considered.

Supervised machine learning approaches have been explored for code and architectural smells detection [25, 26]. However, the authors debated that exiting research using such techniques deal with biased datasets for the training of smells. They stated the need for further research with more realistic datasets to obtain the actual performance of the tool. This leads to the need for further research using unsupervised techniques for smell detection.

The effect of the developer's seniority, frequency of commits, and interval of commits on reducing architectural debts in software were evaluated by Alfayez et al. [27]. The authors determined that seniority and frequency of commits are negatively correlated with reducing architectural debt, whereas the interval of commits is positively correlated. The authors used statistical analysis tests to validate the effects of developer behavior on architectural smells. However, use of unsupervised machine learning for grouping smells was not explored.

Existing tools for detecting architectural smells include *Decor* [28], *Arcan* [29], and *Designite* [30]. Most of these focus on software written in languages other than Java. *AnaConDebt* [31] is used to assess technical debt. Further research is required to include community feedback for the smells detected by those software to prioritize for ATD.

Polomba et al. [32] was able to detect code smell which could not be differentiated via their structures. As a result, they made software suffering from those types of code smells more sustainable by improving detection of smelly code which ultimately set up its removal. The impact of community smells on software was studied based on surveys from software engineers [17]. This motivates us to study how community factors affect sustainability of the software as well.

Using open-source components in design and implementation of research software guarantees its sustainability [33]. The researchers have cited that Mozilla Firefox is a successful and sustainable application, mainly because of its open-source structure. They stated that Mozilla still continues to support, train, and research open-source software [34]. The authors highlighted that based on their years of experience in mentoring open source software projects, there are primarily three areas to consider in order to make an open-source project sustainable and ensure its growth. They are training, peer support, and availability of financial and computational resources. How effective training can be ensured for research-based software is not identified by the authors. Also, they highlighted community support for a largely used software like Mozilla, but how community support can be achieved for a research software which serves a small group of researchers have not been stated.

The importance of studying the experiences of the stewards who developed and used a wide range of open-source software tools to identify and focus on open-source software sustainability has been addressed in [35]. The Hierarchy Data Format (HDF) group has been working with the research community for 30 years, building open-source tools to provide them platforms for data storage, easy access, and analysis. It has analyzed Github and found that nearly 1000 repositories are based on its open-source codes. At the same time several broadly successful open-source systems are centered around HDF. The authors shared their experience of *PyTables* that whenever an existing code is reused or refactored, there is a high probability that it will present unforeseen issues which require correction and addressing, thus reducing the time to market. However, effective refactoring techniques for a fast data sharing software like OneDataShare have not been addressed.

Best practices in software usability and user experience can have a significant effect on software sustainability [36]. By following software usability best practices, failures in the software can be fixed at a lower cost, and performance can be enhanced [37]. The importance of user experience has been emphasized in the academia by designing course works in Human-Computer Interaction (HCI). However, most of the HCI courses are non-major and considered to be esoteric by researchers coming from scientific backgrounds. However, the dynamic nature of scientific software has made user experience an important criterion for its sustainability [36]. As a result, the authors combined heuristic studies, participant-driven interviews and surveys, usability observations, and evaluations to improve user experience of scientific applications. These experiences have been used to develop User Interfaces data exploration and analysis, create workflow models, and design and build data management tools. However, using such tools for a research software that transfers a high volume of data within a desirable time frame has not been addressed.

12.6 Conclusion and Future Work

This chapter identifies and analyzes the impact of human factors in sustainability via a survey. We detected six types of factors and evaluated the impact of those by interviewing experts in the field of software engineering. We proceeded to analyze how software practitioners saw the negative impact of those factors. We provided a questionnaire to the software practitioners to gather their viewpoints related to the adverse effects of the human factors on sustainability. Results show that the "Leadership" and "Communication" factors were rated by the practitioners to have a high impact on sustainability. On the other hand, the factor called *team environment* had less negative impact.

In the future, it may also be helpful to perform a longitudinal study that detects the precision of the survey results by increasing the sample size. Another interesting area would be to extend the list of human factors to include gender inequality and racial discrimination, which also negatively impact sustainability.

References

1. I. of Electrical and E. Engineers. Defining software sustainability. [Online]. https://ieeexplore.ieee.org/Xplore/home.jsp. Accessed 14 Sept 2019

2. Software Sustainability Institute (2016) https://www.software.ac.uk/case-studies. Accessed 10 Sept 2019
3. Albertao F, Xiao J, Tian C, Lu Y, Zhang KQ, Liu C (2010) Measuring the sustainability performance of software projects. In: 2010 IEEE 7th International Conference on E-Business Engineering. IEEE, pp 369–373.
4. Penzenstadler B, Fleischmann A (2011) Teach sustainability in software engineering? In: 2011 24th IEEE-CS Conference on Software Engineering Education and Training (CSEE&T). IEEE, pp 454–458
5. Dagli CH, Kilicay-Ergin N (2008) System of systems architecting. In: System of Systems Engineering: Innovations for the 21st Century. pp 77–100
6. Durdik Z, Klatt B, Koziolek H, Krogmann K, Stammel J, Weiss R (2012) Sustainability guidelines for long-living software systems. In: Software Maintenance (ICSM), 2012 28th IEEE International Conference on. IEEE, pp 517–526
7. Stewart CA, Barnett WK, Wernert EA, Wernert JA, Welch V, Knepper R (2015) Sustained software for cyberinfrastructure: analyses of successful efforts with a focus on nsf-funded software. In: Proceedings of the 1st Workshop on The Science of Cyberinfrastructure: Research, Experience, Applications and Models, ser. SCREAM '15. ACM, New York, NY, pp 63–72. [Online]. http://doi.acm.org/10.1145/2753524.2753533
8. Seacord RC, Elm J, Goethert W, Lewis GA, Plakosh D, Robert J, Wrage L, Lindvall M (2003) Measuring software sustainability. In: International Conference on Software Maintenance (ICSM). IEEE, p 450
9. Chitchyan R, Becker C, Betz S, Duboc L, Penzenstadler B, Seyff N, Venters CC (2016) Sustainability design in requirements engineering: state of practice. In: Proceedings of the 38th International Conference on Software Engineering Companion, ser. ICSE '16. ACM, New York, NY, pp 533–542. [Online]. http://doi.acm.org/10.1145/2889160.2889217
10. URSSI. Developing a pathway to research software sustainability. http://urssi.us/. Accessed 14 Sept 2019
11. Carver JC, Gesing S, Katz DS, Ram K, Weber N (2018) Conceptualization of a us research software sustainability institute (urssi). Comput Sci Eng 20(3):4–9
12. Penzenstadler B, Bauer V, Calero C, Franch X (2012) Sustainability in software engineering: a systematic literature review
13. Penzenstadler B, Raturi A, Richardson D, Tomlinson B (2014) Safety, security, now sustainability: the non-functional requirement for the 21st century. IEEE Softw 1:1
14. Calero C, Bertoa MF, Moraga MÁ (2013) A systematic literature review for software sustainability measures. In: Proceedings of the 2nd International Workshop on Green and Sustainable Software. IEEE Press, pp 46–53
15. Imran A, Kosar T (2019) Software sustainability: a systematic literature review and comprehensive analysis. arXiv preprint arXiv:1910.06109
16. Tockey S (2014) Aspects of software valuation. In: Economics-Driven Software Architecture. Elsevier, pp 37–58
17. Palomba F, Tamburri DAA, Fontana FA, Oliveto R, Zaidman A, Serebrenik A (2018) Beyond technical aspects: How do community smells influence the intensity of code smells?. IEEE Trans Softw Eng
18. de Andrade HS, Almeida E, Crnkovic I (2014) Architectural bad smells in software product lines: an exploratory study. In: Proceedings of the WICSA 2014 Companion Volume. ACM, p 12
19. Zazworka N, Shaw MA, Shull F, Seaman C (2011) Investigating the impact of design debt on software quality. In: Proceedings of the 2nd Workshop on Managing Technical Debt, ser. MTD '11. ACM, New York, NY, pp 17–23. [Online]. http://doi.acm.org/10.1145/1985362.1985366
20. Dillman DA (2011) Mail and Internet surveys: the tailored design method–2007 Update with new Internet, visual, and mixed-mode guide. Wiley

21. Imran A, Nine MS, Guner K, Kosar T (2018) Onedatashare-a vision for cloud-hosted data transfer scheduling and optimization as a service. In: Proceedings of the 8th International Conference on Cloud Computing and Services Science, vol 1
22. Imran A, Kosar T (2020) The impact of auto-refactoring code smells on the resource utilization of cloud software. In: García-Castro R (ed) The 32nd International Conference on Software Engineering and Knowledge Engineering, SEKE 2020, KSIR Virtual Conference Center, USA, 9–19 July 2020. KSI Research Inc., pp 299–304. [Online]. https://doi.org/10.18293/SEKE2020-138
23. Besker T, Martini A, Bosch J (2016) A systematic literature review and a unified model of ATD. In: 2016 42nd Euromicro Conference on Software Engineering and Advanced Applications (SEAA). IEEE, 2016, pp 189–197
24. Martini A, Fontana FA, Biaggi A, Roveda R (2018) Identifying and prioritizing architectural debt through architectural smells: a case study in a large software company. In: European Conference on Software Architecture. Springer, pp 320–335
25. Caram FL, Rodrigues BRDO, Campanelli AS, Parreiras FS (2019) Machine learning techniques for code smells detection: a systematic mapping study. Int J Softw Eng Knowl Eng 29 (02):285–316
26. Fontana FA, Mäntylä MV, Zanoni M, Marino A (2016) Comparing and experimenting machine learning techniques for code smell detection. Empirical Softw Eng 21(3):1143–1191
27. Alfayez R, Behnamghader P, Srisopha K, Boehm B (2018) An exploratory study on the influence of developers in technical debt. In: Proceedings of the 2018 International Conference on Technical Debt, ser. TechDebt '18. ACM, New York, NY, pp 1–10. [Online]. http://doi.acm.org/10.1145/3194164.3194165
28. Moha N, Guéhéneuc Y-G, Le Meur A-F, Duchien L, Tiberghien A (2010) From a domain analysis to the specification and detection of code and design smells. Formal Aspects Comput 22(3–4):345–361
29. Fontana FA, Pigazzini I, Roveda R, Tamburri D, Zanoni M, Di Nitto E (2017) Arcan: a tool for architectural smells detection. In: 2017 IEEE International Conference on Software Architecture Workshops (ICSAW). IEEE, pp 282–285
30. Suryanarayana G, Samarthyam G, Sharma T (2014) Refactoring for software design smells: managing technical debt. Morgan Kaufmann
31. Martini A (2018) Anacondebt: a tool to assess and track technical debt. In: 2018 IEEE/ACM International Conference on Technical Debt (TechDebt). IEEE, pp 55–56
32. Palomba F (2015) Textual analysis for code smell detection. In: Proceedings of the 37th International Conference on Software Engineering – vol 2, ser. ICSE '15. IEEE Press, Piscataway, NJ, pp 769–771. [Online]. http://dl.acm.org/citation.cfm?id=2819009.2819162
33. Cabunoc A (2018) Supporting research software by growing a culture of openness in academia
34. Brown AW, Booch G (2002) Reusing open-source software and practices: the impact of opensource on commercial vendors. In: International Conference on Software Reuse. Springer, pp 123–136
35. Haebermann T (2018) Sustainable open source tools for sharing and understanding data. In: USRRI 1st Workshop on Software Sustainability. USRRI, pp 1561–1570
36. Kitzes J, Turek D, Deniz F (2017) The practice of reproducible research: case studies and lessons from the data-intensive sciences. University of California Press
37. Gilb T, Finzi S (1988) Principles of software engineering management, vol 11. Addison-Wesley, Reading, MA

Chapter 13
Social Sustainability in the e-Health Domain via Personalized and Self-Adaptive Mobile Apps

Eoin Martino Grua, Martina De Sanctis, Ivano Malavolta, Mark Hoogendoorn, and Patricia Lago

Abstract Within software engineering, social sustainability is the dimension of sustainability that focuses on the "support of current and future generations to have the same or greater access to social resources by pursuing social equity." An important domain that strives to achieve social sustainability is e-Health, and more recently e-Health mobile apps.A wealth of e-Health mobile apps is available for many purposes, such as lifestyle improvement and mental coaching. The interventions, prompts, and encouragements of e-Health apps sometimes take context into account (e.g., previous interactions or geographical location of the user), but they still tend to be rigid, e.g., apps use fixed sets of rules or they are not sufficiently tailored toward individuals' needs. Personalization to the different users' characteristics and run-time adaptation to their changing needs and context provide a great opportunity for getting users continuously engaged and active, eventually leading to better physical and mental conditions. This chapter presents a reference architecture for enabling AI-based personalization and self-adaptation of mobile apps for e-Health. The reference architecture makes use of a dedicated goal model and multiple MAPE loops operating at different levels of granularity and for different purposes. The proposed reference architecture is instantiated in the context of a fitness-based mobile application and exemplified through a series of typical usage scenarios extracted from our industrial collaborations.

E. M. Grua (✉) · I. Malavolta · M. Hoogendoorn · P. Lago
Vrije Universiteit Amsterdam (VU), Amsterdam, The Netherlands
e-mail: e.m.grua@vu.nl; i.malavolta@vu.nl; m.hoogendoorn@vu.nl; p.lago@vu.nl

M. De Sanctis
Gran Sasso Science Institute (GSSI), L'Aquila, Italy
e-mail: martina.desanctis@gssi.it

C. Calero et al. (eds.), *Software Sustainability*,
https://doi.org/10.1007/978-3-030-69970-3_13

13.1 Introduction

e-Health mobile apps are designed for assisting end users in tracking and improving their own health-related activities [1]. With a projected market growth of US$102.3 billion by 2023, e-Health apps represent a significant market [2] providing a wide spectrum of services, i.e., life style improvement, mental coaching, sport tracking, and recording of medical data [3]. The unique characteristics of e-Health apps wrt other health-related software systems are that e-Health apps (1) can take advantage of smartphone sensors, (2) can reach an extremely wide audience with low infrastructural investments, and (3) can leverage the intrinsic characteristics of the mobile medium (i.e., being always on, personal, and always carried by the user) for providing timely and in-context services [4].

However, even if the interventions, prompts, and encouragements of current e-Health apps take context into account (e.g., previous interactions or geographical location of the user), they still tend to be *rigid* and not fully tailored to individual users, e.g., by using fixed rule sets or by not considering the unique traits and behavioral characteristics of the user. In this context, we see *personalization* [5] and *self-adaptation* [6–8] as effective instruments for getting users continuously engaged and active, eventually leading to better physical and mental conditions. The addition of intervention tailoring (via personalization and self-adaptation) is a crucial step in addressing the main sustainability concern that e-Health mobile apps want to achieve: *social sustainability*. By providing better interventions, we are not only more likely to have the user interested in maintaining engagement with the app but also help the user achieve better physical and mental conditions by allowing the app to better address the personal needs and by extension the social ones too.

In this work, we combine personalization and software self-adaptation to provide users of mobile e-Health apps with a better, more engaging and effective experience. To this aim, we propose a *reference architecture (RA) that combines data-driven personalization with self-adaptation*. The main design drivers that make the proposed reference architecture unique are:

- The combination of multiple *Monitor–Analyze–Plan–Execute* (MAPE) loops [9] operating at different levels of granularity and for different purposes, e.g., to suggest users the most suitable and timely activities according to their (evolving) health-related characteristics (e.g., active vs. less active), but also to cope with technical aspects (e.g., connectivity hiccups, availability of IoT devices and third-party apps on the user's device) and the characteristics of the physical environment (e.g., indoor vs. outdoor, weather).
- A *dedicated goal model* for representing health-related goals via a descriptive concise language accessible by healthcare professionals (e.g., fitness coaches, psychologists).
- The exploitation of our *online clustering algorithm* for efficiently managing the evolution of the behavior of users as multiple time series evolving over time. This online clustering algorithm has been already extensively tested in a previously

published article [10], showing promising results by doing better than the current state of the art.

The main characteristics of the proposed reference architecture are the following: (1) it caters the personalization of services to specific user preferences (e.g., preferred sport activities); (2) it guarantees the correct functioning of the features via the use of connected IoT devices (e.g., a smart-bracelet) and runtime adaptation strategies; (3) it adapts the services depending on contextual factors such as environmental conditions and weather; (4) it supports a smooth participation of domain experts (e.g., psychologists) in the personalization and self-adaptation processes; and (5) it can be applied in the context of a single e-Health app and by integrating the services of third-party e-Health apps (e.g., already installed sport trackers). All of these characteristics are shown in this work by evaluating the reference architecture and the goal model with fitness coaching scenarios. We want to emphasize how most characteristics have been engineered with the main goal of achieving social sustainability. Possible exceptions are characteristics (2) and (5), which more specifically addresses technical sustainability of the reference architecture. Our emphasis on social sustainability will be further explained and explored throughout the chapter.

Lastly, in a previous study [11] we reported a preliminary version of our Reference Architecture (RA). Here we extend the work by: (1) framing the work in the overall context of social sustainability, (2) document the methodology used to design our RA, (3) report a scenario-based evaluation of our RA, (4) provide a goal model to be used with the RA, (5) a viewpoint definition used to create the view of our RA.

13.2 Background

The notion of *reference architecture* (RA) is borrowed from Volpato et al. [12], who define it as "a special type of software architectures that provide a characterization of software systems functionalities in specific application domains," e.g., SOA for service orientation and AUTOSAR for automotive. In the context of this study, a *self-adaptive software system* is defined as a system that can autonomously handle changes and uncertainties in its environment, the system itself, and its goals [7].

For the definition of *personalization*, we build on that by Fan and Poole [5] and define it as "a process that changes a system to increase its personal relevance to an individual or a category of individuals." Furthermore, to enhance personalization, we use CluStream-GT (CluStream for Growing Time-series) [10]. CluStream-GT was chosen for this RA as it is the state-of-the-art clustering algorithm for time-series data (especially within the health domain). CluStream-GT works in two phases: offline and online. First, the *offline phase* initializes the algorithm with a small initial dataset; this is done either at design time or at the start of runtime. Afterward, during the *online phase* the algorithm clusters the data that is being collected at runtime. Clustering allows the RA to group similar users together, where similarity is determined by the data gathered from the apps. This gives the RA a more sustainable and scalable method of personalization, without requiring to create individual personalization strategies but maintaining a suitable degree of personalization [10, 13]. An example case where clustering can be used to aid personalization is with the use of cluster-based Reinforcement Learning [14].

The methodology used for the design of our RA is the one presented by Angelov et al. [15] (see Fig. 13.1), where the authors present their RA Framework to facilitate software architects in the design of *congruent RAs*, i.e., RAs where the design, context, and goals are explicit and coherent (adapted from [15]).

The RA Framework (or framework for short) consists of two elements: a multidimensional classification space, and a set of predefined RA types (and variants of these types). The former, through the use of strict questions and answers, supports software architects in classifying RAs according to their *context* (**Where?**, **Who?**, and **When?** questions in Fig. 13.1), *goals* (**Why?** in Fig. 13.1), and *design* (**How?** and **What?** in Fig. 13.1) dimensions. The latter consists of specific combinations of values from the multidimensional space. These types, and variants, are used to evaluate the congruence of the RA being designed. If a RA is congruent (i.e. matches a type or variant) it has a greater chance of becoming a success, where, by success the authors mean ". . . the acceptance of the architecture by its stakeholders and its usage in multiple projects" [15]. For each dimension, the authors have defined subdimensions with respective questions and answers. During the design of our RA we have worked with each dimension and, with the use of the framework, classified our RA according to the possible values available for each subdimension. As knowledge of our RA and its components is necessary to understand the design process, we further explain the use of the framework in Sect. 13.7.

In recent years a larger body of work on software engineering and software architecture address sustainability. Sustainability can be divided into four dimensions: technical, economical, environmental, and social [16]. Within this work we present an RA for the e-Health domain with the main goal of better addressing the social dimension of sustainability, whilst the technical contributions of this work include the combination of AI and self-adaptation. In this work we build on the following definition of social sustainability: "focusing on supporting current and future generations to have the same or greater access to social resources by pursuing generational equity. For software-intensive systems, this dimension encompasses the direct support of social communities, as well as the support of activities or processes that indirectly create benefits for such communities" [16].

13.3 Related Work

Several RAs for IoT can be found in the literature [17–20]. In particular, Bauer et al. [19] present several abstract *architectural views* and *perspectives*, which can be differently instantiated. The adaptation of the system's configuration is also envisioned, at an abstract level. IoT-A [18] aims to be easily customized to different needs, and it makes use of *axioms* and *relationships* to define connections among

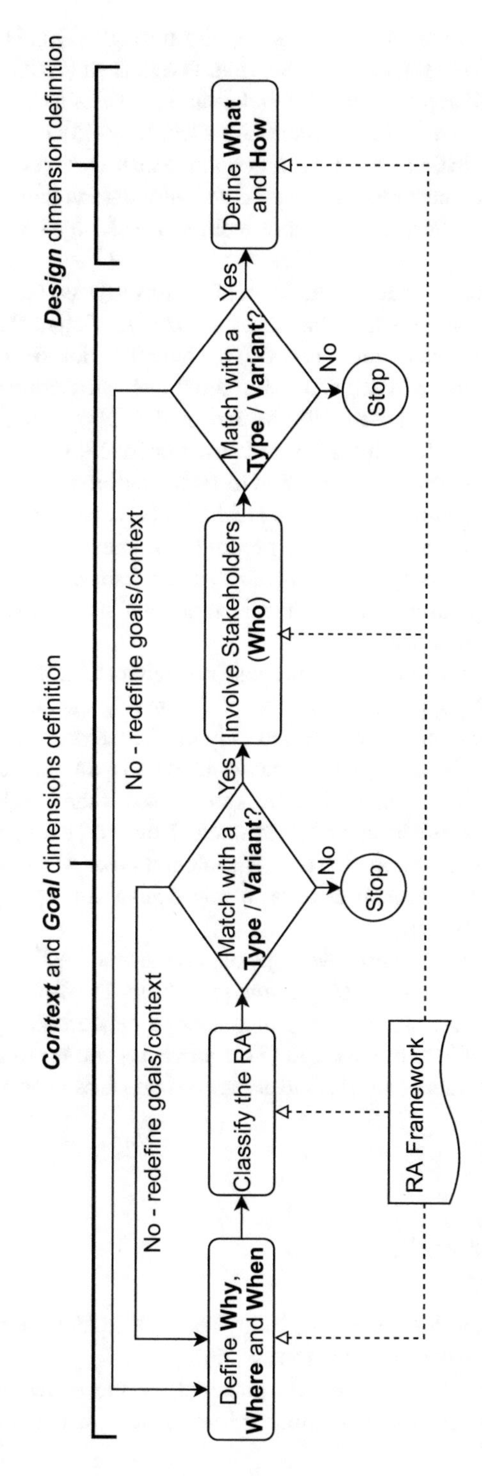

Fig. 13.1 Methodology for the design of our RA [15]

IoT entities. Industrial Internet Reference Architecture (IIRA) [17] is particularly tailored for industrial IoT systems. Web Services Oxigenated (WSO2) [20] presents a layered structure and targets scalability and security aspects too. All of the above RAs are abstract and domain independent. As such, they do not address required features specific to the IoT-based e-Health domain. Moreover, they lack the needed integration with AI for personalization used to tailor interventions to the user's health-related characteristics, an important technique used by the RA to address social sustainability.

Other works providing service-oriented architectures (SOAs) focused on adaptation but neglected user-based personalization. For example, Feljan et al. [21] defined an SOA for planning and execution (SOA-PE) in Cyber Physical Systems (CPS) and Mohalik et al. [22] proposed a MAPE-K autonomic computing framework to manage adaptivity in service-based CPS. Morais et al. [23] present RAH, a RA for IoT-based e-Health apps. RAH has a layered structure, and it provides components for the prevention, monitoring and detection of faults. Different from RAH, our RA explicitly manages the self-adaptation of e-Health mobile apps, both at the user and architectural levels. Mizouni et al. [24] propose a framework for designing and developing context-aware adaptive mobile apps. Their framework lacks other types of adaptation, i.e., adaptation for user personalization and adaptation with other IoT devices—which is possible with our RA.

Lopez and Condori-Fernandez [25] propose an architectural design for an adaptive persuasive mobile app with the goal of improving medication adherence. Accordingly, the adaptation is here focused only on the messages given to the user and lacks the other levels of adaptation (environment adaptation, etc.) that our RA covers. Kim [26] proposes a general RA that can be used when developing adaptive apps and implements a e-Health app as an example. However, being general, the RA lacks the level of detail present in our work, the integration of AI for personalization, and a way for involving domain experts in app design and operation, which is essential in adaptive e-Health.

In summary, to the best of our knowledge, ours is the first RA for e-Health mobile apps that simultaneously supports (1) *personalization* for the different users, **Users** by exploiting users' smart objects and preferences to dynamically get data about, e.g., their mood and daily activities, and (2) *runtime adaptation* to user needs and context in order to keep them engaged and active, so that we can better address social sustainability.

13.4 Reference Architecture

Figure 13.2 shows our RA with the following stakeholders and components. Section 13.8 defines the corresponding viewpoint.

Users provide and generate the data gathered by the e-Health app. At first installation, the users are asked to input information to better understand their

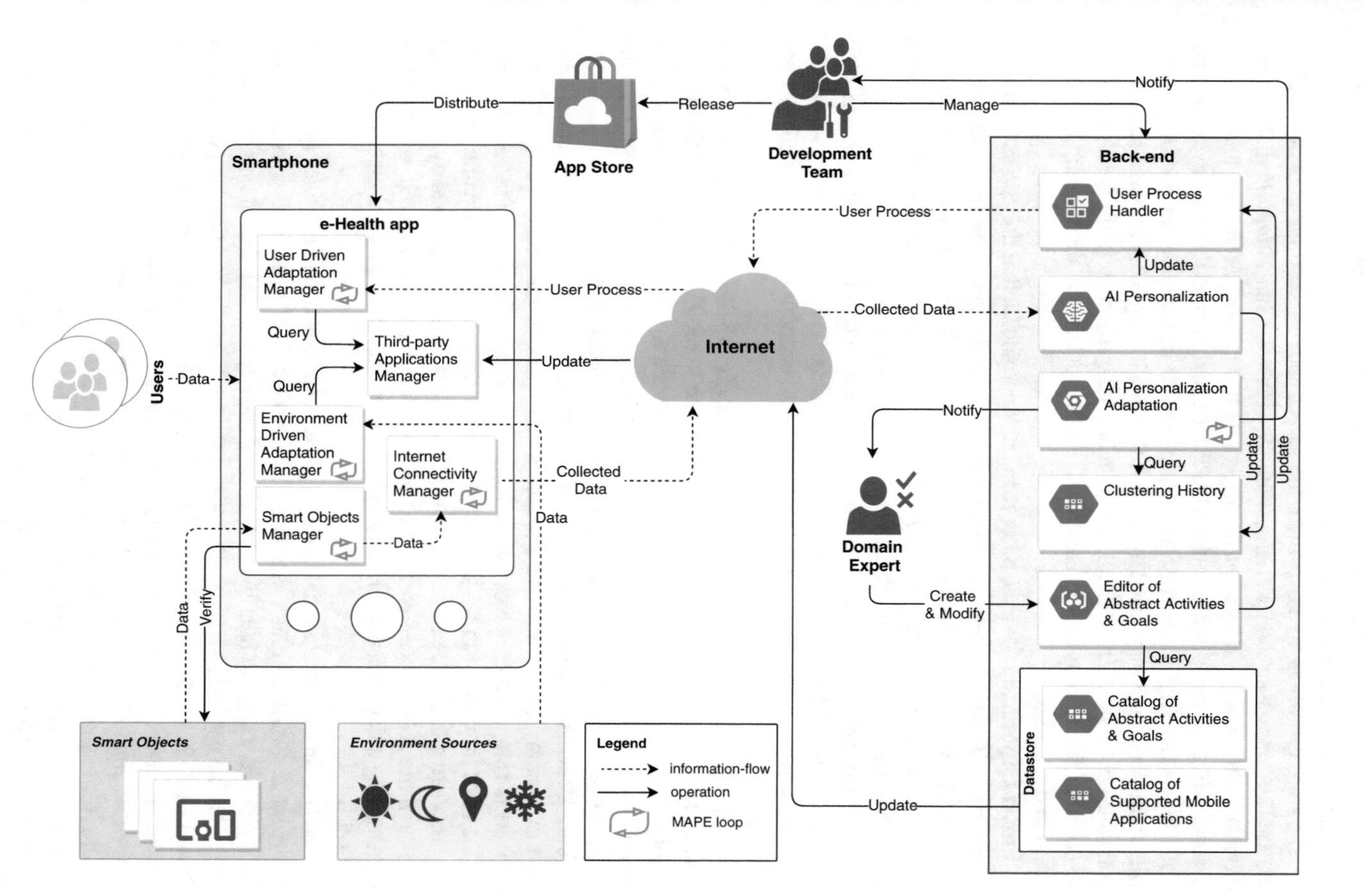

Fig. 13.2 Reference architecture for personalized and self-adaptive e-Health apps

aptitudes. After an initial usage phase and data collection, the system has enough information to assign them to a cluster.

The **smartphone** is the host where the self-adaptive e-Health app is installed. In the mobile app, four components, namely, *User Driven Adaptation Manager*, *Environment Driven Adaptation Manager*, *Smart Objects Manager*, and *Internet Connectivity Manager*, implement a MAPE loop to dynamically perform adaptation. The *Third-party Applications Manager*, in turn, is responsible for communication with third-party apps supported by the RA that can be exploited by the e-Health app both during its nominal execution and when adaptation is performed. It is also responsible for storing user preferences. Further details on these components are given in Sect. 13.5.

Smart Objects are devices, other than the smartphone, that the app can communicate with. They are used to gather additional data about the users as well as augmenting the data collected by the smartphone sensors. For instance, a smartwatch would be used by the app to track the user's heartrate, therefore adding extra information on the real-time performance of the user.

Environment is the physical location of the user, and its measurable properties. It is used by the e-Health app to make runtime adaptations according to its current operational context and to the user's scheduled activities, as described in Sect. 13.5.5.

The back-end of our RA (right-hand side in Fig. 13.2) is Managed by a *Development team*. It additionally exposes an interface to the *Domain Expert* that is also involved in the e-Health app design and operation. The back-end contains the components needed to store the collected user data and to manage the user clusters. It also hosts components supporting the general functioning of the app.

User Process Handler is in charge of sending User Processes to the users; it takes care of sending the same User Process to all users of the same cluster. A *User Process* is composed of one or more *Abstract Activities*. These activities are inspired by the ones introduced in [27], although they differ both in the structure and in the way they are refined, as explained later. An *Abstract Activity* is defined by a vector of one or more *Activity categories* and an associated goal, with each vector entry representing a day of the week. Examples of Abstract Activities are discussed later in Sect. 13.9.

Each *Abstract Activity* is defined by the Domain Expert via the *Editor of Abstract Activities & Goals* and later stored in the *Catalog of Abstract Activities & Goals*. Each Activity category identifies the kind of activity the user should perform. As an example, the user can receive either a *Cardio* or *Strength* Activity category and so should perform an activity of that kind. More precisely, for each user, the Activity categories are converted to *Concrete Activities* at runtime via the use of the *User Driven Adaptation Manager* and based on the user's preferences. For instance, a cardio Activity category can be instantiated into different *Concrete Activities* such as running, swimming, and walking. Moreover, if an *Abstract Activity* is composed of multiple Activity categories, all or some of type Cardio, they can be converted into different *Concrete Activities*. This implies that users who receive the same User

Process will still be likely to have different *Concrete Activities*, therefore personalizing the experience to the individual user (this is further discussed in Sect. 13.5.2).

The goals associated with an *Abstract Activity* are also important for distinguishing between Abstract Activities, besides converting them into *Concrete Activities*. Two Abstract Activities containing the same vector of Activity categories can be different solely based on their associated goal. More details on the goal model are given in Sect. 13.6.

The *User Process Handler* receives Updates from (1) the *AI Personalization* and the *Editor of Abstract Activities & Goals* in order to send User Processes to their associated users. The AI Personalization Updates the *User Process Handler* every time a user moves from one cluster to another, while the *Editor of Abstract Activities & Goals* Updates it every time new clusters are analyzed by the Domain Expert (along with the new associated User Process). These updates guarantee that the *User Process Handler* remains up to date about the User Processes and their associated users.

AI Personalization sends an Update to the *Clustering History* component whenever a change occurs in the clusters. The *AI Personalization* component uses the CluStream-GT algorithm to cluster users into clusters in a real-time and online fashion [10]. It receives the input data from the e-Health app (see Collected Data in Fig. 13.2). More than one instance of CluStream-GT can be running at the same time. In fact, there is one instance per category of data. For example, if the e-Health app is recording both ecological momentary assessment [28] and biometric data, one for the purpose of monitoring **mood** and the other for **fitness**, there will be two running instances of the algorithm.

AI Personalization Adaptation is in charge of monitoring the evolution of clusters and detecting if any change occurs. Examples include the merging of two clusters or the generation of a new one. To do so, it periodically Queries the *Clustering History* database. If one or more new clusters are detected, this component will Notify both the Development Team and the Domain Expert. The Domain Expert will examine the new information and add the appropriate User Process to the *Catalog of Abstract Activities & Goals* via the dedicated editor. In turn, the Development Team is notified just as a precaution so that it can verify if the new cluster is not an anomaly. The specifics of the corresponding MAPE loop are described in Sect. 13.5.1.

The role played by AI via the CluStream-GT algorithm is relevant in our RA as it strongly supports both personalization and self-adaptation, thus guaranteeing a continuous user engagement that is crucial in e-Health apps. Specifically, personalization is achieved by clustering the users based on their preferences and their physical and mental condition. This supports the RA in assigning appropriate *User Processes* to each user, and further adapt them to continuously cope with the current status of the user and by doing so better addressing social sustainability concerns.

Clustering History is a database of all the clusters created by the *AI Personalization* component. For each cluster it keeps all of the composing micro-clusters with all of their contained information.

Editor of Abstract Activities & Goals allows the Domain Expert to create and modify *Abstract Activities* (and their associated goals) and to combine them as User Processes. This is achieved via a web-based interactive UI and the editor's ability to Query the *Catalog of Abstract Activities & Goals*. It is also the editor's responsibility to update the *User Process Handler* if any new User Process has been created and is currently in use.

Catalog of Abstract Activities & Goals is a database of all User Processes that the Domain Expert has created for each unique current and past cluster. When a new cluster is defined, the Domain Expert can assign to it an existing User Process from this catalog, or create a new one and store it.

Catalog of Supported Mobile Applications is a database containing the metadata needed for interacting with supported third-party mobile apps installed on users' devices. This database stores information such as the specific types of Android intents (and their related extra data) needed for launching each third-party app, the data it produces after a tracking session, etc. Indeed, our e-Health app does not provide any specific functionality for executing the activities suggested to the user (e.g., running, swimming); rather, it brings up third-party apps (e.g., Strava[1] for running and cycling, Swim.com[2] for swimming) and collects the data produced by the apps after the user performs the physical activities. The main reasons for this design decision are: (1) we do not want to disrupt the users' habits and preferences in terms of apps used for tracking their activities and, (2) we want to *build on* existing large user bases; we do not want to reinvent the wheel by reimplementing functionalities already supported by development teams with years-long experience.

Whenever the e-Health app evolves by supporting new applications (or no longer supporting certain applications), the *Catalog of Supported Mobile Applications* Updates, through the *Datastore*, the *Third-party Applications Manager*. The *Third-party Applications Manager*'s responsibility is to keep the list of supported mobile apps up to date and provide the corresponding metadata to the *User Driven Adaptation Manager* and the *Environment Driven Adaptation Manager*, when needed.

The e-Health app and back-end communicate via the Internet. Specifically, the communication from the e-Health app to the back-end is REST based and it is performed by the *Internet Connectivity Manager*, which is responsible for sending the Collected Data to the *AI Personalization* component in the back-end. Communication from the back-end to the e-Health app is performed by the *User Process Handler*, which is in charge of sending the User Process to the e-Health app via push notifications.

[1]https://www.strava.com/

[2]https://www.swim.com/

13.5 Components Supporting Self-Adaptation

The RA has five components used for self-adaptation. To accomplish its responsibilities, each of these components implements a MAPE loop.

13.5.1 AI Personalization Adaptation

The main goal of the AI Personalization Adaptation is to keep track of the clusters evolution and to enable the creation of new User Processes. It does it through its MAPE loop depicted in Fig. 13.3.

During its Monitor phase, the AI Personalization Adaptation monitors the macro-clusters. In its Analyze phase it determines if there are changes in the monitored macro-clusters. To do so, the AI Personalization Adaptation periodically queries the Clustering History database. It compares the current clusters with the previously saved ones. If any of the current ones are significantly different, then the AI Personalization Adaptation enters its Plan phase. The Plan phase gathers the IDs of the users and macro-clusters involved in these significant changes. Since this change involves the need for the creation of new User Processes for all of the users belonging to the new clusters, the Domain Expert must be involved in this adaptation. To achieve this we have exploited the type of adaptation described in [29], which considers the involvement of humans in MAPE loops. In particular, in [29] the authors describe various cases in which a human can be part of a MAPE loop. AI Personalization Adaptation falls under what the authors refer to as: 'System Feedback (Proactive/foreground)'. This type of adaptation is initiated by the system which may send information to the human. The human (i.e., Domain Expert) uses this information to execute the adaptation (by creating the new User Processes necessary). To send the needed information to the Domain Expert, AI Personalization Adaptation takes the gathered knowledge from the Plan phase and gives it to Execute. Execute notifies (Fig. 13.2) both the Development Team and the Domain Expert about the detected cluster change(s) and relays the gathered information.

To determine if a cluster is significantly different from another, we use a parameter *delta*. This parameter is set by the Development Team at design time and determines how different the stored information of one cluster has to be from another

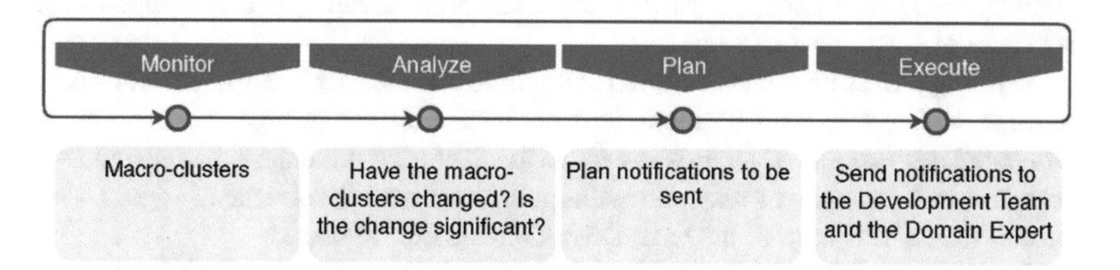

Fig. 13.3 AI Personalization Adaptation MAPE loop

one to identify them as unique. The Development Team is notified as a precaution, to double check the change and verify that no errors occurred.

13.5.2 User Driven Adaptation Manager

The main responsibility of the User Driven Adaptation Manager is to receive the User Process from the back-end and convert the contained Abstract Activities into Concrete Activities. A Concrete Activity represents a specific activity that the user can perform, also with the support of smart objects and/or corresponding mobile apps. As an example, *running* is a concrete activity during which the user can exploit a smart bracelet to monitor their cardio rate as well as a dedicated mobile app to measure the run distance and the estimated burned calories. A Concrete Activity is designed as a class containing multiple attributes that is stored on the smartphone. The attributes are:

- **Selectable**: This is True if the User Driven Adaptation Manager or the Environment Driven Adaptation Manager can choose this Concrete Activity, when dynamically refining Abstract Activities; False otherwise. It is set by the user via the user preferences.
- **Location**: This specifies if the activity is performed indoors or outdoors. This attribute is used by the Environment Driven Adaptation Manager to choose the appropriate Concrete Activity according to weather conditions (see Sect. 13.5.5).
- **Activity category**: This defines what type of category the Concrete Activity falls under; e.g., for a fitness activity, it specifies a cardio or strength training.
- **Recurrence**: This tracks how many times the user has performed the Concrete Activity in the past. It allows the User Driven Adaptation Manager to have a preference ranking system within all the selectable Concrete Activities.

For each user, the Concrete Activities are derived from their preferences stored in the Third-party Applications Manager. During its nominal execution, the User Driven Adaptation Manager is in charge of refining the Abstract Activities in the User Process into Concrete ones. To do this, it queries the Third-party Applications Manager and exploits its knowledge of the Concrete Activities and their attributes. After completing the task, the User Driven Adaptation Manager presents the personalized User Process to the user as a schedule, where each slot in the vector of Activity categories corresponds to a day. Therefore creating a personalized user schedule of Concrete Activities.

Refining a User Process is required every time that the user is assigned with a new process, to keep up with its improvements and/or cluster change. To this aim, a dynamic User Process adaptation is needed to adapt at runtime the personalized user schedule in a transparent way and without a direct user involvement. Figure 13.4 depicts the MAPE loop of the User Driven Adaptation Manager.

Once it accomplishes its main task of refining the User Process, the User Driven Adaptation Manager enters the Monitor phase of its MAPE loop by monitoring the

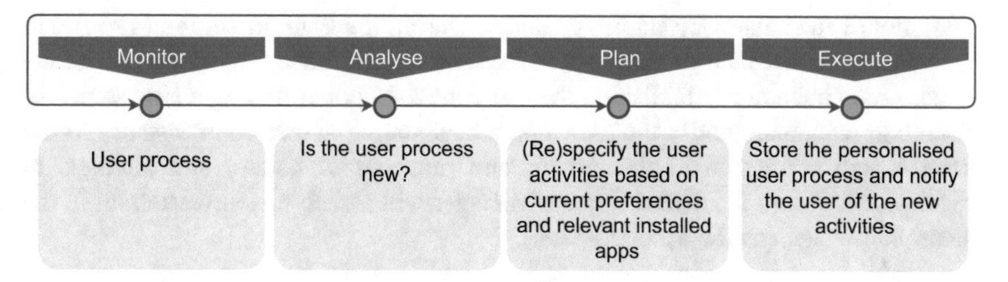

Fig. 13.4 User Driven Adaptation Manager MAPE loop

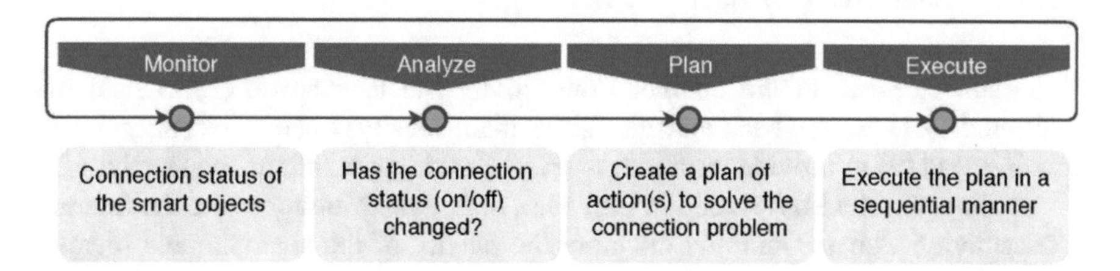

Fig. 13.5 Smart Objects Manager MAPE loop

User Process. The Analyze phase receives the monitored User Process from Monitor. Analyze is now responsible to determine if the user has been assigned a new User Process. If so, the User Driven Adaptation Manager converts the Abstract Activities in this new User Process into Concrete ones, taking into account the user preferences. It makes this conversion by finding suitable Concrete activities during the Plan phase. As all of the Abstract Activities have been matched with a corresponding Concrete activity, the Execute phase makes the conversion, storing this newly created personalized User Process and notifying the user about the new activity schedule.

13.5.3 Smart Objects Manager

This component aims to maintain the connection with the user's smart objects and, if not possible, find alternative sensors to make the e-Health app able to continuously collect user's data and, thus, to perform optimally. To this aim, it implements a MAPE loop, shown in Fig. 13.5, supporting the dynamic adaptation at the architectural level of the smart objects.

The Monitor phase is devoted to the run-time monitoring of the connection status with the smart objects. Connection problems can be due to either the smart objects themselves, which can be out of battery, or to missing Internet, Bluetooth or Bluetooth low energy connectivity. The Analyze phase is in charge of verifying the current connection status (received by Monitor) and see if the connection status with any of the smart objects has changed. During the Plan phase, the MAPE will create a sequential plan of actions that the Execute will have to perform. All of the actions are aimed at reestablishing the lost connection or at finding a new source of data (e.g. reconnect, notify the user, find a new source of data). For instance, if the smart-watch connected to the smartphone runs out of battery and attempts to reconnect to it fail, the Smart Objects Manager will switch to sensors inbuilt in the smartphone (such as the accelerometer).

13.5.4 Internet Connectivity Manager

The main purposes of the Internet Connectivity Manager are to (1) to send the Collected Data to the back-end and store them locally when the connection is missing, and (2) to provide resilience to the e-Health app's Internet connectivity.

As shown in the MAPE loop in Fig. 13.6, during the Monitor phase the Internet Connectivity Manager runtime monitors the quality of the smartphone's Internet connection.

Analyze is then in charge of detecting whether a significant connection quality alteration is taking place. If so, the Internet Connectivity Manager enters the Plan phase and it plans for an alternative. The alternative can include switching the connection type or storing the currently collected data locally on the smartphone. As a new connection can be established, the component sends the data to the back-end to be used by the AI Personalization.

13.5.5 Environment Driven Adaptation Manager

One of the objectives of the e-Health app is keeping users constantly engaged, to ensure that they execute their planned schedule of activities. To this aim, the Environment Driven Adaptation Manager plays an important role, which is essentially supported by its MAPE loop, depicted in Fig. 13.7.

The purpose of this component is to constantly check whether the currently scheduled Concrete Activity best matches the runtime environment (i.e., weather

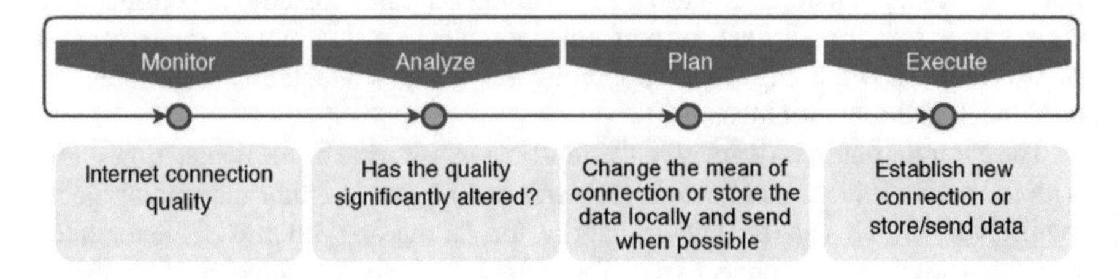

Fig. 13.6 Internet Connectivity Manager MAPE loop

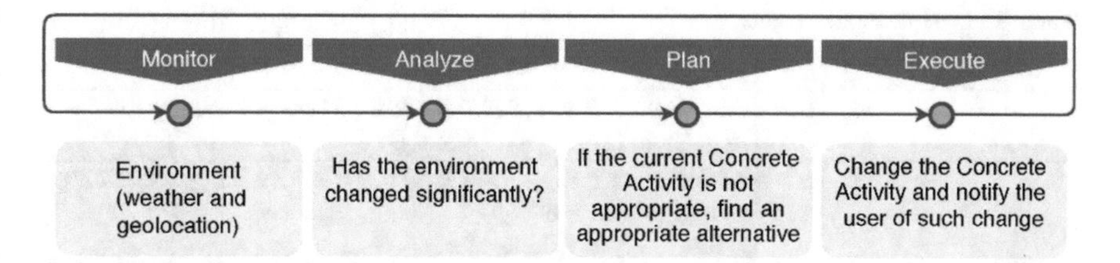

Fig. 13.7 Environment Driven Adaptation Manager MAPE loop

conditions) the user is located in. To do so, the Environment Driven Adaptation Manager monitors in runtime the user's environment. The Monitor phase periodically updates the Analyze phase by sending the environment data. This phase establishes if the environment significantly changed. If so, it triggers the Plan phase that verifies whether the currently planned Concrete Activity is appropriate for the user's environment. If it is not, it finds an appropriate alternative and sends the information to Execute. Execute swaps the planned Concrete Activity with the newly found one and notifies the user of this change.

13.6 Goal Model

Goals have been used in many areas of computer science for a long time. For instance, in AI planning they are used to describe desirable states of the world (e.g., [30]) whereas in goal-oriented requirements engineering (GORE [31]) they are used to model nonfunctional requirements (e.g., [32]). Goals have also been used in self-adaptive systems to express the desired runtime behavior of systems execution [27, 33]. More recently, goals are used to model personal objectives at the user level [34], as done in our work.

As stated before, a User Process is composed of one or more Abstract Activities, each defined as a vector of Activity categories with an associated *goal*. For each cluster, the Domain Expert defines its User Process and corresponding goals, through the *Editor of Abstract Activities & Goals*.

The syntax of our *goal model* is presented in Table 13.1. A goal of an Abstract Activity, namely, G_a, refers to the type of feature that the Abstract Activity represents (e.g., mood, fitness). At the current stage of our work, we have *mood-based goals*, m_g, and *fitness-based goals*, f_g.

A mood-based goal defines as objective a desirable mood that the user should reach, considering their specific pathology. A mood-based goal can be specified in two different ways: as a numerical value belonging to a given discrete range, such as $1, \ldots, n$, or as a string value belonging to a specific string set, such as *[very sad, sad, neutral, happy, very happy]*. This goal type establishes the target mood that users are expected to reach when performing mood-related activities. Specifically, we use the

Table 13.1 Goal model syntax

G_a:	::=	$m_g \mid f_g$
m_g	::=	**one_of** $STRING\_SET$ $?(FREQ) \mid <or \leq or >or \geq or = value$ **in** [1, . . . ; n] $?(FREQ)$
f_g	::=	$INTENSITY_{time}\ ?(FREQ) \mid INTENSITY_{value}\ ?(FREQ) \mid <or \leq or >or \geq or = f_g \mid$ f_g **and** $f_g \mid f_g$ **or** $f_g \mid$ **one_of** $seq\ f_g \mid T \mid \bot$
$INTENSITY_{time}$	::=	$Seconds \mid minutes \mid hours$
$INTENSITY_{value}$	::=	$kcal \mid km \mid step\_count$
$FREQ$	::=	$TIMES$ **per day** $\mid TIMES$ **per week** $\mid TIMES$ **per month**
$TIMES$	::=	[1, . . . ; n] $\forall n \in N$

one_of $STRING\_SET$ construct to allow the Domain Expert to define as goal one mood among the ones listed in the set $STRING\_SET$, for instance in (Eq. 13.1):

$$G_a := m_g \ \textbf{one\_of} \ [neutral, happy, very\ happy] \tag{13.1}$$

When a numerical range is used to describe the user mood, we use relational operators to specify a goal as a value in a subset of the given discrete range. Moreover, for both mood-based goals, the expert can optionally specify the frequency with which the user is asked to register their mood, through the $?\ FREQ$ construct. The frequency can be expressed in terms of $TIMES$ *per day*, *per week*, or *per month*, where $TIMES$ belongs to a discrete range of values, as given in (Eq. 13.2):

$$G_a := m_g \geq 7 \ \textbf{in} \ [1, \ \ldots, 10] \ 3 \ \textbf{per day} \tag{13.2}$$

A mood-based goal m_g *succeeds* if it satisfies the relation expressed by the goal. In the presence of a frequency, instead, the user enters more than one mood. In this case, the mood-based goal succeeds if the average computed among the registered mood satisfies the relation expressed by the goal m_g; it *fails* otherwise.

A fitness-based goal specifies the required *intensity* and *frequency* with which users should perform fitness-related activities. In particular, the goal model provides two constructs to indicate the intensity, namely, $INTENSITY_{time}$ and $INTENSITY_{value}$. The former is used to express the intensity in terms of duration of the activity (e.g., *seconds*, *minutes*, and *hours*). The latter is used to express the intensity in non-time-based terms. Our goal model foresees the use of values such as *kcal*, *km*, and *step_count*. As for mood-based goals, the Domain Expert can optionally specify the frequency with which the user is asked to perform the suggested activities, via the $?\ FREQ$ construct. Relational operators can be used to specify threshold values over intensity-based goals. Moreover, *control-flow constructs*, namely, **and**, **or**, and **one_of**, can also be specified to combine fitness-based goals. These constructs allow us to recursively combine elementary goals, of $INTENSITY_{time}$, $INTENSITY_{value}$, and threshold types, thus to create goals of different complexity. An example is given in (Eq. 13.3):

$$G_a := f_g \geq 1000 \text{ kcal } 1 \textbf{ per day or } f_g > 5 \text{ km} \quad (13.3)$$

A fitness-based goal f_g of type intensity or threshold *succeeds* if the user performs the suggested activities with the required time-based or value-based intensity; it *fails* otherwise. Goals of type **and** and **or** represent combination of goals and they *succeed*, respectively *fail*, as per the rule defined by the involved logical operators. A goal **one_of** *seq* f_g specifies the need for achieving one of the goals in the given sequence. The choice of the goal to target among the available ones can depend on a utility function or a user's choice.

The presented goal model is open and easy to extend. If a new feature different from *mood* and *fitness* is envisaged, it is sufficient to extend the rule related to G_a with a further nonterminal term on the right-hand side of the rule, referring to the new feature, along with one or more associated rules. The ease of use of the goal model, as well as the *Editor of Abstract Activities & Goals* are designed as tools that allow *Domain Experts* to make changes in the tailoring of the app to better meet the interests and needs of the users. This is an important feature of the RA that allows it to better address social sustainability.

13.7 Methodology

As introduced in Sect. 13.2, to design our RA we used the framework and the methodology of Angelov et al. [15]. In Table 13.2 we list all questions for each dimension (i.e., context, goals, and design), with the answers we gave whilst designing our RA and the rationale for each answer.

In the **goal dimension**, the aim of our RA is providing guidelines for the design of personalized and self-adaptive e-Health apps. To the best of our knowledge no RA of this type exists in this domain (**G1**).

In the **context dimension**, our RA is devoted to any organization in the e-Health domain who can benefit from it (**C1**). Particularly, during the design of our RA we have used our collected experience from multiple collaborations with psychologists and e-Health app providers to formulate the requirements needed to be addressed. We were the sole designers of the RA (**C2**). The main objective was to design RA in a way that it can utilize, in the same architecture, relevant techniques needed to achieve both personalization and self-adaptation within this domain (**C3**).

In the **domain dimension**, the main ingredients of our RA are: software components and their connectors, the CluStream-GT algorithm, the MAPE-loops, and the goal model (**D1**). Specifically, the software components and goal model are semi-detailed as they demonstrate implementation feasibility and a clear objective but are not yet implemented. CluStream-GT is detailed as it is previously published and

Table 13.2 RA according to the three dimensions: context, goals, design

Dimension	Values	Rationale
G1: Why is it defined?	Facilitation	Our aim with this RA is to provide guidelines for the design of personalized and self-adaptive e-Health apps.
↓	↓	↓
C1: Where will it be used?	Multiple organizations	Multiple organizations within the e-Health domain.
C2: Who defines it?	Research centers (D),	The RA was designed by the authors who are all researchers.
	User organizations (R), software organizations (R)	Requirements for this RA were derived by collaborations with domain experts and e-Health app providers.
C3: When is it defined?	Preliminary	The algorithms, goal model, and MAPE-loops do not exist in practice yet.
↓	↓	↓
D1: What is described?	Components, algorithms, protocols, etc.	Components, CluStream-GT, MAPE-loops, domain model.
D2: How detailed is it described?	Semi-detailed architecture, detailed algorithms, and aggregated protocols	The goal model and the software components are semi-detailed, CluStream-GT is detailed, and the MAPE-loops are aggregated.
D3: How concrete is it described?	Abstract elements	At the time of design, our RA mainly abstracts from concrete technologies.
D4: How is it represented?	Semi-formal architecture representation and a formal algorithm	The RA is described according to 42010, CluStream-GT is implemented.

Table 13.3 Final match of our RA to one of the five types identified in [15]

	T/V	G1	C1	C2	C3	D1	D2	D3	D4
RA	5.1	X	X	X	X	X	X	X	X

tested work. The MAPE-loops only demonstrate the general communication and are specified at an aggregated level (**D2**). As our RA is described, we mainly abstract from concrete technologies (**D3**); in fact, the majority of the RA is currently presented in a semi-formal manner with the exception of CluStream-GT (**D4**).

In Table 13.3 we present our final match of the RA with respect to the types/variants (T/V) presented by Angelov et al. [15]. In particular, X denotes a match of the architecture values with those in the T/V. As shown, our RA fits one of the architecture variants identified and described by Angelov et al. (specifically variant 5.1); this demonstrates its congruence wrt its context, goals, and design. As stated in [15], if a RA can be classified into one of their identified types it has a better chance of being successful (i.e., "accepted by its stakeholders and used in multiple projects" [15]).

13.8 Viewpoint Definition

This section describes the essential elements of the viewpoint defined to represent Mobile-enabled Self-adaptive Personalized Systems (or MSaPS Viewpoint for short).

We have used it to create the view of our RA for personalized and self-adaptive e-Health Apps as described in Fig. 13.2. It must be noted, however, that the MSaPS Viewpoint is not limited to reference architecture use: one could use it to design specific e-Health mobile-enabled systems, as well as to describe mobile-enabled systems not targeted at e-Health but involving personalization and self-adaptation.

The MSaPS Viewpoint relies on the guidelines provided in the ISO/IEC/IEEE 42010 Standard [35]. Accordingly, after a short description it frames (cf. Table 13.4) the typical stakeholders, their concerns, the meta-model, and the related conforming visual notation. The indication of which stakeholders may have which concerns is further shown in Table 13.5.

13.9 Scenario-Based Evaluation

To evaluate how our RA would cover typical usage scenarios, we used the domain expertise learnt from our industrial collaborations and have defined the example case and associated scenarios described in this section (see Figs. 13.8 and 13.9). For each scenario, we challenged how the RA can be used. Throughout the example we use a hypothetical user named Connor and focus on fitness-based goals.

Scenario 1 (Fig. 13.8a). Connor downloads a fitness app that uses our proposed RA. As a first step, he has to input some preferences about the kind of activities he likes the most, complete a questionnaire used to understand his fitness level and give consent for his data to be tracked and used by the app. The fitness app decides on his first weekly schedule of activities.

This is a default schedule created by the Domain Expert, in accordance with the information provided by Connor. The default schedule, represented as an Abstract Activity, is adapted by the User Driven Adaptation Manager in accordance with Connor's preferences and supported third-party applications. *This scenario highlights how our RA supports both user level adaptation (where the Abstract Activities assigned to Connor are adapted by the User Driven Adaptation Manager) and architecture level adaptation (where the Third-party Applications Manager realizes the Concrete Activities by dynamically integrating the specific apps Connor uses on his mobile device).*

Scenario 2 (Fig. 13.8b). During the first week Connor performs the Concrete Activities assigned to him. This first week is needed by the app to gather enough data from Connor so that the AI Personalization can determine to which macro-cluster Connor belongs. After successfully clustering Connor, the AI Personalization sends an update to the User Process Handler, which is now able to send the appropriate

Table 13.4 Elements of the MSaPS viewpoint

Element	Description
Viewpoint description	This viewpoint captures the essential architectural and contextual elements supporting the design of mobile-enabled self-adaptive and personalized systems
Typical stakeholders	Domain experts, software architects, members development teams, user
Concerns	C1: How to extend a mobile app with personalization and self-adaptation? C2: How to integrate external smart objects and environmental information flows? C3: How to integrate Domain Expert knowledge into the mobile app's personalization? C4: How to integrate third-party apps as part of the mobile app's personalization? C5: What are the components of MAPE loops and how do they interact? C6: Where is the user data stored?
Meta-model	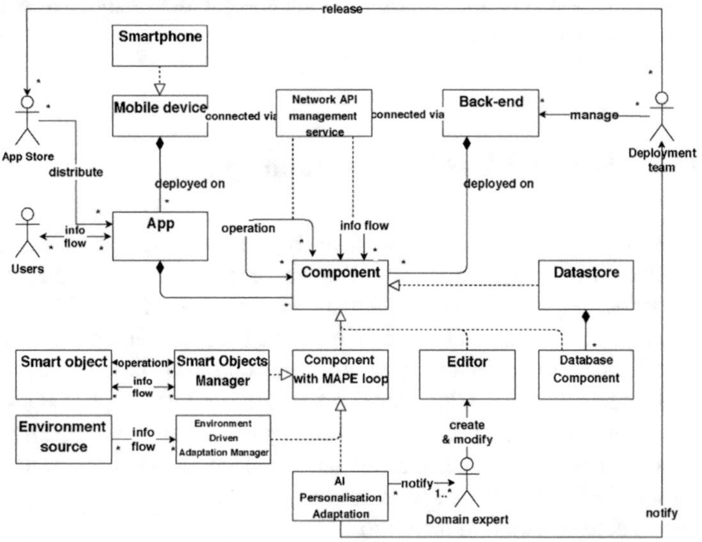 release Smartphone Mobile device connected via Network API management service connected via Back-end manage App Store distribute Deployment team deployed on deployed on Users info flow App operation info flow Component Datastore Smart object operation info flow Smart Objects Manager Component with MAPE loop Editor Database Component Environment source info flow Environment Driven Adaptation Manager create & modify AI Personalisation Adaptation notify 1..* Domain expert notify
Conforming notation	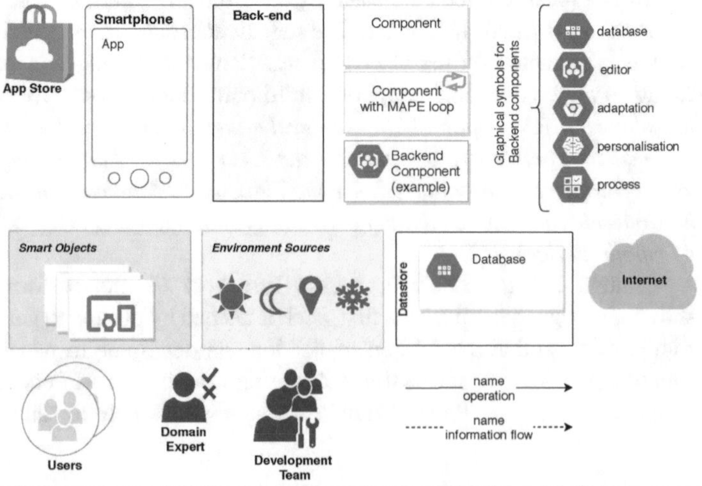 App Store Smartphone App Back-end Component Component with MAPE loop Backend Component (example) Graphical symbols for Backend components database editor adaptation personalisation process Smart Objects Environment Sources Datastore Database Internet Users Domain Expert Development Team name operation name information flow

Table 13.5 Stakeholders and related concerns

Concerns/stakeholders	User	Domain expert	Developer	Software architect
C1: Extend App w/Pers/Adapt			✓	✓
C2: Integrate External Elements			✓	✓
C3: Integrate		✓	✓	✓
C4: Integrate Apps			✓	✓
C5: MAPE Interactions			✓	
C6: User Data	✓			

User Process to Connor. By querying the Third-party Applications Manager, the Abstract Activity is adapted by the User Driven Adaptation Manager into appropriate Concrete Activities. Like with the default schedule, the two *Cardio* entries are converted into running, whilst the newly given *Strength* one is converted into weight lifting. Furthermore, the new goal he receives is more challenging. *This scenario illustrates the same levels of adaptation as scenario 1, completed by the same components. Additionally, the user level adaptation is further personalized by clustering Connor and the User Process Handler sending him his cluster-related User Process.*

Scenario 3 (Fig. 13.8c). On Monday Connor goes running as suggested by the app. Whilst he is running outdoors, both the Wi-Fi and 4G have no connection. The Internet Connectivity Manager detects this and so decides to store the data locally. When Connor gets back home after completing his run, the Wi-Fi connection is reestablished. Aware of this, the Internet Connectivity Manager sends the locally stored Collected Data to the back-end. *This scenario illustrates an architectural level adaptation—performed by the Internet Connectivity Manager by storing the data locally and sending it to the back-end when the Internet connection is reestablished.*

Scenario 4 (Fig. 13.9a). On Wednesday as Connor is in the gym doing the assigned weight training, the connection with the smartwatch is interrupted. The disconnection is detected by the Smart Objects Manager that at runtime reconnects to the smartwatch allowing the app to resume collecting the data about Connor via the smart object. *This scenario describes an example of architectural level adaptation—performed by the Smart Objects Manager when Connor's smartwatch is no longer detected by the app.*

Scenario 5 (Fig. 13.9b). On Friday, the Environment Driven Adaptation Manager detects that the weather forecast predicts rain for the day. As Connor's scheduled Concrete Activity is running, an outdoor activity, the Environment Driven Adaptation Manager needs to make a runtime adaptation. It queries the Third-party Applications Manager for Cardio activities suitable for indoors. As swimming is the best alternative, it switches running with swimming and notifies Connor of the change, saying that the activity will be carried out via the *swim.com* app. *This scenario focuses on both user level adaptation (when the Concrete Activity is adapted by the Environment Driven Adaptation Manager), and architectural level adaptation (when the Third-party Applications Manager accesses the third-party app).*

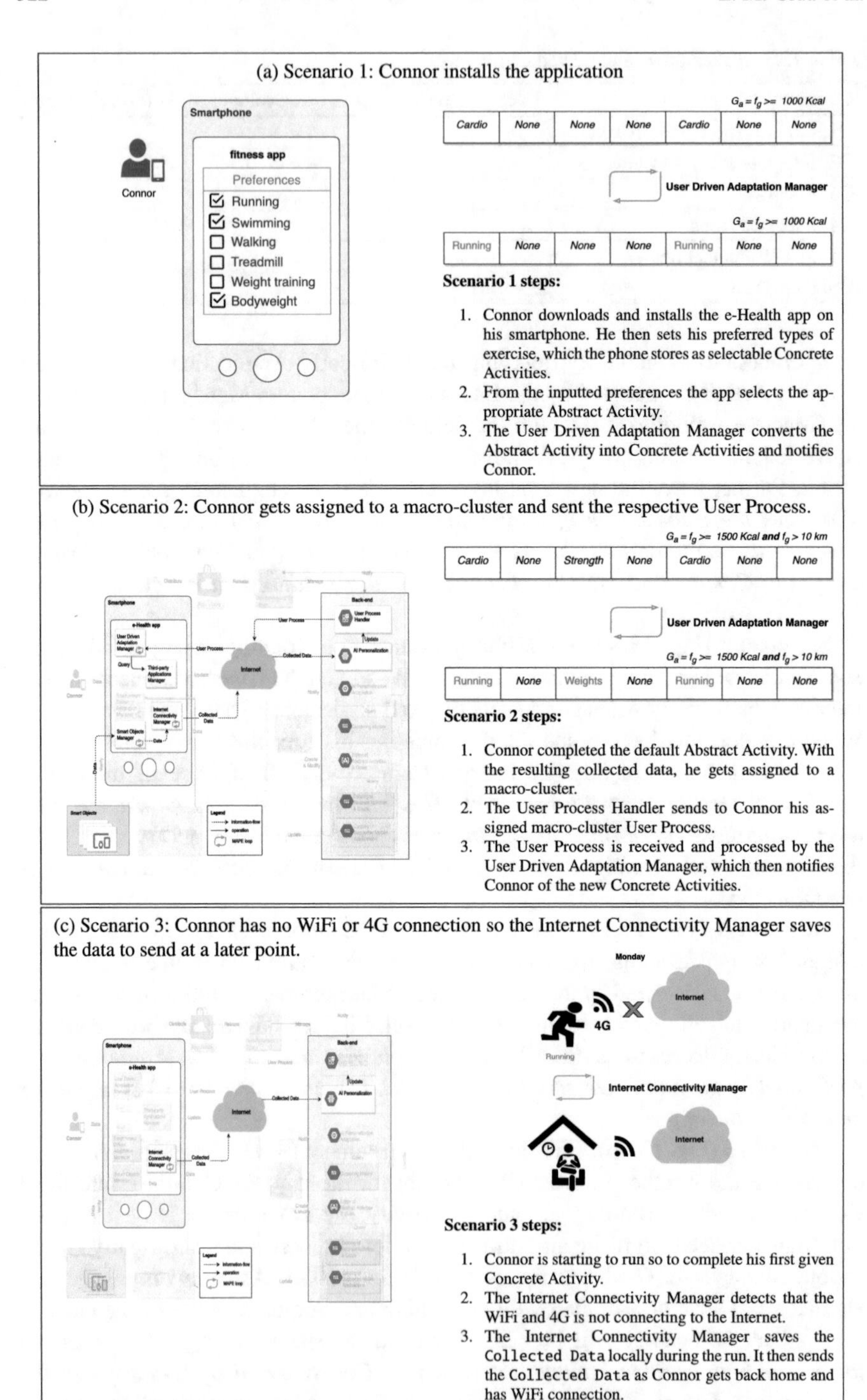

Fig. 13.8 Scenarios 1–3

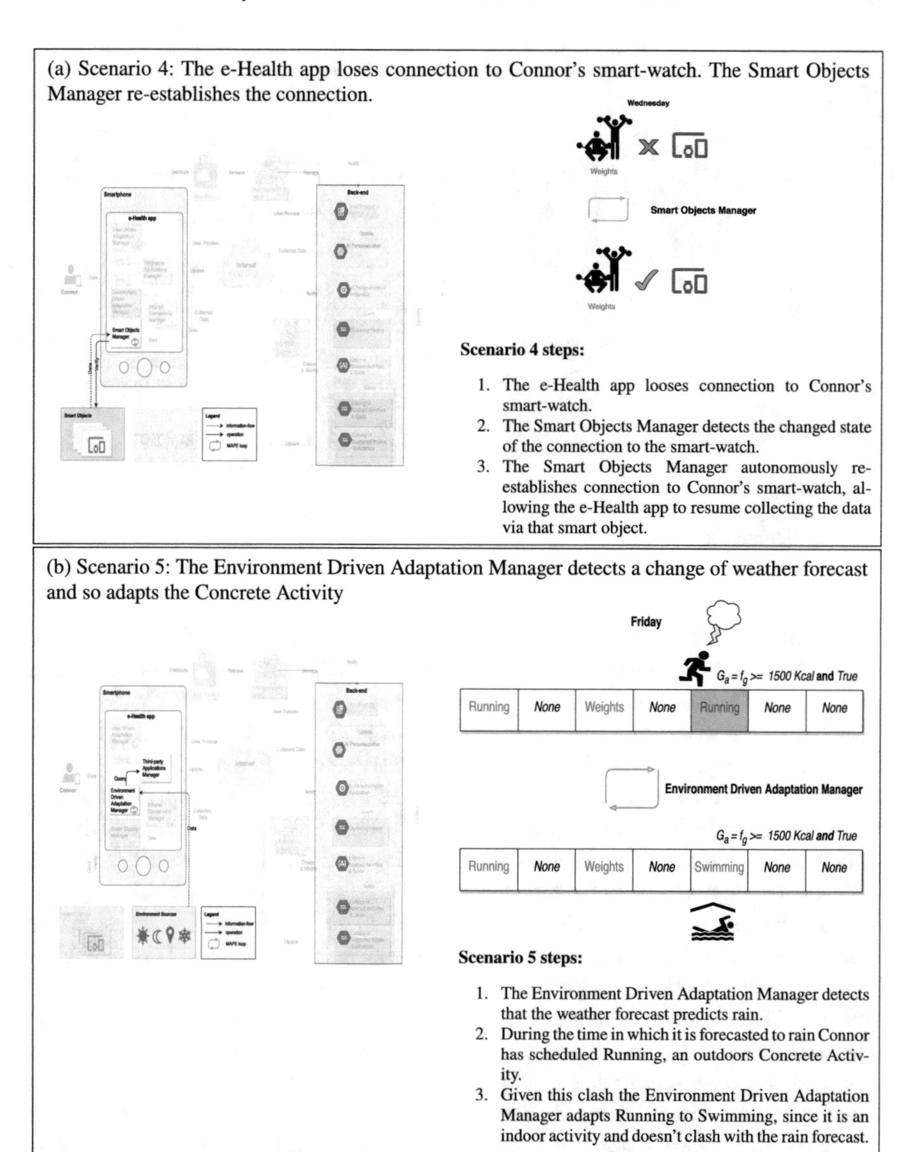

Fig. 13.9 Scenarios 4–6

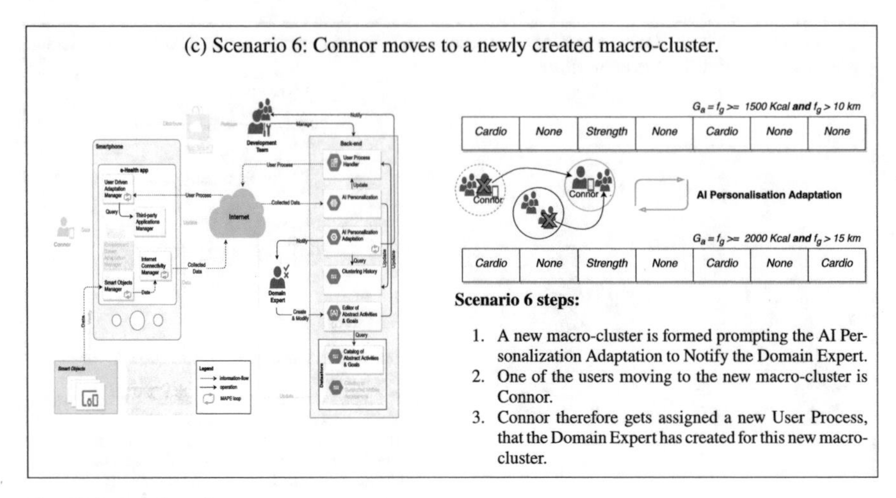

(c) Scenario 6: Connor moves to a newly created macro-cluster.

Fig. 13.9 (continued)

Scenario 6 (Fig. 13.9c). Connor has now finished his second week and has successfully reached his assigned goal. In order to maintain the goal engaging and challenging, Connor's success, along with the success of other users, causes the AI Personalization to create a new macro-cluster for them. As the new macro-cluster is one that has never occurred in the system's history, the AI Personalization Adaptation deems this change significant and so `notifies` the Domain Expert to analyze the new macro-cluster and associate with it a new User Process. The notified Domain Expert makes the new User Process via the web-based Editor of Abstract Activities & Goals. Given the users of the new macro-cluster's success (including Connor), the Domain Expert makes the User Process goal more challenging, increasing the amount of k*cal* to 2000 and the k*m* to 15 (as shown in the figure). This new User Process is sent to the members of the new macro-cluster via the User Process Handler. *This scenario illustrates all three levels of adaptation: the cluster level adaptation to the new macro-cluster performed by the AI Personalization Adaptation, the user level adaptation performed by the User Driven Adaptation Manager when adapting a new User Process, and the architectural level adaptation by associating third-party apps to Concrete Activities performed by the Third-party Applications Manager.*

13.10 Discussion

It is important to note that our RA is *extensible enough to support other domains* beyond fitness and mood. Specifically, the goal model has been designed such that supporting an additional domain can be achieved by adding (1) a new nonterminal term in the root rule G_a and (2) one or more rules describing the goal within the new domain. Also, many of the existing rules (e.g., $FREQ$) are generic enough to be reused by newly added rules. On the client side no changes are required, whereas the only components which may need to be customized to a new application domain are: (1) the Editor of Abstract Activities & Goals, so that it is tailored to the different domain experts and the extended goal model, and (2) the Catalog of Supported Mobile Applications, so that it now describes the interaction points with different third-party apps.

Abstract Activities allow Domain Experts to define *incremental goals* spanning over the duration of the whole User Process. In addition, User Processes are defined at the cluster level (potentially including thousands of users) and can cover large time spans (e.g., weeks or months). Those features make the operation of the RA sustainable from the perspective of Domain Experts, who are not required to frequently intervene for defining new goals or User Processes. Furthermore, these features make the apps adopting our RA socially sustainable on multiple levels. The cluster-level-defined User Processes allow for tailoring to a "community" of similar users, empowering them to achieve a better life. On an individual level, the app fine-tunes the User Processes to better suit the user's needs and interests; this allows the individual user to better achieve their goals both in the immediate and in the systemic (as defined in [36]). Lastly, these features allow for the larger group of users utilizing this RA to reach the same level of health benefits, as the interventions have been specifically tailored for them for this goal.

Through the conversion from Activity Categories to Concrete Activities, which takes place during the dynamic Abstract Activities refinement, we accommodate both *Type-to-Type* adaptation (e.g., from the *Cardio* Activity Category to the *Running* Concrete Activity) and the most common *Type-to-Instance* adaptation (e.g., by using the Strava mobile app as an instance of the *Running* Concrete Activity). Similarly, a *Type-to-Type* adaptation is reported by Calinescu et al. [37] presenting an approach where elements are replaced with other elements providing the *same* functionality but showing a superior quality to deal with changing conditions (e.g., dynamic replacement of service instances in service-based systems). In our approach, however, we go beyond, by replacing activities with others providing *different* functionality to deal with changing conditions. To the best of our knowledge, this adaptation type is uncommon in self-adaptive architectures, despite quite helpful.

The components of the RA running on the smartphone can be deployed in two different ways, each leading to a different business case. Firstly, those components can be integrated into an existing e-Health app (e.g., Endomondo[3] for sports tracking) so as to provide personalization and self-adaptation capabilities to its services. In this case the development team of the app just needs to deploy the client-side components of the RA as a third-party library, suitably integrate the original app with the added library, and launch the server-side components. The

[3]https://www.endomondo.com/

second business case regards the creation of a new meta-app integrating the services of third-party apps, similarly to what apps like IFTTT[4] do. In this case, the meta-app makes an extensive usage of the Third-party Applications Manager component and orchestrates the execution of the other apps already installed on the user device.

Finally, we are aware that our RA is responsible for managing highly sensitive user data, which may raise severe *privacy* concerns. In order to mitigate potential privacy threats, the communication between the mobile app and the back-end is TLS-encrypted and the payload of push notifications is encrypted as well, e.g., by using the Capillary Project [38] for Android apps, which supports state-of-the-art encryption algorithms, such as RSA and Web Push encryption. Eventually, we highlight that, according to the privacy level required by the Development Team, the components running in the back-end can be deployed either on premises or on the Cloud, e.g., by building on public Cloud services like Amazon AWS and executing them behind additional authentication and authorization layers.

13.11 Conclusions and Future Work

In this paper we presented a RA for e-Health mobile apps. Its goal is to combine AI-based personalization and self-adaptation. The RA achieves self-adaptation on three levels: (1) adaptation to the users and their environment, (2) adaptation to smart objects and third-party applications, and (3) adaptation according to the data of the AI-based personalization, ensuring that users receive personalized activities that evolve with the users' runtime changes in behavior. This work emphasizes how personalization and self-adaptation within the e-Health domain can be beneficial in addressing social sustainability. By tailoring user interventions we empower mobile app developers to better help their users in achieving better physical and mental health; this leads to increased support for the community of people who suffer from mental and physical illness and are working on increasing their health. The RA therefore achieves what is defined as the core principal of social sustainability in the realm of software-intensive systems. As future work we are realizing a prototype implementing the RA and designing a controlled experiment to evaluate its effects on user behavior and performance at runtime.

References

1. Williams PAH, McCauley V (2013) A rapidly moving target: conformance with e-health standards for mobile computing. In: 2nd Australian eHealth Informatics and Security Conference

[4]https://ifttt.com/

2. Global Industry Analysts, I (2019) mhealth (mobile health) services – market analysis, trends, and forecasts. https://tinyurl.com/rbvdtc3
3. Paschou M, Sakkopoulos E, Sourla E, Tsakalidis A (2013) Health internet of things: metrics and methods for efficient data transfer. Simul Model Pract Theory 34:186–199
4. Fling B (2009) Mobile design and development: Practical concepts and techniques for creating mobile sites and Web apps. O'Reilly Media, Inc.
5. Fan H, Poole MS (2006) What is personalization? Perspectives on the design and implementation of personalization in information systems. J Organ Comput Electron Comm 16 (3–4):179–202
6. Grua EM, Malavolta I, Lago P (2019) Self-adaptation in mobile apps: a systematic literature study. In: IEEE/ACM 14th International Symposium on Software Engineering for Adaptive and Self-Managing Systems (SEAMS). pp 51–62
7. Weyns D (2017) Software engineering of self-adaptive systems: an organised tour and future challenges. In: Handbook of Software Engineering
8. Yang Z, Li Z, Jin Z, Chen Y (2014) A systematic literature review of requirements modeling and analysis for self-adaptive systems. In: International Working Conference on Requirements Engineering: Foundation for Software Quality. Springer, pp 55–71
9. IBM (2006) An architectural blueprint for autonomic computing. Technical report. IBM
10. Grua EM, Hoogendoorn M, Malavolta I, Lago P, Eiben A (2019) Clustream-GT: Online clustering for personalization in the health domain. In: IEEE/WIC/ACM International Conference on Web Intelligence. ACM, pp 270–275
11. Grua EM, De Sanctis M, Lago P (2020) A reference architecture for personalized and self-adaptive e-health apps. In: Software Architecture: 14th European Conference, ECSA 2020 Tracks and Workshops, L'Aquila, Italy, 14–18 September 2020, Proceedings. Springer, pp 195–209
12. Volpato T, Oliveira BRN, Garcés L, Capilla R, Nakagawa EY (2017) Two perspectives on reference architecture sustainability. In: Proceedings of the 11th European Conference on Software Architecture: Companion. ACM, pp 188–194
13. Kim KJ, Ahn H (2004) Using a clustering genetic algorithm to support customer segmentation for personalized recommender systems. In: International Conference on AI, Simulation, and Planning in High Autonomy Systems. Springer, pp 409–415
14. Grua EM, Hoogendoorn M (2018) Exploring clustering techniques for effective reinforcement learning based personalization for health and wellbeing. In: 2018 IEEE Symposium Series on Computational Intelligence (SSCI). IEEE, pp 813–820
15. Angelov S, Grefen P, Greefhorst D (2012) A framework for analysis and design of software reference architectures. Inf Softw Technol 54(4)
16. Lago P, Verdecchia R, Fernandez NC, Rahmadian E, Sturm J, van Nijnanten T, Bosma R, Debuysscher C, Ricardo P (2020) Designing for sustainability: lessons learned from four industrial projects. In: Environmental Informatics – Sustainability aware digital twins for urban smart environments (EnviroInfo). Springer
17. (2019) The industrial internet of things volume G1: reference architecture. Industrial Internet Consortium. https://bit.ly/2talimM
18. Bassi A, Bauer M, Fiedler M, Kramp T, van Kranenburg R, Lange S, Meissner S (2016) Enabling things to talk: designing IoT solutions with the IoT architectural reference model, 1st edn. Springer
19. Bauer M et al (2013) IoT reference architecture. In: enabling things to talk: designing IoT solutions with the IoT architectural reference model
20. Fremantle P (2015) A reference architecture for the internet of things. WSO2 White paper. https://bit.ly/2RMzCft
21. Feljan AV, Mohalik SK, Jayaraman MB, Badrinath R (2015) SOA-PE: a service-oriented architecture for planning and execution in cyber-physical systems. In: 2015 International Conference on Smart Sensors and Systems (IC-SSS). pp 1–6

22. Mohalik SK, Narendra NC, Badrinath R, Le D (2017) Adaptive service-oriented architectures for cyber physical systems. In: IEEE Symposium on Service-Oriented System Engineering, SOSE. pp 57–62
23. de Morais Barroca Filho I, Junior GSA, Batista TV (2019) Extending and instantiating a software reference architecture for iot-based healthcare applications. In: Int. Conf. on Computational Science and Its Applications. pp 203–218
24. Mizouni R, Matar MA, Al Mahmoud Z, Alzahmi S, Salah A (2014) A framework for context-aware self-adaptive mobile applications SPL. Expert Syst Applic 41(16):7549–7564
25. Lopez FS, Condori-Fernández N (2016) Design of an adaptive persuasive mobile application for stimulating the medication adherence. In: International Conference on Intelligent Technologies for Interactive Entertainment. Springer, pp 99–105
26. Kim HK (2013) Architecture for adaptive mobile applications. Int J Bio-Sci Bio-Technol 5 (5):197–210
27. Bucchiarone A, Lluch-Lafuente A, Marconi A, Pistore M (2009) A formalisation of adaptable pervasive flows. In: WS-FM. pp 61–75
28. Shiffman S, Stone AA, Hufford MR (2008) Ecological momentary assessment. Annu Rev Clin Psychol 4:1–32
29. Gil M, Pelechano V, Fons J, Albert M (2016) Designing the human in the loop of self-adaptive systems. In: International Conference on Ubiquitous Computing and Ambient Intelligence. Springer, pp 437–449
30. Dal Lago U, Pistore M, Traverso P (2002) Planning with a language for extended goals. In: Proceedings of the Eighteenth National Conference on Artificial Intelligence and Fourteenth Conference on Innovative Applications of Artificial Intelligence. pp 447–454
31. Mylopoulos J, Chung L, Nixon BA (1992) Representing and using nonfunctional requirements: a process-oriented approach. IEEE Trans Softw Eng 18(6):483–497
32. Santos M, Gralha C, Goulão M, Araújo J (2018) Increasing the semantic transparency of the KAOS goal model concrete syntax. In: Conceptual Modeling – 37th International Conference, ER. pp 424–439
33. Morandini M, Penserini L, Perini A (2008) Towards goal-oriented development of self-adaptive systems. In: 2008 ICSE Workshop on Software Engineering for Adaptive and Self-Managing Systems, SEAMS. pp 9–16
34. Qian W, Peng X, Wang H, Mylopoulos J, Zheng J, Zhao W (2018) MobiGoal: flexible achievement of personal goals for mobile users. IEEE Trans Serv Comput 11(2):384–398
35. International Organization for Standardization (2011) ISO/IEC/IEEE 42010:2011 – Systems and Software Engineering – Architecture Description. Technical report. International Organization for Standardization (ISO)
36. Lago P (2019) Architecture design decision maps for software sustainability. In: 2019 IEEE/ACM 41st International Conference on Software Engineering: Software Engineering in Society (ICSE-SEIS). IEEE, pp 61–64
37. Calinescu R, Weyns D, Gerasimou S, Iftikhar MU, Habli I, Kelly T (2018) Engineering trustworthy self-adaptive software with dynamic assurance cases. IEEE Trans Softw Eng 44 (11):1039–1069
38. Hogben G, Perera M (2018) Project capillary: end-to-end encryption for push messaging, simplified. https://android-developers.googleblog.com/2018/06/project-capillary-end-to-end-encryption.html?m=1

Chapter 14
Human Sustainability in Software Development

Vijanti Ramautar, Sietse Overbeek, and Sergio España

Abstract Human thriving and outsourcing can go hand in hand. This research aims to outline outsourcing approaches for facilitating human thriving by conducting a semi-systematic literature review. We identified three outsourcing approaches that consider corporate social responsibility: impact sourcing, ethical outsourcing, and Fair Trade Software. The aim of this research is to understand the effect of these approaches on marginalized people, and the benefits and challenges for client organizations. The following main conclusions are drawn. First, impact sourcing provides marginalized people with the opportunity to generate an income, to develop themselves professionally, and to build a social circle. In some cases it can generate harmful impacts such as stress. Second, the benefits of impact sourcing for client organizations compared to traditional outsourcing are reduced costs, reduced employee turnover, improved corporate social responsibility, and new chances for growth. Third, ethical outsourcing protects brand image and can improve stakeholder management. However, the extra investments required may reduce competitiveness. Last, Fair Trade Software is a relatively new model, and therefore the benefits and challenges have yet to be assessed. A potential benefit is capacity building by knowledge transfer and network strengthening. Currently some of the biggest challenges are the lack of audits, caused by a lack of resources, and increasing the adoption rate of this outsourcing model.

14.1 Introduction

Human thriving at work is indicated by the joint experience of vitality and learning at work [1]. Despite the fact that traditional outsourcing often attempts to maximize profits while neglecting human needs, many initiatives show that human thriving and outsourcing can go hand in hand. This research aims to outline outsourcing

V. Ramautar (✉) · S. Overbeek · S. España
Department of Information and Computing Sciences, Utrecht University, Utrecht, The Netherlands
e-mail: v.d.ramautar@uu.nl; s.j.overbeek@uu.nl; s.espana@uu.nl

C. Calero et al. (eds.), *Software Sustainability*,
https://doi.org/10.1007/978-3-030-69970-3_14

approaches for facilitating human thriving. The global outsourcing market is growing, and increasingly work is outsourced to outsourcing suppliers who employ marginalized people [2]. Marginalized people are defined as disadvantaged individuals who have few opportunities for employment [3]. Examples of marginalized people are those who face discrimination, those who are poor, and those who live in rural areas [4]. The education level can vary from no education to a university diploma [5]. Therefore, labor that requires few skills (e.g., entering data) can be outsourced to marginalized people, as well as labor that requires advanced skills (e.g., developing software) [5].

There are many motivations for companies to outsource. Some potential benefits are cost savings, access to new expertise and skills, and the chance to focus on core capabilities [6]. Outsourcing companies, which are often referred to as "client organizations," might not always reap the benefits of outsourcing. Traditional outsourcing, meaning outsourcing practices that do not consider corporate social responsibility, focuses on maximizing profits. High staff turnover and poor marketing effect caused by negative publicity about working conditions can result in an increase of the total costs. Therefore, maximizing profits by means of traditional outsourcing might not be the best strategy [5]. The inability to reach the desired goal has led to the development of more ethically and socially responsible outsourcing approaches. These new approaches implement corporate social responsibility (CSR). CSR is a theory that emphasizes that companies should implement policies and practices toward the good of society [7]. One of these CSR-considering outsourcing approaches is impact sourcing. It focuses on the training and hiring of marginalized people [3]. Another approach is ethical outsourcing, in which work standards are imposed on the outsourcing supplier (sometimes referred to as the providing organization) [5, 8]. These new business models of outsourcing can provide marginalized people numerous benefits: an increase in income, the chance to learn new skills, and an increase in social status [5]. These benefits can contribute to human thriving at work. Client organizations can benefit from incorporating ethically and socially responsible approaches as well. Impact sourcing allows client organizations to maintain similar quality at reduced cost [9]. This is essential since client organizations, even those interested in impact sourcing, generally base their decision-making regarding outsourcing on quality and cost [3, 5, 6, 9]. Therefore, we can conclude that incorporating CSR in outsourcing can be beneficial to both marginalized people and client organizations. Corporate social responsibility is becoming more important in outsourcing [10], and impact sourcing accounted for 12% of the outsourcing market in 2014 [2]. The value that incorporating CSR in outsourcing can provide to both marginalized people and client organizations, as well as the rise of incorporating CSR in outsourcing, warrants an improved and comprehensive understanding of the different ways in which it can benefit marginalized people and client organizations.

To discover how client organizations can successfully implement corporate social responsibility in their information technology outsourcing while enabling human thriving of marginalized people, a semi-systematic literature review is conducted. First, research and study selection criteria are identified. The study quality is then assessed by tracing the findings back to the research method. If this cannot be done or if information on the method is lacking, the study is not included. Basic information on the study is collected and the findings from the included research are coded in NVivo. Three main nodes are defined to code the findings: efficacy, benefits, and challenges. Subnodes are created to group related information from different studies. The findings are then grouped and compared. Lastly, overviews of the key findings, the focus of the research (either marginalized people or clients), and the method of publication of the study are created.

The research performed has provided additional proof and detail on the efficacy for marginalized people and the benefits for client organizations. Moreover, we identified that harmful effects on marginalized people can also exist, in contrast to prior literature [11, 12], which states that outsourcing to marginalized people is a win-win situation for both marginalized people and client organizations. We discovered three outsourcing approaches that consider CSR: impact sourcing, ethical outsourcing, and Fair Trade Software. Each approach is elaborated upon by describing the benefits and challenges. Additionally the efficacy of the first approach is discussed. For the remaining two approaches, no literature on the efficacy was found. In the context of this work, efficacy is part of the Soft Systems Methodology (SSM), which offers a structured way to deal with complex problems that involve different stakeholders [13]. The SSM proposes a set of three variables to measure the performance of transformational methods [13], like impact sourcing and ethical outsourcing. The three variables proposed in the SMM are efficacy, which considers whether or not a result is produced by the method; efficiency, which considers the resources required to produce a result; and effectiveness, which considers the degree to which long-term goals are achieved by the method [13]. An initial survey of impact sourcing and ethical outsourcing literature indicated that efficiency and effectiveness are not reported, and thus, these variables are excluded from this research. To understand if there is a business case for client organizations to invest in impact sourcing, ethical outsourcing, or Fair Trade Software, the benefits and challenges for both approaches are determined.

The following sections will discuss the results from the semi-systematic literature review. Section 14.2 identifies the outsourcing approaches that will be discussed throughout this work. In Sect. 14.3 the efficacy, benefits, and challenges related to impact sourcing are stated. The positive and negative aspects of ethical outsourcing are stated in Sect. 14.4. The notion of Fair Trade Software and barriers for applying it can be found in Sect. 14.5. Finally, the limitations, future research, and conclusion are stated in Sect. 14.6.

14.2 Outsourcing Approaches That Consider CSR

Based on the literature found, we identified three outsourcing approaches that consider CSR.

- *Impact sourcing*, sometimes referred to as "social outsourcing" or "developmental outsourcing," is the act of outsourcing to marginalized people who would otherwise have difficulty finding employment [3]. These marginalized people are typically hired and trained by a social enterprise (i.e., an enterprise that has social aims, as well as business aims) [14]. An important type of impact sourcing is rural outsourcing. This is the case when work is outsourced from urban to rural areas [15]. For client organizations outsourcing to rural areas is often cheaper because the average salary in rural labor pools is typically lower than in urban labor pools. This form of impact sourcing increases the employment opportunities for marginalized people in rural areas [16].
- *Ethical outsourcing*, also referred to as "socially responsible outsourcing," occurs when the client organization imposes minimum social and environmental standards on the organization supplying the outsourced service [5, 8]. Successful implementation of such standards ensures compliance to ethical values and prevents unethical practices, such as child labor, slave wages, and workplace abuse. Setting and pursuing these standards mitigates risks associated with a bad reputation as a result of negative CSR [8]. We differentiate between the terms "ethical outsourcing" and "ethical sourcing." This research only discusses ethical outsourcing, since ethical sourcing also encompasses finding suppliers for goods, in which case sourcing relates more to procurement rather than outsourcing.
- *Fair Trade Software* is a form of software development collaboration with teams from both developing and developed countries with a focus on the transfer of knowledge from the teams from developed countries to the teams from developing countries [17, 18]. It is promoted by the Fair Trade Software Foundation (FTSF), a not-for-profit organization whose main value proposition is to ensure that learning and knowledge transfer processes are put in place to stimulate the knowledge economy of developing countries. One study supports this benefit by observing that software development teams in Kenya were able to learn skills such as project management from more experienced, Western software development teams [19].

The following sections deepen into each of the approaches by explaining them more elaborately and analyzing their effects on marginalized people and client organizations.

14.3 Impact Sourcing: Efficacy, Benefits, and Challenges

This section will discuss the efficacy of impact sourcing for marginalized people and the benefits and challenges for client organizations. The results of the literature study on impact sourcing are summarized in Figs. 14.1, 14.2, and 14.3.

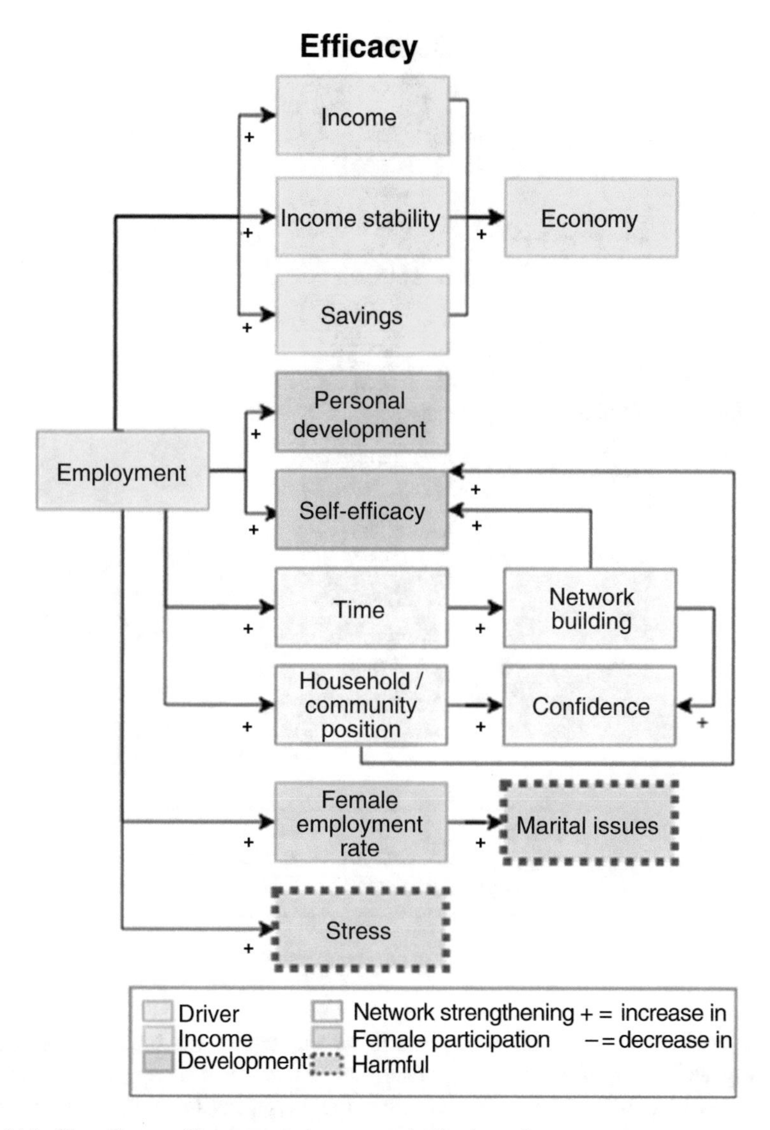

Fig. 14.1 The efficacy of impact sourcing to marginalized people

14.3.1 Efficacy of Impact Sourcing for Marginalized People

Impact sourcing has a positive effect on employment opportunities of marginalized people. Their newfound employment comes with (an increase in) income [14, 20–28], an increase in income stability [14, 23, 28], and an increase in savings [14, 23, 25]. This income is, for instance, spent on education, medical supplies and services, groceries, debt payments, and/or expenses to support family members [23, 24, 27,

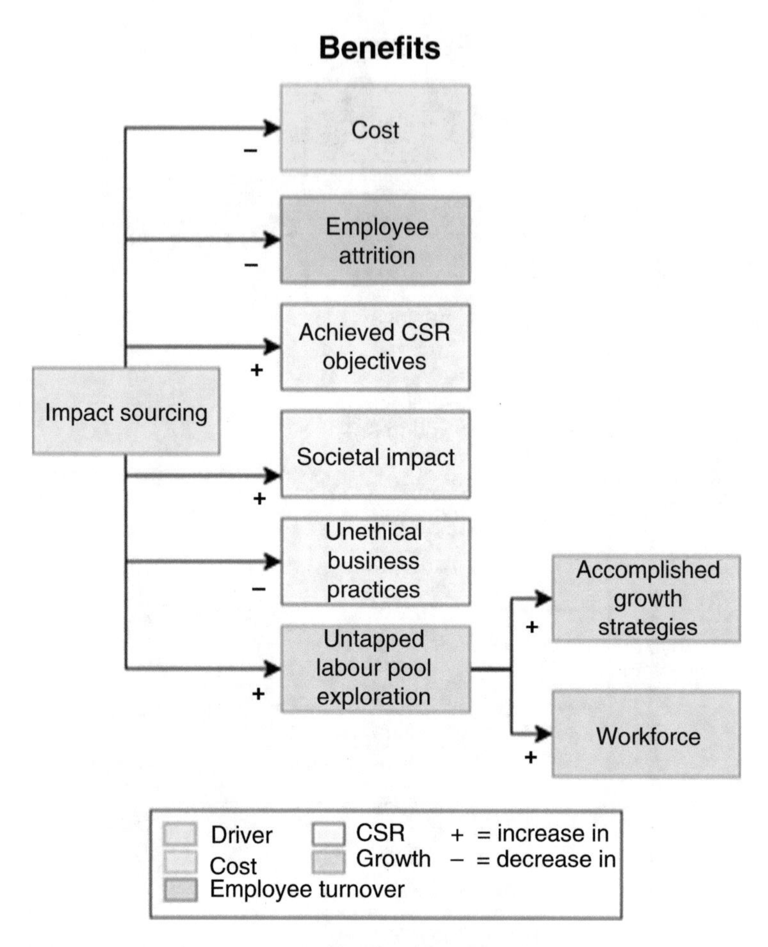

Fig. 14.2 The benefits of impact sourcing for client organizations

28]. An additional benefit of an increase in income can be the improvement of the economy within a community [25], since more money can be spent and invested. Another positive effect of impact sourcing is the possibility for marginalized people to develop themselves and build self-efficacy. Employment affects the self-efficacy of marginalized people mainly through job experience and training given on the job [14, 20–24, 26–30]. Examples of training on the job are ICT training [14], language training [26], and soft skills training [14, 29, 31]. This ability to learn on the job contributes to human thriving at work. Apart from training on the job, employees can spend their (increase in) income on education, to develop more skills and improve job prospects [21, 32]. Income in general is also found to positively affect self-efficacy as it creates financial independence [24, 27]. Gill and Tsai performed a study that focused on the employment of traumatized people. These people received special training to help cope with their shame and lack of sense of self-worth

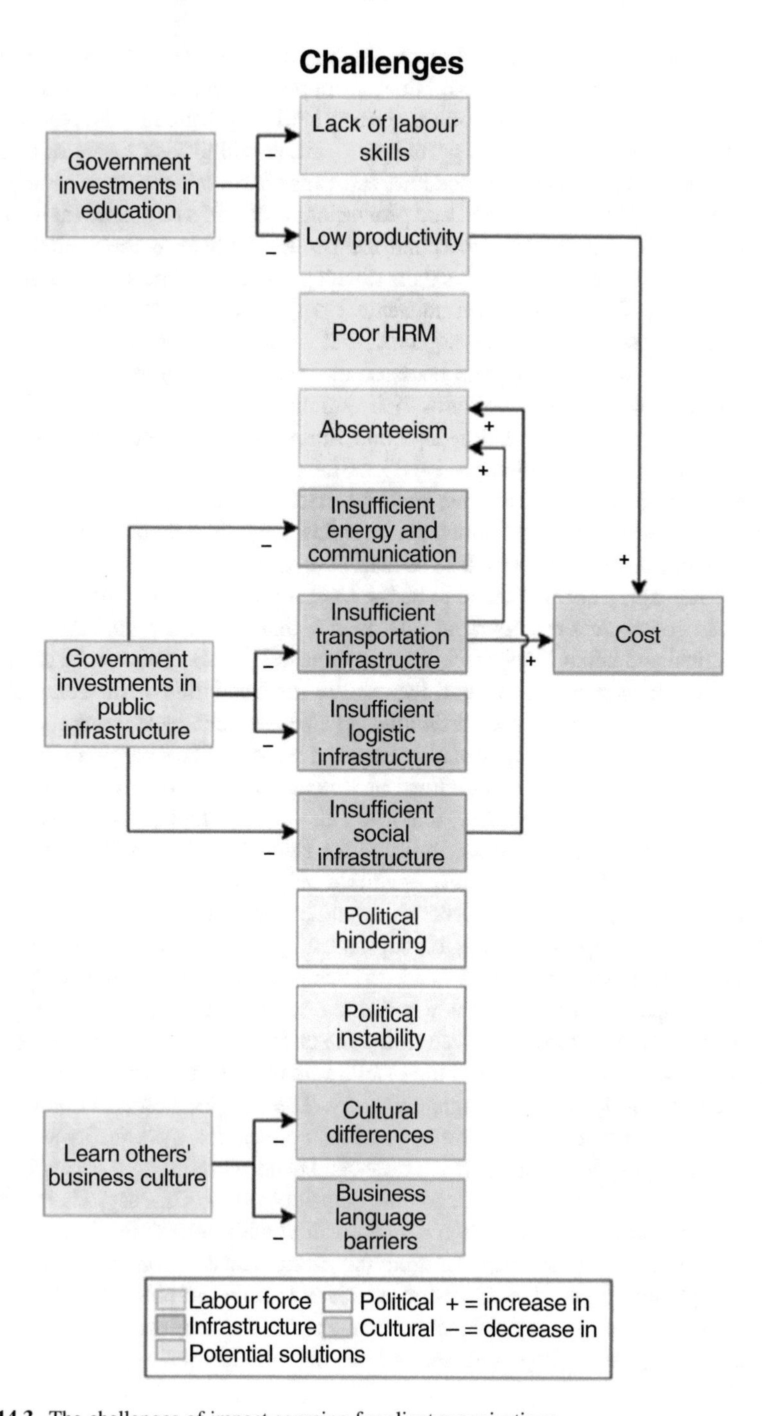

Fig. 14.3 The challenges of impact sourcing for client organizations

[29]. Lacity et al. performed a study focusing on prison inmates. The prison inmates stated that being employed prevented them from performing unlawful conduct [24].

Due to their newly found employment, marginalized people may be perceived as more powerful or receive a more prominent position within their home or community. More convenient working locations can cause them to be able to spend more time at home, and build new social and professional networks or extend their already existing networks. At their job marginalized people are able to form relationships with coworkers, which in turn helps them develop a social and professional network [20, 23, 24, 33]. Within their own household, employees may start to have a more authoritative position due to the fact that they generate income [14, 23, 26]. Employees may also become a source of inspiration or pride for their family members [23, 24] and people within their community [14, 23, 24]. Three studies found that newly formed relationships and improved household positions led to more confidence and self-efficacy [14, 23, 26].

In some cases employment has harmful effects on marginalized people. For instance, when marginalized women start earning more than their husbands or start making money in general, it can lead to marital problems. This is due to the fact that in some cultures men are expected to be the most important or sole income provider [26, 34]. In some cases employment causes an increase in stress and other negative psychological and behavioral effects for marginalized people. This can be caused by inexperience in their new roles and increase in responsibilities. Not being able to effectively communicate these difficulties to management worsens the problems [23, 34, 35]. Sandeep and Ravishankar mention a case where marginalized people struggle with the differences in cultures and values between their own community and the workplace [33]. However, this problem was eventually alleviated by marginalized people introducing their family to their new working environment and to their coworkers. In two cases employees could only work night shifts [34, 35], and this may lead to resistance from other members of the household [35].

The efficacy of impact sourcing to marginalized people is shown in Fig. 14.1. The figure is created based on the factors and effects found in the literature study on the efficacy of impact sourcing. The newly found employment of marginalized people serves as a driver for factors that determine the efficacy. The new job can create an increase in income and saving and the stability thereof. These increases in income-related factors can lead to a stronger economy. The new job can also contribute to personal development and self-efficacy. More convenient working locations can save marginalized people time, which they can spend on building a network. Their new network can help build confidence and contribute to self-efficacy. Their new job can also help marginalized people obtain a position with more authority, which can also contribute to self-efficacy. Employing marginalized women increases the female employment rate; the downside of this is that the number of marital issues can increase due to the employment of women. Another harmful effect can occur when an increase in employment results in an increase in stress.

14.3.2 Benefits of Impact Sourcing for Clients

Impact sourcing can reduce the costs of outsourcing [2, 3, 9, 31, 36, 37]. A study conducted by Everest Group compared the costs between impact sourcing and traditional outsourcing in the same country and found that impact sourcing is cheaper. Cost savings can range wildly from a few percentages up to 40% [2]. This same study compared the costs between an organization impact sourcing and an organization outsourcing to the USA or the UK. The results were cost savings varying from 70% to almost 90% [2]. In impact sourcing, several cost items can turn out to be lower, for example, labor costs, technology costs, operation costs, recruiting costs [9], costs related to employee attrition and turnover, and location costs [2]. When client organizations engage in impact sourcing, outsourcing suppliers were found to have low employee attrition, with four studies citing lower employee attrition in comparison to traditional outsourcing [2, 9, 20, 25]. However, these studies do not provide a clear explanation of why these lower attrition and turnover rates occur. Two studies provide possible explanations: strong family and community ties [2, 25], education opportunities [2], good relationship with the employer [2], and skills matching the job requirements [2].

Impact sourcing can help companies achieve CSR objectives (e.g., increasing supplier diversity) [2, 3, 9] and create a societal impact, since providing work helps increase the livelihood of both marginalized people and the communities they are a part of [2, 5, 9]. It can also prevent unethical business practices, such as employment under poor working conditions [5], and it allows access to a previously untapped labor pool [2, 9]. This can be used to help achieve growth strategies [2, 9] or complement workforce in case of talent shortages [31].

Figure 14.2 shows an overview of the benefits that client organizations can obtain by impact sourcing. In the figure impact sourcing serves as a driver for lower costs, lower employee attrition, more achieved CSR objectives, better societal impact, less unethical business practices, and more access to untapped labor pools. The latter can help organizations accomplish growth strategies and increase their workforce.

14.3.3 Challenges of Impact Sourcing for Clients

Several possible challenges were identified relating to the labor force, namely, not enough skilled labor [9, 15, 23, 30, 38–41], low productivity [15, 39], poor human resource management [15], and absenteeism [3, 23, 42]. Absenteeism can be caused by inadequate transport infrastructure or social services, like day care [3]. Absenteeism can also occur when the income earned from impact sourcing is not the primary source of income [42]. In this case marginalized people may prioritize the activity generation of the primary income over the employment resulting from impact sourcing.

Another challenge can be inadequate public infrastructure. Multiple studies identified energy and telecommunication infrastructure, for instance, electricity and Internet connectivity [9, 15, 30, 38, 43, 44], transportation infrastructure that allows travelling from home to the working place [15, 23, 43, 44], logistic infrastructure (e.g., for importing raw materials) [39], and social infrastructure (e.g., education and day care) [43] as insufficient. This is especially a problem in the more rural areas [43]. When public infrastructure is insufficient, higher costs can be expected to compensate for this deficiency, for example, because transportation has to be provided by the client or impact sourcing service providers (ISSP) [23]. In some places public infrastructure is being improved upon, for instance, in Malaysia [42].

Governments can stimulate or hinder impact sourcing attractiveness. The governments of low-income countries can address challenges for client organizations by enforcing policies. The public infrastructure challenge, for instance, can be reduced if governments invest in electricity supply and telecommunications [15, 23, 25, 42, 45] and government expenditure on education can reduce challenges related to the labor force [23, 43, 45]. Additionally, governments can stimulate outsourcing by providing tax benefits and introducing import/export policies [15, 25]. The political climate can also pose a challenge for client organizations. This is the case when governments fail to stimulate the outsourcing industry and decide to solely focus on regional production or when political instability results in uncertainty regarding the outsourcing industry [9, 15, 46]. Dissimilar cultures and unfamiliarity with the business language [15] or a mutual lack of respect for the different cultures and differences [43] can result in problems. One proposed solution is to exploit any similarities in culture and learn about the other country's business culture [15]. In one study the observation was made that understanding and respecting the culture of the marginalized people played an important role for rural impact sourcers in positioning themselves within the community [25].

Figure 14.3 shows the challenges that client organizations might have to overcome to impact sourcing successfully. It also shows potential solutions for some of these challenges. However, we were not able to identify potential solutions for all challenges. There are challenges related to labor force, infrastructure, politics, and culture. Certain challenges can have a strengthening effect on other challenges. Insufficient transportation and social infrastructure can have a strengthening effect on the level of absenteeism, for instance, because employees cannot go to work by public transport. Insufficient transportation infrastructure can also result in extra costs for the client organizations because they might have to arrange and pay for the transport of their employees.

14.4 Ethical Outsourcing: Benefits and Challenges

In this section the benefits and challenges of ethical outsourcing are discussed. Ethical outsourcing can result in multiple benefits for client organization, such as a positive brand image and better stakeholder interest management. We also discovered that there is a scenario in which ethical outsourcing can result in a competitive disadvantage for client organizations investing in this practice. These benefits and challenges are elaborated upon in the following subsections.

14.4.1 Benefits of Ethical Outsourcing for Clients

Two studies state that ethical outsourcing can create a positive brand image, which subsequently could increase the customer's willingness to purchase products from a particular company [47, 48]. A third study focuses on the absence of ethical outsourcing. In this study some interviewees stated that they would stop purchasing products from a company involved in a scandal [49]. Millennials were interviewed to gain insights on outsourcing scandals concerning *Apple*. Following a scandal, 85% of the millennials said they would continue purchasing Apple products versus 15% saying they would not. The rationale millennials provided to continue using Apple products are product loyalty and lack of competitors with good outsourcing practices. All interviewed millennials stated that they want the unethical practices to be changed though [49]. Two studies discuss the positive effects of ethical outsourcing on stakeholder interest management. Babin and Nicholson argue that newer generations of employees are more concerned with a company's CSR activities [47]. Park and Hollinshead argue that shareholders and senior management have interests in CSR [48].

14.4.2 Challenges of Ethical Outsourcing for Clients

Ang states that client organizations might be hesitant to engage in ethical outsourcing, due to the fact that investments in the provider's CSR capabilities can benefit other clients when they purchase goods and/or services from the same provider. This allows competitors to gain advantage based on the ethical methods of production without additional costs [50]. This scenario is shown in Fig. 14.4. Ang suggests that creating multilateral contract (i.e., contracts that introduce more than two stakeholders) can mitigate the risk of competitive disadvantage caused by ethical sourcing [50].

To summarize, ethical outsourcing can protect the brand, help facilitate stakeholder management, and decrease competitiveness.

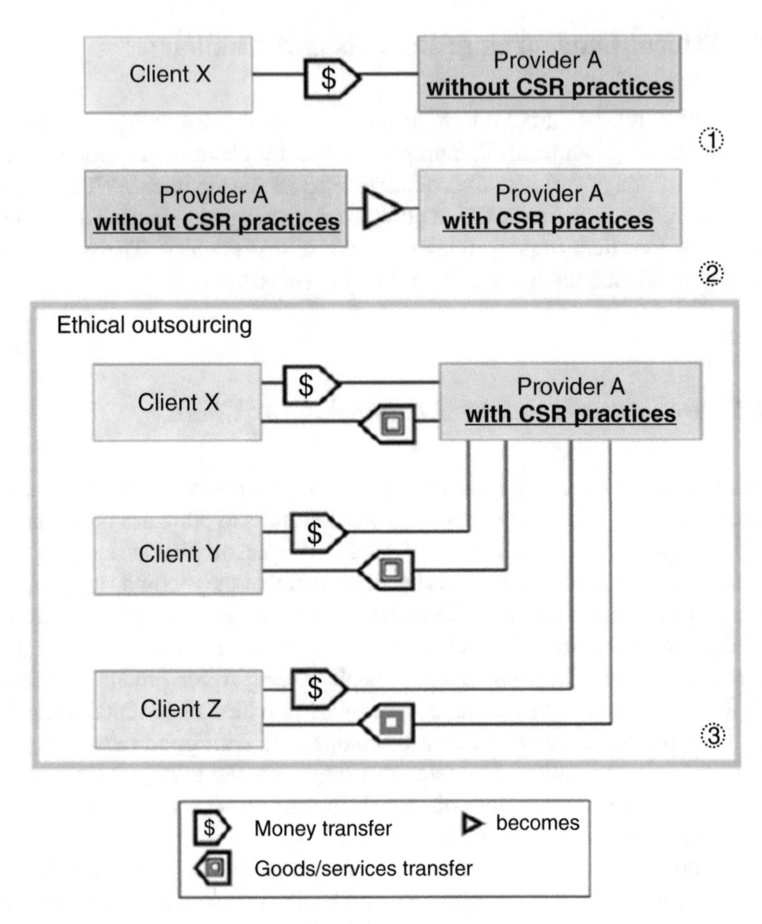

Fig. 14.4 (1) Client X invests in Provider A, so Provider A can incorporate CSR practices. (2) Provider A becomes a provider with CSR practices. (3) Clients X, Y, and Z engage in ethical outsourcing with provider A. Clients Y and Z have not made any CSR-related investments in Provider A, but thanks to the investments made by Client X, they can reap the benefits from ethical outsourcing. Therefore, Clients Y and Z have a competitive advantage over Client X

14.5 Fair Trade Software: Benefits and Challenges

The concept of Fair Trade Software is fairly new, and there are few companies working according to this concept. This section will explain what Fair Trade Software entails, why the model was created, and what the challenges in its application are. Only a few scientific sources discuss Fair Trade Software. We collected more practical insights by means of an interview. Like other Fair Trade models, Fair Trade Software is a movement formed by individuals with a desire to help others in developing countries. Initially formed in 2011 as a form of impact sourcing, over time the model has evolved into its current form. The Fair Trade Software models aim to deliver high-quality and cost-effective software for customers while helping to grow and develop knowledge economies in developing countries [19]. The latter is done by knowledge transfer from teams in developed countries to teams in developing countries. To reach this goal learning and knowledge transfer processes have to be introduced. To further develop the Fair Trade Software model and to stimulate its use, the Fair Trade Software Foundation (FTSF) was founded. The FTSF creates and sustains partnerships between software companies in developed and developing countries, oversees companies using the model, and engages with other stakeholders.

14.5.1 Benefits of Fair Trade Software

There is limited data available to assess the benefits of Fair Trade Software. One of the potential benefits of the Fair Trade Software model, for teams in developing countries, is capacity building. Capacity building is the process by which individuals and organizations obtain, improve, and retain the skills, knowledge, tools, equipment, and other resources needed to do their jobs competently. It allows individuals and organizations to perform at a greater capacity [51]. Since knowledge and skill transfer lays at the basis of the Fair Trade Software model, this presumably is one of the benefits of the model. No literature was found on the benefits for teams in developed countries, but based on an interview with the founder of the Fair Trade Software Foundation, Andy Haxby, we discovered that the benefits for teams in developed countries are motivation and self-satisfaction for individuals and marketing opportunities and brand enhancement for the companies they work for.

14.5.2 Challenges of Fair Trade Software

To develop Fair Trade Software, teams in developed and developing countries have to work together according to the same development method. A key skill that is often missing in developing countries is knowledge on working Agile. Working Agile has proven to be an effective development method. In a study performed by Serrador and Pinto, empirical evidence was collected to prove the success of working Agile. The results mentioned that in 6% of the 1386 cases, the way of working was completely Agile, and in 65% of the cases, there were some Agile elements. This study shows that the higher the level of Agile working (or another iterative approach), the higher the reported project success. The Agile methodology also scored significantly higher on the overall project success, efficiency, and stakeholder success [52]. Budzier and Flyvbjerg found that Agile methods also improve the delivery time of the product [53].

To ensure IT companies in Kenya can reap the benefits of working Agile, the FTSF transfers their knowledge on Agile project management methods [19]. To successfully execute an Agile project, 25–33% of the project team has to be experienced with the Agile methodology [54]. Experienced team members deliver the most added value in Agile projects [55]. So, to successfully communicate and develop products, the team from the developing country needs to collaborate with a team member that is experienced in working Agile. Haxby and Lekhi attest this by stating that it is almost impossible to teach individuals about Agile methods without them being immersed into already existing and experienced teams [19].

In addition to the challenges found in literature, we discovered challenges during the interview with Haxby. The most pressing issues Haxby mentioned are the limited resources within the FTSF. The limited time and people power create a challenge to effectively audit organizations. For instance, if the FSTF helps an organization obtain a grant, the means to test and audit the people and companies involved in the project are too weak. Currently, there is no auditing body. Haxby also states that it is difficult to sell Fair Trade Software, mainly because impact sourcing models are hard to sell. Although FTS is different from impact sourcing, they share many of the same difficulties and challenges. The difficulties and challenges are elaborated upon in Sect. 14.3.3. Additionally, some organizations do not wish to be associated with the Fair Trade brand. This is often the case for organizations that operate in industry sectors typically considered socially or environmentally unsustainable (e.g., the petroleum industry), because the FT brand does not fit with their customer demographic. This is an obstacle in making the model more widely adopted. The last issue Haxby mentioned is the ineffectiveness of networks of responsible enterprises, in relation to supporting Fair Trade Software. Networking events for responsible enterprises, which are supposed to result in more support for models such as Fair Trade Software, often attract people who cannot support these models adequately. Fair Trade Software does not yet mobilize the desired partnerships and resources.

14.5.3 Challenges of Cross-Border Development

In addition to the challenges that relate to Fair Trade Software, specifically we identify four major challenges of cross-border development. These challenges apply for Fair Trade Software as well, since it is a specific form of cross-border development. A common problem of cross-border software development is that developers living and working in different locations sometimes use different software [56]. This can be due to the fact that newer versions of software are not accessible in some countries because of export regulations [57]. If the team members in cross-border locations use different data repositories and these repositories are not compatible with each other, this can lead to problems regarding data transfer [58]. Moreover, developers often do not know each other on a personal level. A good personal and working relationship is essential to the success of a project [59]. Insufficient cooperation because of a lack of personal relationships can have adverse consequences for sharing implicit knowledge and reduce motivation [59–61].

When realizing cross-border software development, the difference between time zones, also known as temporal difference, should be taken into account [62]. It is likely that there are differences in time zones and working hours in the countries involved in a project [56, 60, 63]. Communication is one of the success factors of cross-border software development [64], as well as a good product owner [65]. When developers work and live in different time zones, there might be little overlap in the working hours. The FTSF has mitigated this issue by focusing on collaborations between teams in Europe and Africa. This complicates the use of asynchronous communication technologies (e.g., chat and emails) [66, 67], increasing the chance of miscommunication [68, 69]. Moreover, the use of asynchronous communication technologies increases the response time, causing developers to receive a response only the next day [59]. A delay in response can cause a developer to be unable to continue working [70, 71]. When developers decide to continue working without confirmation or response, it could lead to significant errors in the code [63, 72]. The delay can affect the deadline, and this in turn can have consequences and create frustration [59, 71].

Cultural barriers are one of the most common barriers when it comes to outsourcing and cross-border cooperation in software development [71]. Haxby and Lekhi state that cultural aspects complicated teaching Agile methods in Kenya. The Kenyan education system is very competitive: few assignments are performed in teams and competition among students is encouraged. The culture surrounding individual competition makes it difficult to explain the added value of working Agile. Moreover, miscommunication and/or lack of cultural awareness can cause conflicts among peer and management [67].

Lastly, language can pose a barrier for cross-border development. Usually English is the common language in cross-border development. A developer who is not confident in the English language can have the tendency to choose for an asynchronous communication tool, while synchronous communication tools (e.g., video or teleconference) have prevented misunderstandings [60, 73]. Additionally, it is more probable that native English speakers obtain a higher position, due to their linguistic advantage [60]. When non-native English speakers are skilled in the English language, their fluency is often confused with an understanding of idiomatic expressions [67]. If the non-native speaker is unaware of the actual meaning of the expression, it can result in misunderstandings. Another issue related to linguistics is that people might have different interpretations for similar words [74]. In turn, this can lead to misunderstandings about the meaning of an explicit or implicit message.

14.6 Conclusions and Future Research

To create human sustainability when outsourcing, vitality and learning at work are crucial. Impact sourcing and ethical outsourcing can contribute to human sustainability. In Fair Trade Software especially the learning aspect of human sustainability is emphasized. This research puts the different outsourcing approaches that consider

CSR side by side, describing them and compiling what is known about them. This has revealed that the approaches have similarities and differences, and that there is preliminary evidence that they yield a good impact both on marginalized people and on clients.

The literature postulates that outsourcing using a CSR-considering approach is a win-win situation both for marginalized people and for client organizations. For example, marginalized people may see an increase in income and client organizations may see a reduction in costs. Not surprisingly, incorporating CSR practices is becoming more important in outsourcing [10], and impact sourcing even accounted for 12% of the market in 2014 [2]. Fair Trade Software is a novel concept and little scientific research can be found in this field; therefore, some findings related to Fair Trade Software were derived from practice rather than from literature.

The results of this literature study indicate that the efficacy of impact sourcing can be categorized in four ways, namely: it provides an opportunity for employment, it improves personal and self-efficacy, it can improve existing social relationships and result in new ones, and finally a variety of harmful effects can sometimes occur. No literature was found on ethical outsourcing that relates to efficacy; thus, no conclusions can be drawn on how ethical outsourcing affects the lives of marginalized people.

Impact sourcing provides several benefits for client organizations compared to traditional outsourcing: lower costs, lower employee attrition, and turnover. It helps achieve CSR objectives and societal impact, and finally, it helps achieve growth strategies. Additionally, the quality of the products and services delivered through impact sourcing is of similar quality as that of traditional outsourcing. In order to achieve these benefits of impact sourcing, client organizations have to overcome the following four challenges: productivity and quality of labor force, reliability and quality of public infrastructure, unstable or unfavorable political climate, and finally cultural differences, which may lead to conflict. For ethical outsourcing two benefits and one challenge were found, although supported by few studies. Ethical outsourcing can protect brand image and may improve stakeholder management; however, the extra investments required can cause a competitive disadvantage. With regard to ethical outsourcing, no definitive conclusions can be drawn. We can conclude that impact sourcing is considered beneficial for marginalized people and potentially beneficial for client organizations. Fair Trade Software is a relatively new model and there is little scientific literature mentioning the model. Nonetheless, we were able to identify some potential benefits and challenges. A challenge of Fair Trade Software is that it is difficult to teach teams who are unfamiliar with Agile methodologies about working Agile—without them being part of an experienced team. Additionally, cultural differences can cause employees who are unfamiliar with these methodologies to not understand the added value of working according to an Agile methodology. The limited resources of the FTSF cause them to be unable to audit their members, which might threaten the reputation of Fair Trade Software. Another reputation-related issue is that some organizations do not want to be associated with the Fair Trade brand. Therefore, selling Fair Trade Software to make it widely adapted in the software development landscape is challenging. We also found challenges related to cross-border development in general. The use of different software by teams can cause compatibility issues, differences in time zones can complicate communication, and cultural and linguistic differences can hinder cooperation.

All in all, to discover the efficacy of Fair Trade Software and ethical outsourcing, more research has to be performed. In the case of Fair Trade Software, future research could focus on the impact of Fair Trade Software compared to non-Fair Trade Software, barriers for choosing Fair Trade Software, and methods for guaranteeing that software is produced under fair circumstances. The FTSF is already in the process of engineering a certification method; this could ensure that software is developed according to the Fair Trade Software model. For impact sourcing, more research is necessary on the business case of impact sourcing for client organizations. In particular, more evidence on lower cost benefit and on the observation that quality is similar compared to the output of traditional outsourcing is of importance, since client organizations cited these potential benefits as the most important [3, 5, 6, 9]. Additionally, more research is necessary on ethical outsourcing so that definitive conclusions can be drawn on this research field.

In conclusion, we hope that by contributing a compendium of existing knowledge in the field of impact sourcing, ethical outsourcing, and Fair Trade Software and by delineating new research endeavors, this chapter raises awareness of the importance of these practices as a means to increase the social responsibility of the ICT industry.

Acknowledgments We would like to thank Joost Dijkers for his contributions to the semi-systematic literature review on impact sourcing and ethical outsourcing. We would also like to thank Andy Haxby, Olav Verhoeven, Huseyin Aksoy, Sander Paulus, and Louis Lomans for their contributions on Fair Trade Software and cross-border development.

References

1. Spreitzer G, Porath CL, Gibson CB (2012) Toward human sustainability: how to enable more thriving at work. Organ Dyn 41(2):155–162
2. Everest Group (2014) The case for impact sourcing. Technical report. Everest Group
3. Carmel E, Lacity MC, Doty A (2016) The impact of impact sourcing: framing a research agenda. In: Socially responsible outsourcing. Springer, pp 16–47
4. Balit S (2007) Communication for isolated and marginalized groups. Commun Sustain Dev:101
5. Heeks R (2013) Information technology impact sourcing. Commun ACM 56(12):22–25
6. Lacity MC, Khan SA, Willcocks LP (2009) A review of the it outsourcing literature: insights for practice. J Strateg Inf Syst 18(3):130–146
7. Matten D, Moon J (2008) "implicit" and "explicit" CSR: a conceptual framework for a comparative understanding of corporate social responsibility. Acad Manag Rev 33(2):404–424
8. Roberts S (2003) Supply chain specific? Understanding the patchy success of ethical sourcing initiatives. J Bus Ethics 44(2-3):159–170
9. Bulloch G, Long J (2012) Exploring the value proposition of impact sourcing. Technical report. Accenture and Rockefeller Foundation
10. Nicholson B, Babin R, Lacity MC (2017) Socially responsible outsourcing: global sourcing with social impact. Springer

11. Accenture (2012) Exploring the value proposition from impact sourcing: the buyer's perspective. Technical report
12. Markets MI (2011) Job creation through building the field of impact sourcing. Technical report, Working Paper
13. Checkland P (2000) Soft systems methodology: a thirty year retrospective. Syst Res Behav Sci 17(S1):S11–S58
14. Heeks R, Arun S (2010) Social outsourcing as a development tool: the impact of outsourcing IT services to women's social enterprises in Kerala. J Int Dev J Dev Stud Assoc 22(4):441–454
15. Abbott P (2013) How can African countries advance their outsourcing industries: an overview of possible approaches. Afr J Inf Syst 5(1):2
16. Lacity MC, Carmel E, Rottman J (2011) Rural outsourcing: delivering ITO and BPO services from remote domestic locations. Computer 44(12):55–62
17. Haxby A, van Weperen E (2014) Creating shared value through Fair Trade Software: putting the principle of shared value creation into practice: "Fair Trade Software (FTS); Where Open Source meets Impact Sourcing". Where Open Source meets Impact Sourcing
18. van Nijen S, Espana S, Overbeek S (2018) A method to certify Fair Trade Software practices. B.S. Thesis, Utrecht University
19. Haxby A, Lekhi R (2017) Building capacity in Kenya's ICT market using cross-border scrum teams. In: International Conference on Social Implications of Computers in Developing Countries. Springer, pp 359–366
20. Burgess A, Ravishankar M, Oshri I (2015) Getting impact sourcing right
21. Chertok M, Hockenstein J (2013) Sourcing change: digital work building bridges to professional life. Innov Technol Govern Global 8(1–2):177–187
22. Harji K, Best H, Essien-Lore E, Troup S (2013) Digital jobs: building skills for the future. The Rockefeller Foundation. rockefellerfoundation.org/blog/digital-jobsbuilding-skills-future
23. Kennedy R, Sheth S, London T, Jhaveri E, Kilibarda L (2013) Impact sourcing. Technical report. The Rockefeller Foundation
24. Lacity MC, Rottman JW, Carmel E (2016) Impact sourcing: employing prison inmates to perform digitally enabled business services. In: Socially responsible outsourcing. Springer, pp 138–163
25. Madon S, Ranjini C (2016) The rural BPO Sector in India: encouraging inclusive growth through entrepreneurship. In: Socially responsible outsourcing. Springer, pp 65–80
26. Madon S, Sharanappa S (2013) Social it outsourcing and development: theorising the linkage. Inf Syst J 23(5):381–399
27. Malik F, Nicholson B, Morgan S (2016) Assessing the social development potential of impact sourcing. In: Socially responsible outsourcing. Springer, pp 97–118
28. McKague K, Morshed S, Rahman H (2013) Reducing poverty by employing young women: Hathay Bunano's scalable model for rural production in Bangladesh (innovations case narrative: Hathay Bunano). Innov Technol Govern Global 8(1–2):69–88
29. Gill M, Cordisco Tsai L (2018) Building core skills among adult survivors of human trafficking in a workplace setting in the Philippines. Int Soc Work:538–544
30. Ravi V, Venkatrama Raju D (2013) Rural business process outsourcing in India-opportunities and challenges. Int J Bus Manag Invent 2(8):40–49
31. Herbert IP (2016) How students can combine earning with learning through flexible business process sourcing: a proposition. Technical report. Loughborough University
32. Tinsley E, Agapitova N (2018) Reaching the last mile: social enterprise business models for inclusive development. World Bank
33. Sandeep M, Ravishankar M (2016) Exploring the "impact" in impact sourcing ventures: a sociology of space perspective. In: Socially responsible outsourcing. Springer, pp 48–64
34. Begum KJA (2016) Challenges and opportunities in BPOs–a case study of women working in Bangalore. Skill India Dev Emerging Debates:125–136
35. Deka SJ, Sebastian N (2017) Globalisation and women employees: a study on women employees in BPO industry in India. PhD thesis, Sikkim University

36. Lester DL, Menefee ML, Pestonjee D (2010) An information technology services outsourcing alternative: a business model. J Bus Entrepreneurship 22(1)
37. Lusero C, Taylor AR, Agrawal V (2013) A case review of Xpanxion: a software quality assurance startup. Int J Manag Cases 15(2)
38. Bell B (2015) Creating regional advantage: the emergence of IT-enabled services in Nairobi and Cape Town. PhD thesis, UC Berkeley
39. Fessehaie J, Morris M (2018) Global value chains and sustainable development goals: what role for trade and industrial policies. Inclusive Econ Transf
40. van Gorp D, Brandt M, Kievit H (2015) Assessment of the attractiveness of Bangladesh as an ICT offshoring destination. China-USA Bus Rev 14(2):67–78
41. Keijser C (2016) 10 Changing geographies of service delivery in South Africa. In: Globalisation and services-driven economic growth: perspectives from the global north and south, p 167
42. Ismail SA, Aman A (2018) Impact sourcing initiatives in Malaysia: an insight through porter's diamond framework. In: State-of-the-art theories and empirical evidence. Springer, pp 197–214
43. Allouh A, Maurer R, Walker F, Wilcox Gwynne RH (2017) Designing a socially sustainable impact sourcing model for integrating immigrants in Sweden
44. Wausi A, Mgendi R, Ngwenyi R (2013) Labour market analysis and business process outsourcing in Kenya: poverty reduction through information and digital employment initiative. University of Nairobi, Nairobi. Unpublished Research Report
45. Avasant (2012) Incentives and opportunities for scaling the "impact sourcing" sector
46. Babin R, Myers P (2015) Social responsibility trends and perceptions in global it outsourcing. In: Proceedings of the Conference on Information Systems Applied Research ISSN, vol 2167, p 1508
47. Babin R, Nicholson B (2009) How green is my outsourcer-environmental responsibility in global it outsourcing. In: ICIS 2009 Proceedings, p 83
48. Park KM, Hollinshead G (2011) Logics and limits in ethical outsourcing and offshoring in the global financial services industry. Compet Chang 15(3):177–195
49. Mboga J (2017) A critical analysis of ethical outsourcing using comparative case examination and eliciting consumer millennials perspectives. Eur J Econ Financ Res
50. Ang YS (2015) Ethical outsourcing and the act of acting together. In: Empowering organizations through corporate social responsibility. IGI Global, pp 113–130
51. Potter C, Brough R (2004) Systemic capacity building: a hierarchy of needs. Health Policy Plan 19(5):336–345
52. Serrador P, Pinto JK (2015) Does Agile work?—A quantitative analysis of agile project success. Int J Proj Manag 33(5):1040–1051
53. Budzier A, Flyvbjerg B (2013) Making sense of the impact and importance of outliers in project management through the use of power laws. In: Proceedings of IRNOP (International Research Network on Organizing by Projects), Oslo, vol 11
54. Dorairaj S, Noble J, Malik P (2012) Knowledge management in distributed Agile software development. In: 2012 Agile Conference. IEEE, pp 64–73
55. Lindvall M, Basili V, Boehm B, Costa P, Dangle K, Shull F, Tesoriero R, Williams L, Zelkowitz M (2002) Empirical findings in Agile methods. In: Conference on extreme programming and agile methods. Springer, pp 197–207
56. Herbsleb JD, Moitra D (2001) Global software development. IEEE Softw 18(2):16–20
57. Battin RD, Crocker R, Kreidler J, Subramanian K (2001) Leveraging resources in global software development. IEEE Softw 18(2):70–77
58. Bhat JM, Gupta M, Murthy SN (2006) Overcoming requirements engineering challenges: lessons from offshore outsourcing. IEEE Softw 23(5):38–44
59. Holmstrom H, Conchúir EÓ, Agerfalk J, Fitzgerald B (2006) Global software development challenges: a case study on temporal, geographical and socio-cultural distance. In: 2006 IEEE International Conference on Global Software Engineering (ICGSE'06). IEEE, pp 3–11
60. Noll J, Beecham S, Richardson I (2011) Global software development and collaboration: barriers and solutions. ACM Inroads 1(3):66–78

61. Phalnikar R, Deshpande V, Joshi S (2009) Applying agile principles for distributed software development. In: 2009 International Conference on Advanced Computer Control. IEEE, pp 535–539
62. Carmel E (1999) Global software teams: collaborating across borders and time zones. Prentice Hall PTR
63. Kiel L (2003) Experiences in distributed development: a case study. In: Proceedings of International Workshop on Global Software Development at ICSE, Oregon, USA
64. Perry DE, Staudenmayer NA, Votta LG (1994) People, organizations, and process improvement. IEEE Softw 11(4):36–45
65. Bass JM, Haxby A (2019) Tailoring product ownership in large-scale agile projects: managing scale, distance, and governance. IEEE Softw 36(2):58–63
66. Hossain E, Bannerman PL, Jeffery DR (2011) Scrum practices in global software development: a research framework. In: International Conference on Product Focused Software Process Improvement. Springer, pp 88–102
67. Rao MT (2004) Key issues for global it sourcing: country and individual factors. Inf Syst Manag 21(3):16–21
68. Carmel E, Agarwal R (2001) Tactical approaches for alleviating distance in global software development. IEEE Softw 18(2):22–29
69. Damian DE, Zowghi D (2002) The impact of stakeholders' geographical distribution on managing requirements in a multi-site organization. In: Proceedings IEEE Joint International Conference on Requirements Engineering. IEEE, pp 319–328
70. Boland D, Fitzgerald B (2004) Transitioning from a co-located to a globally-distributed software development team: a case study at Analog Devices Inc. In: 3rd Workshop on Global Software Development. IET
71. Khan K, Zafar AA, Alnuem MA, Khan H (2012) Investigation of time delay factors in global software development. World Acad Sci Eng Technol 63:380–388
72. Braun A, Dutoit AH, Brügge B (2003) A software architecture for knowledge acquisition and retrieval for global distributed teams. In: International Workshop on Global Software Development, International Conference on Software Engineering, Portland, OR
73. Niinimaki T, Piri A, Lassenius C (2009) Factors affecting audio and text-based communication media choice in global software development projects. In: 2009 Fourth IEEE International Conference on Global Software Engineering. IEEE, pp 153–162
74. Agerfalk PJ, Fitzgerald B, Holmstrom Olsson H, Lings B, Lundell B, Ó Conchúir E (2005) A framework for considering opportunities and threats in distributed software development. In: Proceedings of the International Workshop on Distributed Software Development

Chapter 15
The Importance of Software Sustainability in the CSR of Software Companies

Mª Ángeles Moraga, Ignacio García-Rodríguez de Guzmán, Félix García, and Coral Calero

Abstract Organizations around the world, as well as their stakeholders, are becoming increasingly aware of the need for, and the benefits of, socially responsible behavior, and sustainability is a core aspect of this. Given the presence of software systems in most companies and almost every aspect of modern-day life, the promotion of the environmental aspects of software systems is a key factor in sustainable development, and any company aspiring to be considered as a first-class corporate citizen should provide for it in their CSR.

This chapter aims to ascertain how well the policies of companies that develop software are aligned with Software Sustainability, as well as to give recommendations on including specific actions in their CSR to promote Software Sustainability.

The CSR policies of the ten biggest software companies have been studied, identifying a list of actions that the software industry should include in their CSR. In order to do this, different meetings were held among researchers. As a result, a list of actions specific to Software Sustainability that the software industry should be including in their CSR has been proposed. Moreover, we have analyzed the CSR of a Spanish software company, obtaining that the percentage of coverage in respect of the actions defined is 40%. The dimension with more actions is the human dimension, where the percentage of coverage is above 90%. Regarding the economic and environmental dimensions, the company took into consideration 36% and 13% of the actions, respectively. These resulted in a D level of Software Sustainability (possible values: A–E). Based on these results, we have suggested some actions to be implemented in order to improve the industry's Software Sustainability level.

M. Á. Moraga (✉) · I. G.-R. de Guzmán · F. García · C. Calero
Alarcos Research Group, Institute of Technologies and Information Systems, University of Castilla-La Mancha (UCLM), Ciudad Real, Spain
e-mail: MariaAngeles.Moraga@uclm.es; Ignacio.GRodriguez@uclm.es; Felix.Garcia@uclm.es; Coral.Calero@uclm.es

C. Calero et al. (eds.), *Software Sustainability*,
https://doi.org/10.1007/978-3-030-69970-3_15

15.1 Introduction

Organizations around the world, as well as their stakeholders, are becoming increasingly aware of the need for, and the benefits of, socially responsible behavior, and sustainability is a core aspect of it. Indeed, sustainability has increasingly become more important to businesses and must be tackled if we are to successfully develop sustainable societies [1]. By means of sustainable development, the needs of the present are fulfilled without compromising the ability of future generations to meet their own needs [2]. To achieve this aim, sustainable development must satisfy the requirements of three dimensions: society, the economy, and the environment [3]. A business that fails to include sustainable development as one of its top priorities could receive considerable public criticism and subsequently lose market legitimacy [4]. Therefore, sustainable business models (SBMs) are not just a passing fancy but are a field in their own right [5], and commercial organizations have begun to redesign their business models on the basis of sustainability, treating sustainable development as a new source of innovation, a new opportunity to cut costs, and a new mechanism for gaining competitive advantages [4]. All of this can be brought together under the umbrella concept of "strategic sustainability" [6]. When pursuing strategic sustainability, technology is doubly important, as noted by [4]: on one hand, because it helps organizations to tackle environmental issues (using web conferences, repositories, and so on) and, on the other, because technology itself is often responsible for major environmental degradation (e.g., due to the amounts of energy consumed by the engineering processes used to manufacture products). This mixed role that technology plays places organizations under tremendously conflicting types of pressure. Internally, they are under pressure to transform existing engineering processes into ones that are more environmentally friendly, while externally they are expected to design new products that improve the sustainability of society at large.

While sustainability is a standardized practice in several engineering disciplines, there is currently no such awareness within the software engineering community, as stated in [7]. It is of fundamental importance that such awareness be promoted in the software industry by championing "sustainable software"—that is, software whose direct and indirect negative impact resulting from its development, deployment, and usage is either minimal or has a positive effect on sustainable development with respect to the economy, society, humanity, and the environment [8]. But going a step further, the whole software development process could be supported, with sustainable software engineering being defined as "the art of defining and developing software products in such a way that the negative and positive impacts on sustainability that result from and/or are expected to result from the software product over its whole life cycle are continuously assessed, documented, and optimized" [9]. There are several areas in which software sustainability needs to be applied: software systems, software products, web applications, data centers, etc.

We therefore consider it to be of prime importance to find out the impact of software sustainability in (1) the companies that develop software, (2) those who buy it, and (3) the people who use it. From an organizational perspective, an essential reference document for analyzing how software sustainability is tackled is the corporate social responsibility (CSR) document. The objective of social responsibility is to contribute to sustainable development [2], and organizations are now subject to greater scrutiny by their various stakeholders than ever before. CSR has to "meet the needs of a firm's direct and indirect stakeholders (such as shareholders, employees, clients, pressure groups, communities, etc.) without compromising its ability to meet the needs of future stakeholders as well" [10]. Indeed, this much is stated by Friedman [11]:

> people today expect (and demand) more of business than simply that they maximize their profits without coming to grief by some violation of law. Consumers want and expect attributes from what they buy—quality, safety, value. Employees want more than a paycheck. Communities want the company to be a good corporate citizen and hire from the community, provide employees with a living wage, not pollute and to pay its fair share of taxes and support the community.

Therefore, the perception and reality of an organization's performance as regards social responsibility can influence, among other things, its competitive advantage, its reputation, and its ability to attract and retain workers or members, customers, clients, or users; it may also have an impact on the maintenance of employees' morale, commitment, and productivity and affect the view of investors, owners, donors, sponsors, and the financial community as well as the organization's relationship with companies, governments, the media, suppliers, peers, customers, and the community in which it operates. According to the results of the study by [12], disseminating companies' CSR results in improved brand value, and publishing complete sustainability reports comes over as a matter of importance for companies. Nave and Ferreira state that sustainability emerges as an increasing concern for those companies which have focused on reducing the impact that their activities have on the environment, while also implementing activities with social and economic dimensions [13]. As a consequence, some related works deal with the management of corporate sustainability in CSR, such as the theoretical model put forward by Butler [14] which deals with integrating Green IS (information systems) into the normal operations of a company, aligning these Green IS with the firm's CSR. Another example is the paper by Baumgartner [1], which proposes a framework for corporate sustainability management that sets out the different tasks and action levels for the transition of a company toward becoming "sustainable."

In summary, given the presence of software systems in most industries and in almost every aspect of current-day life, the promotion of the environmental aspects of software systems is a key factor in sustainable development, and any company aspiring to be regarded as a first-class corporate citizen should provide for it in their CSR. This chapter aims to ascertain how well the policies of companies that develop software are aligned with software sustainability, as well as to give recommendations on including specific actions in their CSR to promote software sustainability. In our quest to fulfill this goal, we will study the CSR policies of the ten biggest software companies, identifying a list of actions that the software industry should include into their CSR. Finally, we have analyzed the CSR of a specific software company and have suggested some actions to improve it.

The remainder of this chapter is organized as follows: the next section will present the software companies selected for our study, along with an analysis of their respective CSR documents. Section 15.3 will show the specific sustainability actions for the software industry and its companies to take, together with indicators to select the ones most suitable for a given company. Section 15.4 presents the improvements we recommend for the CSR of a Spanish company we investigated, and finally, Sect. 15.5 will set out our conclusions and outline future work.

15.2 Overview of the CSR in Software Industries

15.2.1 Software Companies: A Representative Selection

To find out if, and to what degree, software companies are concerned about the environmental aspects of the sustainability of the software they develop, we analyzed the CSR of several leading software companies. These were chosen based upon the list of the top companies suggested in [6, 15]:

1. Apple
2. Microsoft
3. IBM
4. Oracle
5. SAP
6. Symantec Corp
7. EMC
8. Hewlett-Packard
9. VMware
10. CA Technologies

Next, we studied their CSR statements in depth from the point of view of software sustainability.

15.2.2 Analyzing the CSR Software Sustainability Actions in Software Companies: Work Method

To analyze the degree of awareness on the part of the selected companies as regards the role of software sustainability in their CSR policies, we followed a specifically defined method. The review was carried out by examining the sustainability actions of each respective company, as included in the CSR information available on their corporate websites.

The template we built in order to collect from the companies' CSRs data about their actions on software sustainability includes the following sections:

- Category, which includes the general categories used to group together related actions, such as "People," "Planet," etc.
- Subcategory, to cover specific categories for the actions, for example, "Empowering communities" or "Empowering employees," which fall under the general category "People."
- Action, which covers the specific actions carried out in the context of the CSR, for example, the action "Employee feedback counts," which allows workers to participate in anonymous polls that serve to improve their work conditions. Special care was taken to fill in the specific actions of all the companies and to guarantee that each action included had a similar granularity level to the others, thereby avoiding any validity threats to the comparative study.
- Sustainability dimension, used to classify a given action according to the particular dimension or dimensions in which it is applicable. The dimensions are environmental, human, and economic. It should be noted that the environmental dimension can in turn be divided into Green Software and Green Hardware (this latter subdimension is out of our scope).

Having prepared the template and the inclusion and exclusion criteria, the research method was defined and agreed on by all researchers. In the first step, each researcher (the four authors of this chapter) was responsible for filling in the templates of two to three companies. To do so, the weblink to the CSR of each company, the empty template, and the inclusion and exclusion criteria were used. The output of this step was a first set of completed templates (one per company).

Then, in the second step, each researcher reviewed the templates that had been filled in by the other researchers, with the aim of ensuring that all the relevant information was included and classified appropriately into categories, subcategories, actions, and dimensions. The inputs for this step were the links to the CSRs of the companies, the inclusion and exclusion criteria, and the completed templates. The output was four sets of completed and reviewed templates.

The following step consisted of a meeting to discuss the differences identified between the four reviewed templates for each company, and to resolve any discrepancies. There were no actions which did not obtain the full consensus of the participating researchers. The input for this step consisted of the CSR links of each company, the completed and reviewed templates, and the inclusion and exclusion criteria. The output was a list of agreed-on software-related sustainability actions, taken from the verbatim CSR statements of each company.

Finally, and considering those sustainability actions, the researchers met to propose a list of actions that could be valuable for software companies from the point of view of sustainability. In addition, researchers evaluated every action in the final list to provide (1) the *added value* that the inclusion of this action could provide to the company and (2) an approximate *complexity level* that the implementation of the action in the company would require. Section 15.3 presents the outcome of the process described above.

15.2.3 Analyzing the Companies' CSR from the Point of View of Software Sustainability

As previously mentioned, the first step was the classification of the actions, more information of which can be found in [16]. The next step was to analyze the corporate social responsibility document of each selected company, with a view to determining whether software sustainability aspects had been taken into account. In the following subsections, we will present the results obtained from a top-down perspective, as illustrated in Fig. 15.1.

15.2.3.1 Analysis of Software Sustainability Actions

In this analysis a comparison was carried out between the software sustainability actions and other actions of the company. As can be observed in Fig. 15.2, the majority of the actions are intended to address aspects not related to sustainability. We see this as a clear demonstration of the relatively low importance that the companies give to the issue of software sustainability.

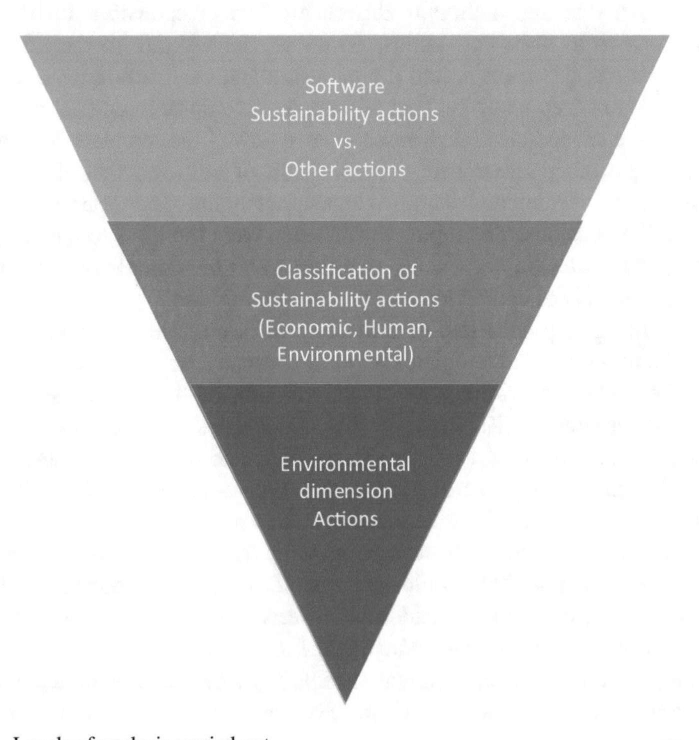

Fig. 15.1 Levels of analysis carried out

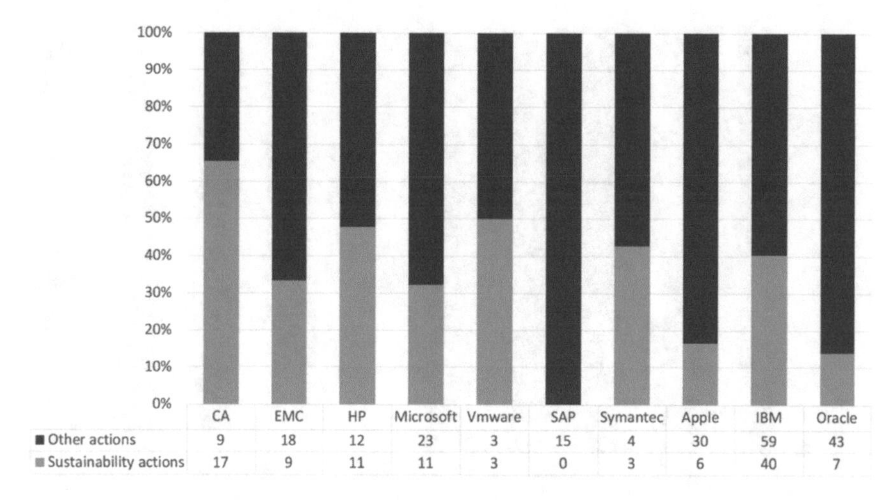

Fig. 15.2 Comparison between sustainability-oriented actions and other kinds of actions

Table 15.1 Percentage of activities devoted to the different dimensions of sustainability per company

Company	Human	Economic	Environmental
CA	14.81%	25.00%	13.51%
EMC	9.26%	0.00%	10.81%
HP	9.26%	6.25%	13.51%
Microsoft	14.81%	6.25%	5.41%
VMware	3.70%	0.00%	2.70%
Symantec Corp	3.70%	6.25%	0.00%
Apple	9.26%	0.00%	2.70%
IBM	33.33%	37.50%	43.24%
Oracle	1.85%	18.75%	8.11%

15.2.3.2 Analysis of Software Sustainability Actions

Software sustainability actions can be related variously to the human, economic, and environmental dimensions of the company's actions.

To give a more detailed breakdown of each of those activities listed in the CSR documents of the different companies that are oriented toward software sustainability, we classified them into three dimensions: human, economic, and environmental. Table 15.1 presents a view of the relative effort each company makes, with respect to the others, as regards these software sustainability dimensions. For each dimension, the percentage figure of each company has been calculated as a mean between the number of actions of the company and the total number of actions proposed by all companies in that dimension. Table 15.1 thereby attempts to represent the relative importance given by each company to each software sustainability dimension, according to the number of actions proposed by them. It should be noted that, although the CSR documents of all ten companies were analyzed, Table 15.1 shows only those companies with software sustainability actions.

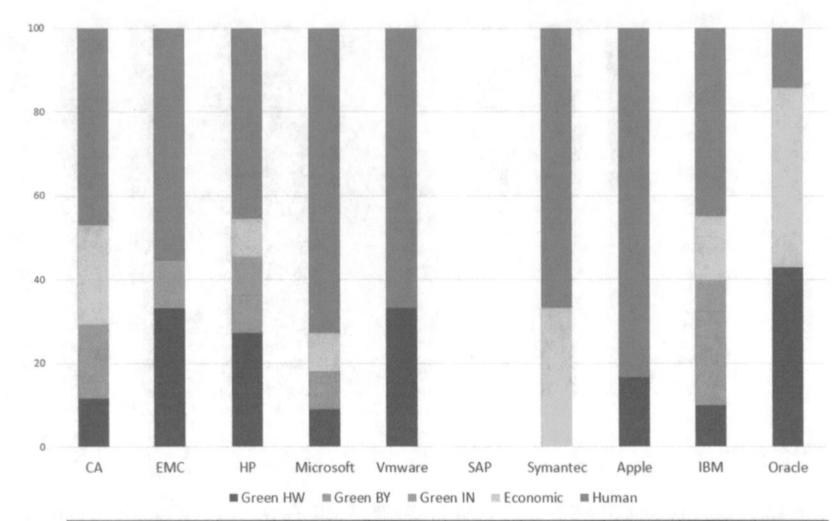

	CA	EMC	HP	Microsoft	Vmware	SAP	Symantec	Apple	IBM	Oracle
Green HW	2	3	3	1	1	0	0	1	4	3
Green BY	2	0	1	1	0	0	0	0	9	0
Green IN	1	1	1	0	0	0	0	0	3	0
Economic	4	0	1	1	0	0	1	0	6	3
Human	8	5	5	8	2	0	2	5	18	1

Fig. 15.3 Distribution of the activities of the CSR per company

As already mentioned, we have classified the environmental dimension actions as Green Hardware (GH) and Green Software—divided further between Green-IN Software (GI) and Green-BY Software (GB). The distribution of the software sustainability actions is depicted in graph form in Fig. 15.3.

From the previous analysis of each of the software sustainability dimensions, we can conclude that two companies are more aware of software sustainability than the others: IBM and CA.

We can thus conclude that, from the perspective of software sustainability actions, IBM and CA are the most balanced and are the companies that propose the most actions. The rest of the firms display different behavior but, in general, have most of their actions classified into only one of the dimensions, with a few actions in the other two.

15.2.3.3 Environmental Dimension Actions

Focusing our analysis on the environmental dimension, Tables 15.2, 15.3, and 15.4 present a more detailed view of the actions for both categories in this dimension, particularly as regards Green Software and Green Hardware.

Table 15.2 shows the percentage of activities based on Green Software, demonstrating the percentage of actions in each company as compared to the number of actions proposed by all companies. At first sight the table reveals something that is

Table 15.2 Percentage of environmental (software-related) activities in each company's CSR

Company	Green software
CA	15.79%
EMC	5.26%
HP	10.53%
Microsoft	5.26%
VMware	0.00%
SAP	0.00%
Symantec Corp	0.00%
Apple	0.00%
IBM	63.16%
Oracle	0.00%

Table 15.3 Percentage of environmental (Green-IN and Green-BY) activities in each company's CSR

Company	Green-IN	Green-BY
CA	16.67%	15.38%
EMC	16.67%	0.00%
HP	16.67%	7.69%
Microsoft	0.00%	7.69%
VMware	0.00%	0.00%
SAP	0.00%	0.00%
Symantec Corp	0.00%	0.00%
Apple	0.00%	0.00%
IBM	50.00%	69.23%
Oracle	0.00%	0.00%

Table 15.4 Percentage of environmental (hardware) activities in each company's CSR

Company	Green hardware
CA	11.11%
EMC	16.67%
HP	16.67%
Microsoft	5.56%
VMware	5.56%
SAP	0.00%
Symantec Corp	0.00%
Apple	5.56%
IBM	22.22%
Oracle	16.67%

highly significant: while only five companies implement actions that are categorized as Green Software, the same number of companies is unaware of the importance of providing such type of actions. A more detailed view of the results of this table may be observed in Table 15.3: this presents a fine-grained view of Green Software, which is made up of Green-IN and Green-BY actions.

These results are in sharp contrast to the actions belonging to the category of Green Hardware (Table 15.4), where eight companies propose actions. It is also

Table 15.5 Percentage of software (Green-IN and Green-BY) and hardware efforts for each company

	Green-IN	Green-BY	GREEN HW
CA	20.00%	40.00%	40.00%
EMC	25.00%	0.00%	75.00%
HP	20.00%	20.00%	60.00%
Microsoft	0.00%	50.00%	50.00%
VMware	0.00%	0.00%	100.00%
Apple	0.00%	0.00%	100.00%
IBM	18.75%	56.25%	25.00%
Oracle	0.00%	0.00%	100.00%

striking that with respect to the Green Hardware category, all the percentages are somewhat more balanced than those in the Green Software category. This remarkable finding stems from the fact that in current IT infrastructures, hardware resources have been the focus of continuous optimizations, in a quest to save energy and reduce their carbon footprint.

From both analyses we can determine that the conclusion reached by Calero and Piattini [17] from their research on software sustainability can be applied also to the software industry in general: i.e., that there is more awareness of the need for Green Hardware than for Green Software.

To conclude our analysis, Table 15.5 presents a comparison between the percentages of actions (categorized into Green-IN, Green-BY, and Green Hardware) for the eight companies whose CSR documents include actions in the environmental dimension. These data are especially useful to show the extent and distribution of the efforts of the different companies in these three categories.

As far as the companies listed in Table 15.5 are concerned, only four of them propose actions for Green-BY Software and another four propose actions for Green-IN Software; the percentage is very low. All the companies shown present actions for Green Hardware, allowing us to confirm our prior observation that the element to which companies give more importance is Green Hardware, rather than Green Software.

Finally, the actions of Green-IN represent the lowest percentages. This may be due to a lack of knowledge about the impact of software on the environment, but it may also be due to a lack of actions to reduce this impact.

15.3 Specific Actions for Software Industries

Based upon our analysis of the actions of these leading software companies' CSRs, a set of actions has been chosen that we believe are particularly interesting for software companies, from the point of view of sustainability.

In Tables 15.6, 15.7, and 15.8, these actions are shown, grouped together according to the sustainability dimension to which they belong. In each table the following information is included:

Table 15.6 Actions proposed to improve the human dimension

ID	Action	Added value (1–3)	Complexity (1–3)
	Employees must be encouraged to be successful in their jobs and to be innovative. To this end:		
H1	Employees must be supported to improve their skills and acquire the ability to work in a different, innovative, and open-minded way	1	1
H2	Employees must be encouraged to propose and implement solutions that are innovative	2	1
H3	At the organizational level, a culture will be fostered that facilitates employees and partners providing the necessary feedback for the transformation of both the processes and the business itself	3	2
	Ethics and rights:		
H4	(a) Create a work environment that respects the personal circumstances of each employee and allows them to manage their work responsibilities, reconciling these with their personal lives. (b) Provide our employees with principles, guidelines, directives, and tools that enable them to effectively manage their work	2	2
H5	Nondiscriminatory policies will be included to make the company a safe place to work	2	2
H6	An open work environment characterized by trust, mutual respect, and empathy is created, promoting leadership guided by ethics and integrity	2	2
H7	Human rights, including the right to privacy and freedom of expression, must be respected and upheld	3	2
H8	A good relationship between all company personnel should be encouraged and good communication between organizational levels should be promoted	2	1
	Women and technology:		
H9	The company must be committed to the professional advancement of its employees and encourage their access to leadership positions within the company	3	2
	Development of training programs for the acquisition of knowledge and skills:		
H10	The company must offer training programs to improve the skills and abilities of its employees, fostering a positive culture for both the employee and the organization	2	2
	Protecting people by:		
H11	Setting standards that are rigorous to protect employees and the planet during the software development process	3	3
H12	Rigorous standards should be established to make the organization's facilities safe	3	3
H13	Occupational risk review, assessment, and awareness programs, covering both the physical and mental health of employees, should be implemented to ensure the health and well-being of employees	3	3

Table 15.7 Actions proposed to improve the economic dimension

ID	Action	Added value (1–3)	Complexity (1–3)
E1	Sustainability must be part of the organization's business model	3	3
E2	Employees should preferably use video conferencing or similar technologies in their communications, travelling only when essential	1	1
E3	Policies that support software business continuity should be encouraged	3	3
E4	The use of energy-efficient technologies should be considered within the business model	3	3
E5	The security and privacy of business and customer data must be ensured to avoid excessive costs due to threats to data	2	1
E6	Customers must be provided with secure solutions with full connectivity and availability	3	2
E7	GDPR (or current country-specific legislation) must be implemented in all the organization's contracts to ensure compliance	3	3
E8	A policy must be defined to manage potential operational, legislative, and financial risks affecting business continuity, and technological support must be provided to carry out this management	3	2
E9	Customers must be provided with IT solutions that optimize resources, minimizing both unnecessary expenses and energy	3	3
E10	It is necessary to analyze and monitor compliance with the economic forecasts of software projects, identifying the reasons for deviations and applying actions to correct them, where necessary	2	2
E11	It is necessary to carry out the digital transformation of the company, using software solutions that support the business model by providing the necessary levels of security	3	3

- Action: Short description of the action.
- Added Value: Effort required to be implemented in the company and the value provided. To facilitate understanding of what is meant by "value" and "effort," a scale has been adopted where "1" is the lowest (easiest and least valuable, respectively) and "3" is the highest (more complex and more valuable, respectively).
- Complexity: Indicates how difficult it is to implement the action.

15.4 Analyzing and Improving the CSR of a Specific Company

Having defined the actions specific to software sustainability that the software industry should be including in their CSR, we now turn to analyze the CSR of a specific medium-sized company in Spain. It is a consulting company which is

Table 15.8 Actions proposed to improve the environmental dimension

ID	Action	Added value (1–3)	Complexity (1–3)
M1	Wherever possible, the resources needed for software development will be reduced and reused	2	2
M2	It is necessary to reduce the KW/h required by each software functionality	3	3
M3	Mechanisms should be defined to qualify software products with respect to energy-saving criteria	3	1
M4	Energy efficiency of software products must be defined and implemented as a corporate objective	3	3
M5	Regular monitoring is needed, through a process of accurate and rigorous collection of software energy consumption in all facilities	2	1
M6	Unnecessary energy expenditure should be avoided by using shared infrastructures for software development and execution	1	1
M7	The environmental footprint of software companies' DPCs must be reduced by using the state-of-the-art energy efficiency and cooling technologies	2	1
M8	Operations associated with software development must be persuaded to use efficient technologies (cloud computing, service virtualization, parallelization, SaaS, infrastructure, etc.)	2	2
M9	Software solutions for energy saving should be used in all company facilities	1	1
M10	Any software features that promote sustainability throughout its life cycle should be evaluated and improved	3	2
M11	The use of reporting systems for the sustainability management of software products is recommended	1	1
M12	Environmental aspects (energy efficiency, sustainability, etc.) should be incorporated into all stages of the life cycle prior to the operation of the software	3	2
	Regarding the process:		
M13	It is necessary to develop and keep the process assets needed for software development updated	2	2
	Regarding the requirements:		
M14	The environmental sustainability requirements (green requirements) of the software must be selected, analyzed, specified, validated, and managed throughout its life cycle	2	1
	Regarding design:		
M15	Green software requirements must be analyzed to obtain an internal description of the software structure that serves as a basis for its construction	3	2
M16	Solutions that promote green software must be provided when defining its architecture (organization into components and their relationships)	3	2

(continued)

Table 15.8 (continued)

ID	Action	Added value (1–3)	Complexity (1–3)
M17	Design decisions must be analyzed and any consequent corrective actions that impact on the green requirements must be carried out	3	2
M18	Design constraints related to green software must be managed and supported	2	1
	Regarding construction of the software:		
M19	Functional software must be created to meet green requirements through a combination of coding, verification, unit testing, integration testing, and debugging	2	1
M20	Construction approaches and technologies that support green requirements must be selected	3	2
	Regarding testing:		
M21	The software must be dynamically verified to meet green requirements through a finite set of test cases, appropriately selected	2	1
M22	Problems with any green requirements identified during testing should be verified as being satisfactorily resolved	2	1
	Regarding maintenance:		
M23	Software maintenance must be performed to ensure compliance with green requirements	1	2

specialized in Oracle technology and which carries out different types of projects ranging from digital transformation to business analytics, data management to security. Although the company has several locations throughout Spain, it is not normal practice for its developers to relocate among these various locations. One of the hallmarks of this company is a concern for the quality of life of its workers (from the perspective of the work environment).

The company's CSR is based on three basic foundations: the company, the people, and the planet. Integrated in their CSR, they have a specific program which addresses four fundamental topics, seen as being complementary to the day-to-day work of their employees: personal well-being, solidarity, teamwork, and ecological focus. This program is derived from the philosophy and the way of living of the people who are part of the company. With regard to their ideal of solidarity, it should be noted that the company collaborates with different associations and holds events to support them. As regards well-being they propose, for example, that employees have at least one healthy breakfast per month (fruit) or that they engage in a "wellness month." This activity consists in adopting healthy lifestyle habits, in four areas (physical, mental, spiritual, and emotional). To foster teamwork, they organize different recreational activities outside of the working day. In this way the employees get to know each other better and can connect on a personal level and not only professionally. Regarding their ecological focus, the company's objective is to contribute to the improvement of our environment. As part of their campaigns to

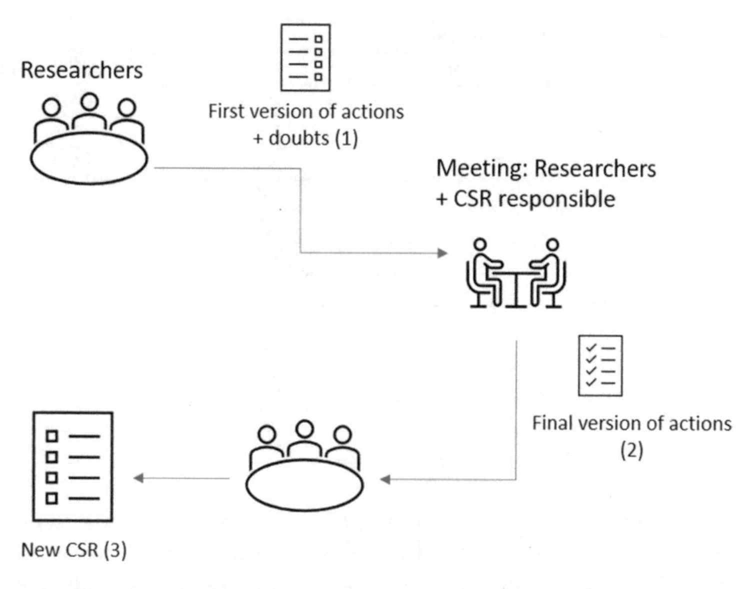

Fig. 15.4 Process to analyze and improve the CSR of the company

raise employees' awareness of the environmental importance and impact of their actions, last year, they, for example, removed all plastic cups and cutlery and replaced them with glass cups and wooden cutlery. The company has achieved the certification of the *HappyIndex AtWork*, which is based on the evaluations given by the employees and which recognizes them as a company in which the workforce feels happy and motivated. Lastly, we can add that they have drawn up their own code of ethics which is based on the principles of honesty, integrity, and respect.

With these special characteristics of the company in mind, we decided to analyze their CSR. Their concern for their employees and their well-being, for the planet, and so on was clear, but would they have also considered aspects related to software sustainability within the CSR?

The process presented in Fig. 15.4 explains the different steps which were carried out. Firstly, the researchers analyzed different documents which were provided by the company: specifically the CSR, the code of ethics, a special program to translate some general actions of the CSR into concrete actions, and a final report. With the aim of selecting the list of actions to be proposed to the company, we compared these documents with our list of proposed actions based on software sustainability (see Tables 15.6, 15.7, and 15.8). As a result, a set of actions was selected—Step 1 in Fig. 15.4. However, since the actions of the CSR are general in nature, the researchers had doubts as to some of these actions (i.e., they were not sure whether these actions were to be taken into account or not). To clear up these doubts, a meeting was held among researchers and the company in question, as a result of which the final set of actions that would be proposed to the company was obtained—Step 2 in Fig. 15.4.

In Table 15.9, the final list of included actions is shown.

Table 15.9 Actions included in the CSR

Dimension		
Economic	Human	Environmental
E2	H1	M1
E5	H2	M6
E7	H3	M8
E11	H4	
	H5	
	H6	
	H7	
	H8	
	H9	
	H11	
	H12	
	H13	

Once the actions were identified, a study of the coverage was made. In Fig. 15.5, the percentage of coverage in respect to the actions defined by us is shown. As can be seen, some 40% of the actions have been considered: a value which, although good as a starting point, can be improved.

Taking into account the total number of actions that we had defined for each dimension, we obtained the following coverage graphic (see Fig. 15.6). As can be noted, the dimension with more actions is the human dimension, where the percentage of coverage is above 90%. In the economic dimension, the company took into consideration 36% of the actions. The worst result is in the environmental dimension, in which only 13% of the actions were considered.

What stands out from these investigations is that, despite their avowed concern for environmental issues, the company's percentage of coverage in the environmental dimension is very low. Consequently, special attention was paid to this problem, and interviews were carried out with the aim of detecting whether there had been any misunderstanding. We concluded, however, that although the company is aware of the impact that their daily actions have on the environment, such as the use of

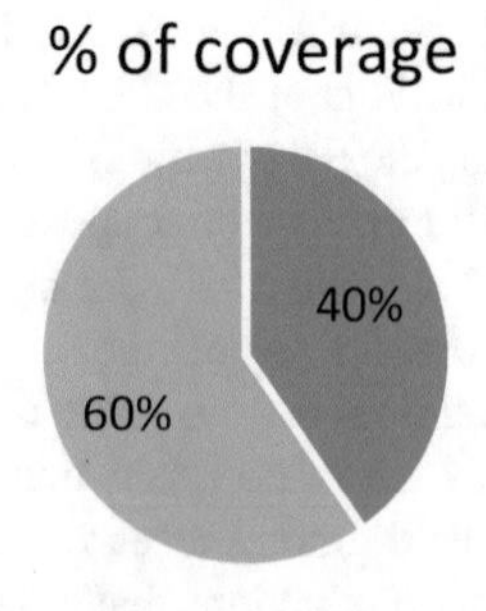

Fig. 15.5 Percentage of coverage considering the total number of actions

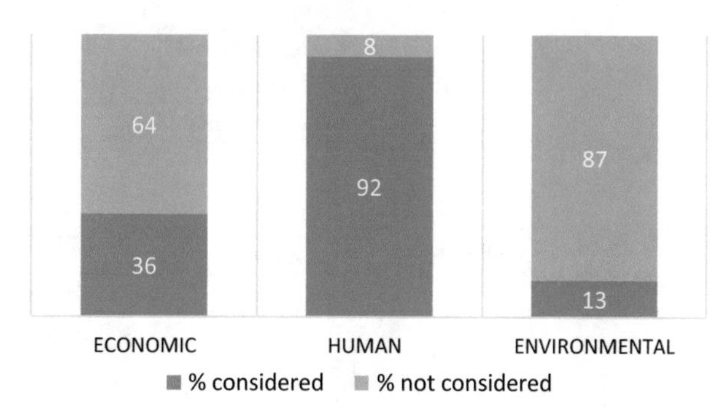

Fig. 15.6 Percentage of coverage for each dimension

Table 15.10 Label according to the % of coverage

% of coverage	Label for the dimension
81–100	A
61–80	B
41–60	C
21–40	D
0–20	E

plastics, not having the refrigerator open for a long time, etc., they had not considered the damaging impact of software on the environment, and this in spite of the fact that their work is focused on software development.

In addition, we have defined a label system that assigns to each company a value ranging from A to E. This value is assigned depending on the level achieved on each dimension, according to the percentage of coverage shown in Table 15.10. In order to assess it, first off, it is necessary to determine the label of each dimension according to Table 15.10.

In order to assess the final software sustainability level, the labels are converted into numbers (the correspondences are shown in Table 15.11), and the software sustainability level (SS) can be calculated following the next formula:

$$SS = (\%ofcoverageEconomic + \%ofcoverageHuman + \%ofcoverageEnvironmental)/3 \quad (15.1)$$

And, finally, we calculate the Software Sustainability label using the information in Table 15.10.

Bearing all this in mind, the analyzed company has the values shown in Table 15.11, concluding that its software sustainability level is C.

With the aim of improving the software sustainability level of the company, a list of actions to be included for the next version of the CSR was suggested by the researchers to the company—see Step 3 in Fig. 15.1. The criteria for the selection of these actions were twofold. The first is the difficulty of applying the action, and the

Table 15.11 Labels for the company

Dimension	Label
Economic dimension	D
Human dimension	A
Environmental dimension	E
Software sustainability	C

Table 15.12 Recommended actions to be included

Dimension		
Economic	Human	Environmental
E6	H10	M3
E8		M5
E10		M7
		M9
		M14
		M18
		M19
		M21
		M22

second is the benefit that the company could obtain from implementing the action (as shown in the two right-hand columns in each of Tables 15.6, 15.7, and 15.8). The list of the actions finally chosen is shown in Table 15.12.

As a future work, the company should include these actions into their CSR. Once some or all of them are included, a new analysis should be done to determine whether their software sustainability level has improved or not.

15.5 Conclusions and Future Work

It is essential that companies consider certain basic aspects of software sustainability within their CSR. We have carried out an initial analysis of the CSRs of leading software-related companies with the aims of (1) analyzing whether they take software sustainability into account and (2) determining an initial set of actions that should be included in future versions of their CSRs.

Subsequently, we have applied this process to a Spanish company which collaborated with us so that we could check the degree of coverage of the proposed actions and recommend subsequent improvements to its CSR, based on the analysis carried out. The new version of the CSR will need to be further revised and refined to include more actions in a gradual way.

As to future work, we want to extend the study with more companies so as to refine and complete the proposed actions, as well as to corroborate their applicability.

The results of this work are not static, but rather require an annual review in order to ensure that the vision of actions toward sustainability is both realistic and reflects how this issue evolves over time.

Acknowledgments This work has been funded by the BIZDEVOPS-GLOBAL project (Ministerio de Ciencia, Innovación y Universidades, y Fondo Europeo de Desarrollo Regional FEDER, RTI2018-098309-B-C31) and by the SOS and TESTIMO projects (Consejería de Educación, Cultura y Deportes de la Dirección General de Universidades, Investigación e Innovación de la JCCM, SBPLY/17/180501/000364 and SBPLY/17/180501/000503, respectively).

References

1. Baumgartner R (2014) Managing corporate sustainability and CSR: a conceptual framework combining values, strategies and instruments contributing to sustainable development. Corp Soc Responsib Environ Manag 21. https://doi.org/10.1002/csr.1336
2. ISO26000 (2010) Guidance on social responsibility
3. UN (1987) Report of the World Commission on environment and development: our common future. In: United Nations Conference on Environment and Development
4. Du W, Pan SL, Zuo M (2013) How to balance sustainability and profitability in technology organisations: an ambidextrous perspective. IEEE Trans Eng Manag 60(2):366–385. https://doi.org/10.1109/TEM.2012.2206113
5. Lüdeke-Freund F, Dembek K (2017) Sustainable business model research and practice: emerging field or passing fancy? J Clean Prod 168:1668–1678
6. Sroufe R, Sarkis J (eds) (2007) Strategic sustainability: the state of the art in corporate environmental management systems. Greenleaf Publishing, Sheffield
7. Penzenstadler B, Bauer V, Calero C, Franch X (2012) Sustainability in software engineering: a systematic literature review for building up a knowledge base. In: 16th International Conference on Evaluation and Assessment in Software Engineering (EASE 2012)
8. Dick M, Naumann S, Kuhn N (2010) A model and selected instances of green and sustainable software. Springer, Berlin, pp 248–259
9. Dick M, Naumann S (2010) Enhancing software engineering processes towards sustainable software product design. In: Greve K, Cremers AB (eds) 24th International Conference on Informatics for Environmental Protection. pp. 706–715
10. Dyllick T, Hockerts K (2002) Beyond the business case for corporate sustainability. Bus Strateg Environ 11:130–141. https://doi.org/10.1002/bse.323
11. Friedman J (2013) Milton Friedman was wrong about corporate social responsibility. http://www.huffingtonpost.com/john-friedman/milton-friedman-was-wrong_b_3417866.html. Accessed 2017
12. Alcaide M, De La Poza E, Guadalajara N (2019) The impact of corporate social responsibility transparency on the financial performance, brand value, and sustainability level of IT companies. Corp Soc Responsib Environ Manag 27:642–654. https://doi.org/10.1002/csr.1829
13. Nave A, Ferreira J (2019) Corporate social responsibility strategies: past research and future challenges. Corp Soc Responsib Environ Manag 26(4):885–901
14. Butler T (2011) Compliance with institutional imperatives on environmental sustainability: building theory on the role of Green IS. J Strateg Inf Syst 20(1):6–26. https://doi.org/10.1016/j.jsis.2010.09.006

15. Seth S (2017) World's top 10 software companies. https://www.investopedia.com/articles/personal-finance/121714/worlds-top-10-software-companies.asp
16. Calero C, García-Rodríguez de Guzmán I, Moraga M, García F (2019) Is software sustainability considered in the CSR of software industry? Int J Sustain Dev World Ecol 26(5):439–459
17. Calero C, Piattini M (2017) Puzzling out software sustainability. Sustain Comput J Sustain Comput Inf Syst 16:117–124
19. Stoller K (2017) The world's largest tech companies 2017: Apple and Samsung Lead, Facebook Rises. https://www.forbes.com/sites/kristinstoller/2017/05/24/the-worlds-largest-tech-companies-2017-apple-and-samsung-lead-facebook-rises/#274f0e72d140. Accessed Nov 2017

Chapter 16
Sustainability ArchDebts: An Economics-Driven Approach for Evaluating Sustainable Requirements

Bendra Ojameruaye and Rami Bahsoon

Abstract Sustainability refers to the ability of an architecture to achieve its goals and continue to deliver value on technical, environmental, social, and/or economic dimensions. Given the increased awareness of the need to conserve resources and be more sustainable, users are becoming more reluctant to support design decisions that are overdesigned or underperforming. There is a need for an efficient requirement evaluation framework that ensures that optimal software performance is achieved at a minimal cost. The goal of this chapter is to develop an objective decision-support framework for reasoning about sustainability requirements in relation to architecture decisions under uncertainty. We propose an economics-driven architectural evaluation method which extends Cost Benefit Analysis Method (CBAM) and integrates principles of modern portfolio theory to address the risks when linking sustainability requirements to architectural design decisions. The method aims at identifying portfolios of architecture design decisions which are more promising for adding/delivering value while reducing risk on the sustainability dimensions, and it quantifies the sustainability debt of these decisions. The results show that the method can make the value, cost, and risks of architectural design decisions and sustainability requirements explicit.

16.1 Introduction

Sustainable development can be defined as "the ability to meet the needs of the present without compromising the ability of future generations to satisfy their own needs" [1, 2]. Systems should be environmentally, socially, economically, and technically sustainable. The ability of a system to continue to deliver value on these dimensions over time determines the level of the software's sustainability. The processes of requirement engineering and architectural design are interleaved and intertwined, encompassing both technical and value-based (cost, value, and risk) decision making. Linking technical decisions to value-based reasoning facilitates

B. Ojameruaye (✉) · R. Bahsoon
University of Birmingham, Birmingham, UK
e-mail: bendra.o@gmail.com; r.bahsoon@cs.bham.ac.uk

C. Calero et al. (eds.), *Software Sustainability*,
https://doi.org/10.1007/978-3-030-69970-3_16

better understanding of the risks and likely value added, when engineering requirements and architecting under uncertainty. Economics-driven software engineering has acknowledged that integrating value-based theories with technical decision making can yield measurable improvements in development cost, time, and risk management.

An architecture is a set of design decisions that intends to meet all of the functional requirements while optimizing the desired quality attributes [3]. The problems in developing architectures are often due to an incomplete knowledge of the system and its environment and capacities. It can be also due to poor estimates or misperceptions of system requirements of the system environment and external conditions. While the process of architectural design implies a fit between the requirements, system conditions, and constraints, incomplete information and uncertainty may increase the cost of the architecture, introduce risks, alter its value, and influence the extent to which it can evolve and sustain over time. We generally neglect to consider that many of the design problem parameters are not completely known or deterministic, but they tend be estimated or stochastic. As a result, we often fail to recognize that our final design may not have completely met the actual requirements.

Technical debt is, per Brown et al., "a situation in which long-term code quality is traded for short-term gain" [4]. Technical debt in the requirement phase of system design is different from that in the implementation phases. It is incurred when requirements are prioritized, which do not deliver the most value to the customer [5]. Technical debt in the implementation refers to choices, to sacrifice quality for short-term gain [4]. To mitigate the risks of not meeting the system's goal and failure, software architects tend to overdesign the components of the system and build in redundancies rather than deal with the problem of uncertainty of parameters. While architectures have become increasingly sophisticated to meet the requirements, architectural design decisions are often still based on estimates and subjective judgment. The harm with this approach is that the final design is not usually optimized for value creation and technical debt reduction. In previous work [6], we introduced the concept of compliance debt. We had presented the hypothesis that one way to understand compliance debt, in relation to goals and obstacles, is to characterize it as the gap between what can be achieved with the available resources and the hypothesized "ideal" environments, where the goals are successfully achieved. In this chapter, we introduce the concept of sustainability debt as another form of technical debt, which provides a metric and quantifies the gap between the level of sustainability that will be achieved with a specific architecture and an ideal environment where the sustainability requirements are completely achieved.

Consider the case of architecting for sustainability. Given the increased awareness for the need to conserve resources and to be more sustainable, users are becoming more reluctant to pay the "premium" that results in architectures which are overdesigned. They are also reluctant to support design decisions which underperform on sustainability dimensions. These drivers have created a demand for an efficient requirement evaluation framework that ensures that optimal software performance is achieved at a minimal cost. In many cases, only some requirements actually determine and shape a software architecture; these requirements are called architecturally significant requirements (ASRs). Architecturally significant requirements are those requirements that play an important role in determining the architecture of the system [7]. Such requirements require special attention. They can be explicitly or implicitly architecturally significant, and, in this chapter, we posit that sustainability requirements can be considered as ASRs. Identifying and tracking technical debt on the architecture over time can provide the decision maker with insights into the extent to which the solution continues to be sustainable or tends to hedge behind the optimal value.

This work proposes an economics-driven architectural evaluation method which extends the Cost-Benefit Analysis Method (CBAM) and integrates the principles of modern portfolio theory to evaluate software architectures for sustainability. The extension acknowledges that architectural design decisions are based on estimates and thus are subject to varying degrees of risks, uncertainty, and technical debt. It hopes to identify architectural design decisions which minimize costs, reduce risk, and maximize value on the identified sustainability dimensions—technical, environmental, social, economic, and/or human. The analysis can provide insights on the extent to which architectural design decisions overperform/underperform from the ideal values—depicted as debt. As architectural design decisions are responsible for fulfilling the system's goals, their contribution to sustainability needs to be evaluated in relation to value creation. We hope to support the reasoning about value concerning the impact of alternatives on sustainability goals.

We build on the concept of debt in requirements [6] as a form of technical debt at the requirement level, which is the result of neglecting nonfunctional compliance when engineering the requirements of software. We propose a value-driven solution, which builds on the CBAM by using portfolio-based thinking and technical debt analysis to systematically manage sustainability requirements and the way they can be met by the architecture. In this context, we hypothesize that some requirements when met by the architecture may introduce debt that needs to be managed for creating value and mitigating risks. In finance, a portfolio denotes a collection of assets (investments) by an investor, usually used as a strategy for minimizing risks and maximizing returns [7]. The goal of portfolio thinking is to select the combination of assets using a formal mathematical procedure that can minimize risks for an expected level of return on investment while accounting for uncertainty in the real world. We use portfolio thinking to make the links between costs, benefits, risks, and technical debt explicit for requirement engineers.

16.2 Background

Research on three major areas—sustainability in software, architectural evaluation methods, and portfolio theory—is related to our work.

16.2.1 Sustainability and Goal-Oriented Requirements

The term sustainable system can be interpreted as either the system is sustainable or the system's purpose is to support sustainability goals [3]. A sustainable system is energy efficient, minimizes the environmental impact of the processes it supports, and has a positive impact on the other dimensions including economic, social, human, and technical sustainability [3, 8].

Sustainability requirements need to be aligned with prevailing regulations to control compliance, to add value, and to meet the needs of the business. Sustainability thinking for software is:

- Reasoning about sustainability as a value-seeking process
- Reasoning about sustainability in the presence of uncertainty and dynamism
- Reasoning about sustainability in heterogeneous and scalable environment

Our working hypothesis is that software sustainability can be considered as a nonfunctional requirement and a composite attribute that is affected by other quality attributes such as reliability, availability, and flexibility to change. Nonfunctional requirements do not have simple true or false satisfaction criteria; rather their level of satisfaction can vary [9]. Although sustainability requirements are crucial to the business economic goals, they do not have a clear link to revenue generation. Henceforth, the benefits and returns of sustainability investments are difficult to comprehend and visualize. To analyze sustainability requirements, we employ parallels in defining the objectives and elaborating the goals and risk analysis techniques to identify relevant debts. We argue that making software sustainable is ultimately an investment activity that requires value-driven decision making—about selecting the right decision options or alternatives based on their impacts on different sustainability perspectives, while handling the trade-offs associated with those decision and quantifying the risks associated with wider issues.

16.2.2 Architectural Evaluation

The architecture of the system is the first design artifact that addresses the quality goals of the system such as security, reliability, usability, modifiability, stability, and real-time performance. Architectural evaluation is an activity for developing an assessment of an architecture against the quality goals. Our method builds on the previous methods for dealing with the uncertainty in software architecture decisions, notably the Cost-Benefit Analysis Method (CBAM) [10, 11].

The CBAM is an architecture-driven method for analyzing the costs, benefits, and schedule implications of architectural decisions. It consists of the following steps:

- Choosing scenarios and architectural strategies (ASs)
- Assessing quality attribute (QA) benefits
- Quantifying the architectural strategies

- Costs and schedule implications
- Calculating desirability
- Making decisions

Upon completion of the evaluation using the CBAM, the stakeholders are guided to determine a set of architectural strategies that address their highest priority scenarios. These chosen strategies furthermore represent the optimal set of architectural investments. They are optimal based upon considerations of the benefits, costs, and schedule implications, within the constraints of the elicited uncertainty of these judgments and the willingness of the stakeholders to withstand the risk implied by uncertainty.

To quantify the benefits of architectural strategies, the stakeholders are asked to rank each architectural strategy (AS) in terms of its contribution to each quality attribute of −1 to +1. A +1 means that this AS has a substantial positive effect on the QA (e.g., an AS under consideration might have a substantial positive effect on performance) and −1 means the opposite. Each AS can be assigned a computed benefit score from −100 to +100. The CBAM doesn't provide a way to determine the cost; it considers that cost determination is a well-established component of software engineering and is outside its scope. The benefits and scores result in the ability to calculate desirability metrics for each architectural strategy. The magnitude of desirability can range from 0 to 100.

16.2.3 Portfolio Management and Requirements

Modern portfolio theory [12] was introduced in 1952 by Harry Markowitz. Its goal is to select the combination of assets using a formal mathematical procedure that can minimize risks for an expected level of return on investment while accounting for uncertainty in the real world. In finance, a portfolio denotes a collection of weighed compositions of assets (investments) by an investor, usually used as a strategy for minimizing risks and maximizing returns. Portfolio theory attempts to show the benefits of holding a diversified portfolio of risky assets rather than assets selected individually. The theory can also assist in determining the optimal strategy for diversification of assets to minimize risks and maximize returns. This can be linked to the process of analyzing obstacles to sustainability, where analysts make decisions on which obstacles should be resolved given a certain amount of resources with minimum risks.

In modern portfolio theory, the risk of a portfolio R_P is determined by the individual risks associated with each asset R_1, the weight of each asset in the portfolio W_1, and the correlations between the assets P_{IJ}. These correlation coefficients range from −1 to +1 and 0 indicates no relationship between the items.

The link between the selection of requirements and market value using portfolio has been first explored by [13]. They proposed a market-driven, systematic, and more objective approach to supplement the selection of requirements, which accounts for uncertainty and incomplete knowledge in the real world using portfolio reasoning [13].

Sivzattian and Nuseibeh [14] elaborated on portfolio analysis to the software decision problem. Sivzattian and Nuseibeh [14] presented an approach for selecting and prioritizing requirements based on portfolio analysis.

Asundi et al. [1] applied portfolio theory to the CBAM to combine architectural strategies with the aim of combining uncorrelated or negatively correlated AS to reduce uncertainty. Our use of portfolio is different: We identify an architecture as an optimal portfolio of architectural strategies. We employ the analysis on the goal and architectural levels. We explicitly look at linking sustainability goals to architectural strategies, risk, returns, and debt.

16.2.4 Related Work

In this section, we provide a review of software architecture evaluation methods that are related to our work.

The Architecture Trade-off Analysis Method (ATAM) [15] does not only reveal how well an architecture satisfies the particular quality goals but also provides insight into how these goals interact with each other—how they trade off against each other [16]. The ATAM is a scenario-based architectural evaluation method. The Software Architecture Analysis Method (SAAM) [17] elicits stakeholders' input to identify explicitly the quality goals that the architecture is intended to satisfy. Unlike the ATAM, which operates around a broad collection of quality attributes, the SAAM concentrates on the attributes for modifiability, variability (suitable for product line), and achievement of functionality. The Cost-Benefit Analysis Method (CBAM) [10] is a method for analyzing the costs, benefits, and schedule implications of architectural decisions. It builds upon the ATAM to model the costs and benefits of architectural design decisions and to provide means of optimizing such decisions.

Our work builds on the CBAM. While the CBAM can be used for evaluating quality attributes, it does not explicitly address sustainability. Also, although it ranks architectural strategies based on their return on investments, it does not explicitly address identifying optimal candidate architectures.

However, some approaches have considered identifying the optimal candidate architecture. GuideArch [18] guides architectural decision making such as ranking of the architectures and finding the optimal candidates, under uncertainty using fuzzy logic. Letier [19] presents an evaluation method that allows describing uncertainty about the impact of alternatives on stakeholders' goals. The method calculates the consequences of uncertainty through Monte Carlo simulation and shortlists candidate architectures based on the expected costs, benefits, and risks. It also assessed the value of obtaining additional information before deciding.

Although none of these approaches explicitly address sustainability [1, 14], they can be viewed as complementary to our approach as they could be used to shortlist architectures and identify optimal architectures based on value.

16.3 The Problem

Sustainability needs to be aligned with the systems and business goals as well as the prevailing regulations, to add value, and to meet the needs of the business. With recent developments and demands for sustainable software, some software systems have a few sustainability requirements to which it should comply with to retain or improve its value. However, the stakeholders are interested in maximizing returns of the systems while minimizing risks and costs of development.

A key to a good alignment between the business domain and sustainability requirements is to keep the focus on value at both the requirement and architectural level (architecturally significant requirements). Requirements and architectural decisions need to add value, and this value can be characterized by maximizing benefits while minimizing risks at a reasonable cost.

The objective is naturally to aid requirement analysis and decision making by using an evaluation method that quantifies and visualizes value. In this context, a lot of work has already been done in requirement analysis and software evaluation method. Some architectural evaluation methods based on cost-benefit assessments [10, 17, 20] already exist. However, these methods are generally applied at the architectural level, once the requirements have been selected and specified. This results in a gap between understanding the value of sustainability requirements and the impact on the architecture.

Therefore, this work is focused on the requirement level, linking architectural decisions with requirement analysis, instead of being only focused on architectures.

Existing methods are generally focused on finding the optimal architecture. However, they fail to consider that the optimal solution may be unnecessary and unaffordable or not deliver the best value to the customer. This leads to a gap referred to as a debt between the optimal solution identified by the evaluation method and the actual solution required with respect to the relevant decision space. This lack of identification and quantification of debt prevents its understanding and management (trade-offs), which may lead to poor decision making.

The problem we address is how to evaluate sustainability goals in architectures quantitatively under uncertainty. We look at how the value of the architecture is computed for higher-level sustainability requirements given the refinement questions and metrics for sustainability-related variables. We also look at quantifying the debt that may be incurred in any suboptimal architecture.

16.3.1 Requirements for the Model

We started our investigation of the problem by undertaking a survey of portfolio management practices in software and other industries [8, 21]. We included portfolio management in our survey because we believe a portfolio viewpoint is needed to manage risks [15]. As a result of this investigation, we identified a number of requirements that any useful model ought to address. These requirements, which include a definition of the two categories of risk our model aims to address, are:

- Linking technical decisions to value (cost, value, and risk).
- Evaluating sustainability as an architecturally significant requirement.
- Evaluating technical debt in the requirement phase of system design.
- Evaluating the impact of sustainability on architecture selection decisions.
- Evaluating the impact of uncertainty in cost estimates as a risk.
- Evaluating the impact of obstacles to operationalizing goals as a measure of risk: While goals capture the desired objectives, obstacles [3] to these goals capture undesirable properties that may obstruct those sustainability goals.

16.3.2 Requirements and Value Component Relationship Model

Value breaks down into three components: risks, costs, and benefits. A typical value-based approach begins with the identification of the decision parameters, followed by the identification and analysis of risks attached to those decisions. After the risk assessment process, a cost-benefit analysis is done to quantify the costs and the benefits of the decisions. In the context of requirement/architectural evaluation, at the end of the process, the value of an architecture plays a significant role in determining how well it meets the initial requirements. Therefore, to systematically understand the value of an architecture, it is necessary to explicate the relationships between the sustainability requirement and the value components.

To systematically inquire about the value components expressed (or missing) in sustainability requirement descriptions, we extend the Common Criteria model [16] to also include sustainability requirements. The resulting model, as shown in Fig. 16.1, leads to the conceptualization of a sustainability requirement in terms of its related value components.

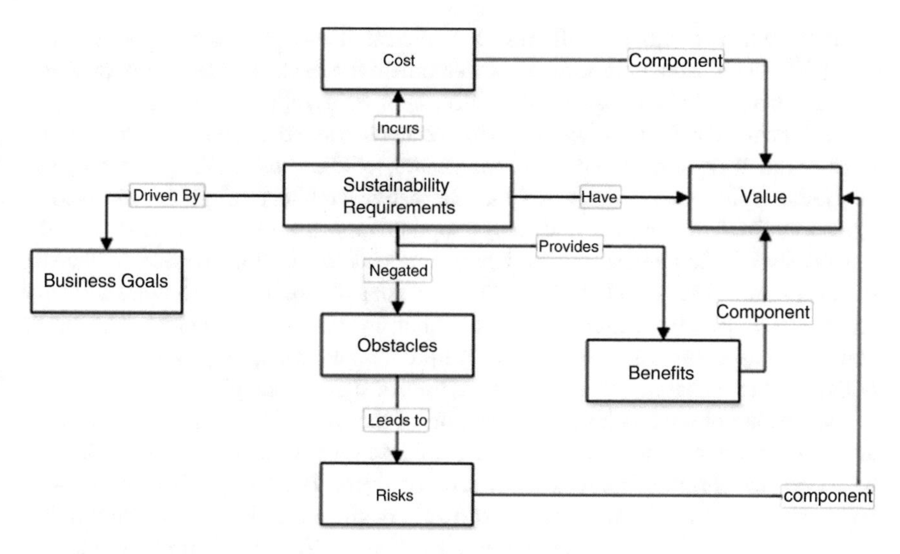

Fig. 16.1 Sustainability requirements and value component relationship model

16.4 Sustainability: A Technical Debt Perspective

The goal of the software architecture process is to create a detailed recipe of design decisions for developing software that will satisfy the functional, quality, and business requirements. Consequently, achieving this goal necessitates active decision analysis of the goals and architectural strategies (ASs) in order to ensure stakeholders' satisfaction and optimal investment decisions under uncertainty. Architectural portfolio management involves decision making aimed at creating a mix of ASs that returns maximum value for the resources invested. This leads to critical decisions involving the selection of an optimum mix of requirements aimed at satisfying the necessary goals.

Unlike previous work, we introduce a new dimension of using sustainability debt as a decision factor for elaborating and managing requirements through architectural evaluation. We incorporate the debt analysis at the requirement analysis and early architectural decision levels. Technical debt can inform the architectural evaluation process and the decision for investing in a specific architecture at the early stages of the development life cycle. Our objective is to avoid inappropriate selection of an architecture that is not value driven and debt aware. The key principle here is to tackle and manage the increased and unjustified debt, which can be associated with the selection of and consequently the inappropriate architecture candidates. We assume that debt can vary with different architectures and each architecture can deliver its own trade-offs for risk, value, cost, and debt reduction. We advocate a predictive approach for anticipating and managing debt at the requirement and early architectural evaluation stages. A predictive approach can be applied during the early stages of the engineering process to predict the debt, its impact, when it will be incurred, when it will pay off, and the interest if any. Classical approaches to managing technical debt in software development life cycle tend to be retrospective. Unlike retrospective approaches, predictive approaches allow planning.

Technical debt in sustainability requirements can be traced back to requirements—the way requirements are engineered, elicited, selected, prioritized, and analyzed. Debt can be linked to the alternative architectural strategies used to operationalized requirements, their appropriateness, the resources used, and the risk involved. The interest on this debt can be characterized as the rate of increase in this distance. Uncertainty about whether or not a decision is appropriate will have an associated penalty, which may incur a debt. In this sense, technical debt can be considered as a particular type of risk; the problem of managing sustainability debt boils down to managing risk and making informed decisions [22].

Our methodology combines the strengths of the CBAM [7] with portfolio theory and technical debt. Our motivation for combining the aspects of these methods is based on the observable strengths in each of them. The CBAM is a quantitative approach for economic modeling of software engineering decisions, which builds upon the Architecture Trade-off Analysis Method (ATAM) [1]. It provides the cost and benefit of different architectural candidates. On the other hand, the goal of modern portfolio theory is to select the combination of assets using a formal mathematical procedure that can minimize risks for an expected level of return on investment while accounting for uncertainty in the real world.

Given a set of design alternatives, these alternatives have varying prospects under uncertainty, which in turn affects the overall value of the architecture. Requirement engineers and architects should be able to classify their requirements and design options at any given time. Beyond that, they should also understand how these options affect the architecture's value. Portfolio thinking can help us manage alternatives more effectively in different ways. It allows us to estimate the value of an architecture, before the refinement process actually ends, as well as assess each design alternative as the development process progresses.

The set of requirements and design alternatives to be used can be called the option space. This space is defined by metrics that help address issues such as selecting an alternative. These metrics include the standard net present value and a measure of the uncertainties.

The steps in our methodology are designed to support developers and analysts in systematically estimating the value of prioritized or selected architecturally significant requirements and architectures. The resulting insights help to establish the link among sustainability requirements from various perspectives in the context of system operation and understand the true value of requirements in the presence of risk, cost, and uncertainty. Our value and risk-aware approach consists of the following basic steps (Table 16.1).

Table 16.1 The value and risk-aware method

Step	Description
1	Elicit and prioritize the goals, cost, and the desired sustainability threshold for these goals
2	Develop architectural strategies (ASs) for the goals, elicit their parameter values, and define the architectural decision model
3	Determine architectural strategies' (ASs) impact on sustainability
4	Elicit sustainability goals, design decisions, obstacle, and risk analysis
5	Determine the expected benefit of each design option
6	Analyzing the costs, benefits, risks, and value using portfolio thinking
7	Identify the optimal architecture, calculate debt, and rank other architectures

16.4.1 Step 1: Elicit and Prioritize the Goals, Cost, and the Desired Sustainability Threshold for These Goals

The first step of the CBAM consists of identifying the decisions to be taken together with their architectural strategies (ASs), defining the goals against which to evaluate the decisions. We use existing goal-oriented requirement engineering methods [23] to elicit, elaborate, and refine the goal model. These ASs have parameters, and each parameter is assigned a value.

16.4.2 Step 2: Develop Architectural Strategies for the Goals, Elicit Their Parameter Values, and Define the Architectural Decision Model

The decisions to be taken together with their architectural strategies (ASs), defining the goals against which to evaluate the decisions, are identified using a goal-oriented requirement engineering method. Based on the requirements implied by the elicited goals, ASs or design options are developed and the parameter values elicited. Following the CBAM principles, we associate utility with the AS in meeting different sustainability requirements and the level of satisfaction. This consists of eliciting the values of the contribution of the AS to the goal for all parameters in the architectural models. The CBAM is however not constrained to the use of any particular elicitation technique. For each AS and goal pair, the impact that the AS is expected to have on that goal is then estimated.

16.4.3 Step 3: Determine Architectural Strategies' Impact on Sustainability

Once the options have been developed and their parameter values elicited, its relationship to sustainability goals such as technical and economic sustainability is determined. The degree, to which each of these attributes is expected, varies from domain to domain. The sustainability parameters [8] used in the framework proposed in this chapter are as follows (the evaluation benefiting from these parameters is long term and strategic in nature):

- **Human sustainability [H]** parameter checks the extent to which the AS being evaluated is likely to contribute and support the individual benefiting from the system over time. As an example, consider the case of the architecture of a wearable medical device supporting an elderly individual. Parts of human sustainability cover privacy, safety, security, and usability. In addition, there is a strong focus on personal health and well-being, which still needs to be made explicit in the requirements. Within software development, an example is the number of people hours involved in the development and implementation. Software is developed and managed by people. Therefore, it is fundamental to consider the value aspects related to human capital while taking product management decisions.
- **Economic sustainability [EC]** parameter checks the extent to which the AS being evaluated is economically viable and likely to economically sustain over time. In this context, the AS is treated as an asset that needs to be maintained over time and be monitored for its operational cost, its risks, and its ability to deliver value and unlock its potentials during the lifetime of the system. The economic sustainability is taken care of in terms of budget constraints and costs as well as market requirements and long-term business objectives that get translated or broken down into requirements for the system under consideration. Within a software development context, the value aspects related to economic sustainability need to be considered while taking decisions.
- **Environmental sustainability [EN]** parameter checks the extent to which the AS being evaluated is likely to improve human welfare by reducing the carbon footprint, protecting natural resources, and being energy aware. In particular, the evaluation covers the production, operation, maintenance, and evolution lifetime of the software system. The requirements with regard to resource flows, including energy, can be elicited and analyzed by the life cycle analysis (LCA). Other aspects are efficiency and time constraints. When building a software system, these dimensions appear in the form of things like cost of energy, efficient algorithms, and development processes.
- **Technical sustainability [T]** parameter checks if the AS being evaluated is necessary to achieve technical sustainability. Technical sustainability has the central objective of long-time usage of systems and their adequate evolution with changing surrounding conditions and respective requirements.

Penzenstadler and Venters et al. [3, 24] proposed that sustainability should be considered as a measure of a system's availability, integrity, maintainability, and reliability. We view technical sustainability as the degree to which a system is maintainable, reliable, usable, and portable.

- **Social sustainability [S]** parameter checks the extent to which the AS being evaluated is likely to sustain and contribute to society. This parameter evaluates the ability of the AS to maintain social capital and preserve societal communities and their solidarity in the way they perceive the benefits of the AS over time. In particular, the AS is evaluated for its likely positive/negative effects on society. This analysis becomes more important once we develop the architectures of systems of systems (SoS), ultra-large-scale architectures, architectures for smart and federated cities, and social network systems, among others. Part of the social sustainability dimension considers political, organizational, or regulatory requirements [8].

We use the Software Value Map [15] which offers a unified view of value to evaluate the impact of these dimensions based on three value perspectives: the financial, the customer, and the internal business process.

- The **financial perspective** looks at the aspects of the implementation decision that affect short- and long-term financial goals. Some of the most common financial measures are the cost of implementation, earned value analysis, and profit margins.
- The **customer perspective** looks at the value proposition of the software to the customers. Measures that are selected for the customer perspective on sustainability should calculate the value that is delivered to the customer with respect to the perceived time, quality, performance and service, and cost [15].
- The **internal process perspective** is concerned with the processes that maintain and encourage sustainability. Quality, cycle time, productivity, and cost are some aspects where performance value can be measured [17] (Table 16.2).

We look at the interrelationship between the impacts on different value aspects. This is necessary because selection decisions based on single-dimension impacts could obscure other impacts that might cause equal or greater damage. The interrelationships can exist as the following effects:

- A positive impact on one value aspect might have a positive impact on one or more sustainability dimensions.
- A negative impact on one value aspect might have a negative impact on one or more sustainability dimensions.
- A positive impact on one value aspect might have a negative impact on one or more sustainability dimensions and vice versa.

The next step is to compute the AS that has a likely impact on the sustainability dimensions. The sustainability value of the AS is a utility function of the utility values on H, S, EC, EN, and T. The parameters are assigned integer values from 1 to

Table 16.2 Sustainability dimensions, value perspectives, and value indicators

	Financial	Customer	Internal process
Individual	The financial implications of user satisfaction and retention	Sustained perceived value to the user (e.g., perceived satisfaction, convenience, etc.)	Implications on resources including human, time, etc. in relation to development, maintenance and evolution, etc.
Economic	Total development, maintenance, and evolution costs, among others	Cost and value in supporting user's specific concerns (e.g., can relate to users' specific functional and nonfunctional requirements)	Cost, value, and risks for developing, maintaining, and monitoring sustainability
Environmental	The financial implication of complying and/or noncompliance with environmental sustainability. Market losses or gains relative to market sustainability	Environmental impact due to energy usage and CO_2 emissions	Meeting market and sustainability requirements and their cost. Risk and value implications
Technical	Total development, maintenance, and evolution costs, among others	Cost and value in supporting functional and nonfunctional requirements	The value of maintaining the system over time
Social		Value from incentives such as cost and complementary value through network of users and technologies	Development team management

5. Sp is the sustainability impact parameter and can assume integer values from 1 to 5. These parameters are calculated as follows:

$$\mathrm{Sp} = (\alpha\mathrm{H} + \beta\mathrm{S} + \gamma\mathrm{En} + \delta\mathrm{Ec} + \epsilon\mathrm{T}) \quad (16.1)$$

where α, β, γ, δ, and ε are the relative weights assigned to the different parameter values based on the relative importance of the system.

Usually qualitative metrics may be extended by quantifying or replacing the value such as high, medium, or low with numerical weights. Although this principle is often used in practice, problems such as lack of precision (for precise requirements) and lack of clear criteria to validate the estimation still exist as this technique is subjective. For simplicity, we propose a quantitative measurement metric that considers the impact of the goal on different attributes based on subjective criteria such as the stakeholders' experience. The parameter values ranging from 1 to 3 are assigned depending on whether the impact is high, medium, or low. For the valuation to be accurate, it is important to estimate the parameters accurately. In this chapter, our focus is more on developing a framework for quantitative analysis system rather than model parameterization. The decision on which level the parameters are investigated depends on the scope of the system under analysis.

16.4.4 Step 4: Elicit Sustainability Goals, Design Decisions, Obstacle, and Risk Analysis

Architectural design decisions should take into consideration the risks associated with each architectural decision and the resulting candidate architecture.

We compute the sustainability value and the corresponding risk values of the AS. We take a goal-oriented requirement engineering (GORE) [23] approach for specifying and refining **H**, **S**, **EC**, **EN**, and **T**.

We use the obstacle analysis [23] of GORE for identifying the risks on sustainability in conjunction with the AS. In particular, various ASs correspond to the agents operationalizing the goals in GORE terms. The presence of obstacles is an indication of risks. Risks refer to the potential damage that can occur if that obstacle is not resolved. When identifying the risks of the AS, each strategy should not be considered in isolation. A risk to one AS might negate other AS. The use of GORE provides decision makers with a systematic and traceability method for reasoning about architectures in conjunction with sustainability goals and risks. The process of goal refinement and elaboration benefits from the fundamentals of GORE; its analysis and evaluation can be iterative and continuous. The approach can inform the decision for further refinements for sustainability. The metric can also inform the desirable stopping criteria for the refinements and analysis processes for accepting or rejecting the AS.

Risks can be formally modeled using two properties—the criticality and the likelihood of the risks occurring. The criticality of a risk indicates how bad the consequence of the risk is likely to be. It also depends on its impact on higher-level goals and the relative importance of those goals. Every risk incurs a cost for its mitigation (mostly resource consumption), and this affects the risk's criticality. In the proposed method, criticality is computed on a five-point scale as shown in Table 16.3.

Table 16.3 Criticality values of risks

Criticality	Interpretation
Very high (5)	The risk has a high impact on multiple higher-level goals
High (4)	The risk has a high impact on a single higher-level goal
Medium (3)	The risk produces a fair contribution to goal negation
Low (2)	The risk has a low impact on multiple higher-level goals
Very low (1)	The risk has a low impact on a single higher-level goal. Its impact can be insignificant

Likelihood denotes the probability that the risk will occur. In our framework, we calculate the likelihood of a risk occurring using historical data (e.g., figures corresponding to T, EC, etc.) and on the basis of the value of evidence that increases and reduces the likelihood. The likelihood is defined quantitatively and can take the following values: *almost certain (5), likely (4), occasionally (3), rare (2), and unlikely (1)*. It can be determined by past occurrences, which represents the previous incidents that have occurred due to the risk.

$$R_v = \log_2(L_o * C_o) \tag{16.2}$$

The reason for taking the logarithm to base 2 is to normalize the result within a scale of 0–5. Every risk has a risk factor. R_v is the value of the risk, L_O is the likelihood that the risk will occur, and C_O is the criticality. Another lightweight technique to support this process consists of using the standard risk analysis matrix in which the likelihood and criticality are estimated.

16.4.5 Step 5: Determine the Expected Benefit of Each Design Option

We use the CBAM principles here, and we track the benefits on goals/scenarios along with the sustainability goals. The benefit or the effectiveness of an AS is an assessment of how well the option helps in achieving the sustainability goal. To quantify the benefits, we associate utility with aspects of the option such as its satisfaction of the sustainability and quality attributes. Based on the information collected, the total sustainability benefit of each AS can be calculated across goals by summing the impact on the sustainability attributes associated with each goal (weighted by the importance of the goal and the attribute). This can be calculated using CBAM (Eqs. 16.3 and 16.4).

16.4.6 Step 6: Analyzing the Costs, Benefits, Risks, and Value Using Portfolio Thinking

We use the CBAM principles here and we track the benefits on goals/scenarios along with the sustainability goals. The benefit or the effectiveness of an AS is an assessment of how well the option helps in achieving the sustainability goal. To quantify the benefits, we associate utility with aspects of the option such as its satisfaction of the sustainability and quality attributes. Based on the information collected, the total sustainability benefit of each AS can be calculated across goals by summing the impact on the sustainability attributes associated with each goal (weighted by the importance of the goal and the attribute). This can be calculated using CBAM (Eqs. 16.3 and 16.4).

$$\text{Benefit}(\text{AS}) = (\text{contribScore})\,(\text{Qattrib}) \tag{16.3}$$

$$\text{Return}(\text{AS}) = \text{Benefit}/\text{Cost} \tag{16.4}$$

We now analyze the costs, benefits, and risks associated with alternative decisions to create an architectural design using portfolio theory. Since sustainability requirements do not have simple true or false satisfaction criteria, but are satisfied up to a level [25], we can determine how well the goal needs to be achieved with the available resources. With the measurements of the value of the individual options as described above, all input information for the portfolio approach is ready. We can start making decisions using the portfolio approach. Each AS has a risk value R_1, a cost P_1, and the weight of the risk W_1. Based on these values, we can then decide on how many instances of the risks to the goal need to be resolved so that the global risk of the goal being obstructed is reduced.

One can see that an architecture has similarities to the financial market. However, we use portfolio theory to identify an architecture as an optimal portfolio of architectural strategies with a total value that minimizes risks and maximizes return.

The total cost of an architecture is

$$C = \sum_{i=1}^{m} C_1 \tag{16.5}$$

And weight W is

$$W = C_1 / \left(\sum_{i=1}^{m} C_1 \right) \tag{16.6}$$

Where P_1 is the return of the architectural strategy and W_1 is the weight assigned to that strategy, the expected return E_p of an architecture is

$$\sum_{i=1}^{n} E_\text{p} = 1 \tag{16.7}$$

$$E_\text{p} = \sum_{n=1}^{m} W_1 (B_1/C_1) \tag{16.8}$$

$$E_{\mathrm{p}} = \sum_{n=1}^{m} W_1(W_1 P_1) \tag{16.9}$$

Where R_1 is the local risk of the architectural strategy and W_1 is the weight assigned to that strategy, the global risk of an architecture is

$$R_{\mathrm{p}} = \sum_{n=1}^{m} \sqrt{W_1^2 R_1^2} \tag{16.10}$$

16.4.7 Step 7: Identify the Optimal Architecture, Calculate Debt, and Rank Other Architectures

Evaluating and selecting the best architecture based on the requirements is a core activity in the requirement elaboration process. We evaluate the architecture by considering the amount of sustainability debt that each architecture may incur as the deciding factor.

The next step is to shortlist valid and optimal architectures. An architecture is valid, if it satisfies the relevant constraints such as cost, risk, schedule, or sustainability score threshold. The default is to shortlist valid candidate architectures that maximize the expected net benefits and minimize risks on dimensions related to sustainability. Identifying the optimum architecture is a core activity in the evaluation process. An architecture is optimal for a given problem if it achieves minimum risk and maximizes value while satisfying critical constraints. This is defined as a portfolio optimization problem to minimize risk and maximize return subject to cost constraints. Where E_{R} is the expected return and P_{d} is the standard deviation of the portfolio, we define this term of Sharpe ratio as:

$$\text{Maximise } E_R / P_d \tag{16.11}$$

This can also be used for comparison and benchmarking of other architectural portfolios. The architecture with a high Sharpe ratio is the one that achieves the best combination of value and risk in the presence of uncertainty.

The ability to find the optimal solution is complemented with the ability to rank architecture candidate portfolios. The architecture is a portfolio of ASs, where each AS has its risk and benefits on goals and the sustainability dimensions. We rank the architecture (i.e., portfolio for our case) by considering the amount of debt that each architecture may incur as the deciding factor. We calculate and assign the sustainability debt of each alternative. We evaluate the different architectures by considering the amount of debt that each resolution tactic may incur as the deciding factor. From our earlier explanation of the debt using the technical debt metaphor in as the gap between what can be achieved with the available resources and the hypothesized "ideal" environments where the goals are successfully achieved, we formulate the value of the debt as:

$$T_D = I_{RT} - R_T \tag{16.12}$$

where T_D is the debt, I_{RT} is the Sharpe ratio of an "optimal" architecture, and R_T is the Sharpe ratio cost of the selected architecture. The ideal value is context dependent and is application and business dependent. T_D for any other tactic is calculated as the gap between the value of architecture k (I_{RT}) and the value of the architecture in question. The architectures can then be ranked from best (low T_D) to worst (high T_D).

This approach models technical debt as a function of requirement trade-offs. It presents the optimal solutions at a given point in time in terms of specific scenarios. This approach is similar to technical debt in the implementation: Even though we may implicitly know that a particular architecture is not optimal in the long term, presumably we choose this architecture precisely because it works at that point in time. It is only when time passes that debt begins to accrue as the sustainability and business needs begin to favor the optimal solution more than the actual solution. With respect to "interest" on the debt, which we loosely translate as the cost of neglecting the debt, this is the rate of change between the current solution and the current state of the requirements.

Debt can be attributed to the inability of the architecture to partially or fully meet the changes in the requirements and the environment. We claim that a sustainable system tends to add value with time and as it evolves and the added value is relative to the stakeholders involved. Debt can be relative to the scenarios. It could be discussed from the perspective of the end users and the perspective of the provider (system developer). When T_D is discussed in conjunction with and relative to all the scenarios of interest, it can give an indication of the overall system's sustainability. T_D at the requirement level can be attributed to different situations such as:

1. The architecture does not fully meet the requirements.
2. An overdesign of the architecture so the potentials of the architecture following composition are not fully utilized and the operational cost tends to exceed that of the generated benefits.
3. The absence of publicly available historical data for analyzing the impact of decisions that may lead to poor and quick decisions that add a value in the short term but can introduce long-term debt in situation where improving the system is unavoidable.

16.5 Evaluation

We apply the proposed method to the case study from the literature [18], referred to as Emergency Deployment System (EDS), to motivate, describe, and evaluate our research. The objective of this case study is to validate the method and demonstrate the derivation of variables and functions for sustainability evaluation from the system's goals. Our choice stems from the adequate level of complexity involved, while taking into account the demands for the need for sustainable architecture and the contradictory needs of the different stakeholders. For the simplicity of exposition, we look at a sustainability scenario which needs to be considered when architecting the software system. In particular, the AS and the architecture that compose these strategies need to preserve the reliability of the system, minimize power consumption, minimize costs, and maximize response time to ensure technical, environmental, and economic sustainability.

The evaluation hopes to demonstrate the extent to which the method can systematically evaluate architectural strategies for sustainability and select a portfolio of ASs which have the potential for reducing cost and risk and adding value on the sustainability dimensions. The approach was modeled as a multi-objective optimization problem in which the utility, risk, and development cost functions formed the three objectives. The case study relied on the following hypothesis:

- The framework is applicable to large-scale, industrial projects.
- The framework provides a better guide to addressing sustainability requirements than conventional requirement models.
- The technique allows one to systematically identify a set of risks, some of which may be missed if no systematic technique is applied, and to link each risk to sustainability to the system goals it may impact.

The hypothesis can be considered true if the following objectives are met:

- Demonstrates the capability of the framework to describe, model, and analyze sustainability requirements of software systems at design time systematically both at the requirement and architectural levels
- Demonstrates how the results achieved using the framework offer a better guide at addressing sustainability requirement
- Demonstrates the use of sustainability debt as a metric and measure of risk

In order to examine the validity of the hypothesis through these objectives, a qualitative analysis of the use case models is conducted. The case study was selected from emergency response domain as the systems in this domain need to be sustainable to accommodate uncertain conditions, increase efficiency, and reduce cost with minimum risk to people.

16.5.1 The Problem

The software to be designed is an Emergency Deployment System whose purpose is to support the deployment of personnel in emergency response. For the case study, the original design team applied GuideArch, an approach which supports design decisions under uncertainty [18].

16.5.2 Design Decision Evaluation

16.5.2.1 Step 1: Elicit and Prioritize the Goals, Cost, and the Desired Sustainability Threshold for These Goals

The original design team elicited priority values for each of the seven goals, presented in Table 16.4. Goals G_1, G_2, G_3, and G_5 have sustainability implications, which should be factored into the evaluation of the design. The strategic requirements related to the sustainability requirements are listed in Table 16.4.

16.5.2.2 Step 2: Develop Architectural Strategies for the Goals, Elicit Their Parameter Values, and Define the Architectural Decision Model

The original team had also defined models for computing the impact of the AS on the goals using a three-point estimate for each parameter: Since the model has 25 possible ASs which could be induced by the alternatives and 7 goals, we have 25×7 (175) parameters. As we are interested in evaluating the sustainability of the ASs, our architectural model would look at the contribution of an AS to a goal in terms of sustainability. We thus define the goal evaluation function as the sum of the contributions of each AS composing an architecture and its corresponding sustainability impact.

Table 16.4 Sustainability requirements for EDS

1.	Minimize cost
2.	Minimize battery usage
3.	Maximize reliability
4.	Minimize response time
5.	Minimize ramp-up time
6.	Minimize development time
7.	Minimize deployment time

16.5.2.3 Step 3: Determine Architectural Strategies' Impact on Sustainability

Once the AS has been identified and their parameter values are elicited, we determine the likely impact an AS can have on sustainability goals, including technical and economic sustainability. Since our work is concerned with sustainability goals, it was necessary for us to design a characterization for sustainability which we use to elicit sustainability-specific questions for evaluation.

Given an AS, we are interested in understanding its impact on sustainability. We estimate the impact of the AS on sustainability goals using the already elicited parameters. Since goals G_1, G_2, G_3, and G_5 are likely to impact sustainability, we use the impact parameters to quantify the impact. For example, battery usage can be linked to environment, and technical sustainability as an architecture with a high battery usage may lead to low availability; the solution may not be technically sustainable in the context of emergency response systems. Also, a high battery usage means an increased carbon footprint, and this inversely affects environmental sustainability. Similarly, reliability and response time figures could be linked to technical and human sustainability, where any degradation in reliability may have negative consequences on the system and human, threatening the system's sustainability. Increased development and deployment times could be linked to increased carbon footprint, which can be linked to environmental sustainability. We compute the impact values using a scale of 1–5 utilizing a three-point estimate for each parameter (most likely, lowest, and highest values). Table 16.3 shows the result of 4 ASs out of the 25 ASs identified.

After the impact parameters are determined, we compute the relative weights of each sustainability goal, to reflect their relative concerns. For this case, we illustrate the computation on the technical sustainability. The total impact of an AS on sustainability is described using Eq. (16.1) (Table 16.5).

16.5.2.4 Step 4: Elicit Sustainability Goals, Design Decisions, Obstacle, and Risk Analysis

After computing the architectural options, sustainability impact values, the risks to these ASs, and the corresponding risk values need to be modeled. The model used for the original case study had no definition of risk. We model the risks of each design decision alternative in terms of loss probability, criticality, and its effect on sustainability on a scale of 1–5.

Table 16.5 Summary of requirements for the case study

Requirements	Alternatives	Related quality/sustainability goals
Location finding	1. GPS 2. Radio transmission	1. Minimize cost 2. Minimize battery usage 3. Maximize reliability 4. Minimize response time 5. Minimize ramp-up time 6. Minimize development time 7. Minimize deployment time
Hardware platform	1. Nexus 1 (HTC) 2. Droid (Motorola)	
File sharing package	1. File manager 2. In-house	
Report synchronization	1. Explicit 2. Implicit	
Chat protocol	1. Openfire 2. In-house	
Map access	1. On demand (Google) 2. Cached (Google server) 3. Preloaded (ESRI)	
Connectivity	1. Wi-Fi 2. LG on Nexus 3. 3G on droid 4. Bluetooth	
Database	1. MySQL 2. SQLite	
Architectural style	1. Peer to peer 2. Client-server 3. Push based	
Data exchange format	1. XML 2. Compressed XML 3. Unformed data	

16.5.2.5 Step 5: Determine the Expected Benefit of Each Design Option

Based on the information collected, the total sustainability benefits of each AS can be calculated across goals by summing the impact on the sustainability goals associated with each goal (weighted by the importance of the goal and the attribute). This is calculated using the CBAM's equations (Eqs. 16.3 and 16.4).

16.5.2.6 Step 6: Analyzing the Costs, Benefits, and Risks Using Portfolio Thinking

The next step consists of analyzing the costs, benefits, and risks associated with alternative ASs to create an architectural portfolio which optimize for risks related to sustainability goals using portfolio thinking. After calculating the return and risks associated with each AS, the next step is to determine the overall effectiveness of a candidate architecture, which is liken to a portfolio of architectural strategies. The overall effectiveness of an architecture is a function of the total expected return and the global risk of that portfolio. Table 16.4 shows the result for one architecture.

16.5.2.7 Step 7: Identify the Optimal Architecture, Calculate Debt, and Rank Other Architectures

We shortlisted valid architectures that maximize the expected net benefits and minimize risks on dimensions related to sustainability. An architecture is valid, if it satisfies the relevant constraints such as cost, risk, schedule, or the sustainability score threshold. The overall effectiveness of an architecture is a function of the total expected return and the global risk of that portfolio. Table 16.6 shows the analysis and result of a candidate architecture, from selecting specific AS.

One of the requirements elicited by the original design team in EDS was to keep the cost of a single handheld device below $1000. For this example, we assume that the optimal architecture is the architecture with the highest Sharpe ratio while satisfying the cost constraint of $1000.

For the case study, there was a total of 6912 alternatives, and after the constraint was applied, we had about 2730 valid architectures and the optimal architecture had a cost of $968.94, an expected return of 11%, and a global risk of 68%. The whole method was implemented in Matlab to speed up the decision making.

Using Eq. (16.12), we ranked the other architectures from best (low T_D) to worst (high T_D).This can also be used for comparison and benchmarking of other candidate architectural portfolios. The architecture with a low debt Sharpe ratio is the one that achieves the best combination of value and risk in the presence of uncertainty. The ideal architecture is context dependent.

16.5.3 Findings

After modeling the different combinations of architectural strategies using portfolio thinking with the cost constraint, one optimal architectural design portfolio is selected. While some of the findings from the analysis may appear unsubtle, they

Table 16.6 Candidate architecture along with the results of the analysis

Goals	Candidate AS	Result of the analysis	
Location finding	Radio transmission	Ep (expected return)	10%
Hardware platform	Nexus 1 (HTC)	Rp (global risks)	77%
File sharing package	In-house	Cost	756.78 pound
Report synchronization	Explicit	2[4]*Sharpe ratio	2[4]*12%
Chat protocol	Openfire		
Map access	Preloaded (ESRI)		
Connectivity	Wi-Fi		
Database	MySQL		
Architectural style	Push based		
Data exchange format	Unformed data		

were not obvious prior to the evaluation of the architecture using the method and given the possible valid combinations for constructing a portfolio.

The main objective of the approach is to improve sustainability by maximizing the returns of architectural strategies and minimizing the risks associated with the architecture achieving the system's goals through a portfolio. The debt metric provides insights on the significance of an architecture in mitigating risks given the resources in hand. This is calculated as the gap between the Sharpe values of the architecture in question relative to the ideal architecture achieving the system's goals for resolving this obstacle. As investing in the ideal architecture is not always affordable, the metric is an expression for the risks tolerated if a different architecture is chosen. This analysis provides analysts and architects with an objective tool to assess and rethink their investment decisions in architecture using the expected returns, global risk, and Sharpe ratio of portfolio value assuming the validity and accuracy of the parameters' probability distributions. The use of the debt metric had made both the short-term and long-term risks visible in the evaluation and selection process. In addition, the results should improve the architect's confidence in the fitness of the proposed architecture toward meeting critical requirements.

The data reported in the case study are limited. We used the data from the literature where the case study was initially presented. We used subjective values, which were derived by consulting similar projects in the literature. While this may lead to subjective results, the estimates are useful in providing insights. The evaluation and the quantification of values we used were intentionally oversimplified to focus on demonstrating the technique and reasoning.

The technique presented in this chapter does not consider the interrelationship between the impacts on different value aspects.

The evaluation presented in this section is limited primarily by the assumed nature of the sustainability requirements. We assumed these requirements are visible to software engineers and that their impacts are easily quantifiable.

16.5.4 Threats to Validity

This case study was based on a study reported in the literature. The objective for carrying out the case study was to provide feedback on the method. For this reason, we relied on the parameters elicited by the initial case study and made some simplified assumptions. This imposed a few threats to their validity. These threats considered based on Wohlin [26] are construct, internal, and external validity threats.

1. Construct Validity: A threat to construct validity is that the case study was the data used for the study. The data was assumed a fair representation of the actual characteristics of the system to be built as it was elicited from stakeholders with expert knowledge. In reality, this cannot be guaranteed. In addition, the analysis was performed by a PhD candidate, who may be biased to produce a positive result. This is considered a threat to the *construct validity* of the study. This risk

was reduced by comparing the results of this study to the published results of the original study.

2. Internal Validity: This study does not account for all goals and architectural strategies that may affect the system's sustainability.
3. External Validity: A threat to external validity of the study was that the architecture had already been evaluated in the literature; hence, we had an idea of what the optimal architecture should be. In the analysis of a new system, deriving the parameters for the models may not be as straightforward or easy.

16.6 Discussions

Using the framework proposed by [27], we compared our evaluation technique with plain CBAM as shown in Table 16.7 The results indicate that the debt-aware and portfolio-based approach is a more optimized and holistic approach for evaluating architectures for sustainability. In particular, our method, which is based on the concept of portfolio theory for finding the optimal architecture with risks on the sustainability goals, offers a better coverage at optimizing architectural investment decisions and ranking candidate architectures based on debt. At the moment, the maturity stage rating of our method (i.e., inception) makes it hard to arrive at strong claims about its generalization. Consequently, more examples, case studies, and empirical investigations are required to further validate the method.

Table 16.7 Comparison of evaluation methods

Components	CBAM	Our technique	GuideArch
Maturity stage	Refinement	Inception	Refinement
Process support	Comprehensive		Comprehensive
Method's goals	Cost-based analysis	Cost, risk, debt-based analysis	Analysis under uncertainty
Quality attributes	Multiple attributes	Sustainability attributes	Multiple attributes
Applicable project stage	Before implementation	Before implementation	Before implementation
Evaluation approaches	Elicitation of values from stakeholders and tool-based analysis	Elicitation of values from stakeholders, historical data, and economic tool-based analysis	Elicitation of values from stakeholders, fuzzy numbers, and tool-based analysis
Tool support	Available	Partially available	Available
Stakeholders involved	All major stakeholders	All major stakeholders	All major stakeholders
Ranking	Use return on investment	Uses the Sharpe ratio	Uses the normalized value of the architecture

Several aspects of the method contribute to making the method we propose easy to use. For instance, the wide use of guidelines and templates to explain each phase of the method facilitates the application of the method by users without previous experience in evaluating the requirements at the architectural level. The method is based on intuitive concepts organized in a structured manner, which contributes to its repeatability. The users of the method do not need to understand the details of the decision theory behind the method. On the other hand, they must have a comprehensive understanding of the organization's objectives and knowledge about the domain. Given that we have personally conducted the case studies, it is difficult to have a conclusive validation of the overall ease of use of the method. As criticism, eliciting the numeric values for the portfolio analysis is difficult. The evaluation team preferred to use a qualitative scale ranging from low, medium, and high. As a result, it is necessary for further evaluation to test the suitability of the prioritization in large selection projects.

It was observed the method covers all major steps necessary to conduct a disciplined evaluation process. The phases provide a clear, structured process to guide the acquisition and analysis of relevant information. The benefit of the method is that it provides guidance on how to manage risk by quantifying the debt associated with different candidate architectures. We evaluated the method using some general qualitative characteristics including simplicity of use, openness, and comprehensiveness:

1. A notable desirable feature of the method is its flexibility and openness; the method does not define rigorous ways for estimating its parameters, conducting its steps, and confirming specific actions to execute. The nature of the decisions made when applying the technique varies from one project to another, with the addressed problem, and across domains. As a result, the effectiveness of its application is subject to the context in which the model is applied.
2. The method is open; it could be easily integrated to complement existing evaluation methods, with the objective of explicit evaluation for compliance while taking a value-based perspective.

Using our method takes more time than the CBAM as more trade-offs are involved in the analysis. In an environment where numerous components are involved, our method may require much more time to get the results. The criticality of the architecture or application domain may guide the solution and how this trade-off can be prioritized, ignored, or empowered.

Portfolio theory is a well-accepted concept for diversifying risks; it is well grounded in theory. The framework presented here, although useful, has its limitations. Analyzing the portfolio depends on identifying the threats and estimating their likelihood. This approach assumes sufficient awareness and experience of sustainability goals, which can be subjective and suffer from inherent limitations for completeness and accuracy. Furthermore, it assumes that stakeholders are confident enough to anticipate the probabilities and the likely risks involved. Nevertheless, anticipating risks is rather a subjective exercise, which can be biased to the perspective and the experience of the stakeholders involved. Consequently, due to the different variables that might be estimated in a subjective way, this approach can only provide a best-case/good-enough portfolio rather than optimal portfolio.

The requirements tend to change over time. Though the current study does not explicitly cater to change and evolution in sustainability goals, the approach assumes that the considered requirements provide baseline for realizing sustainability at that specific time. However, the same process can be reiterated with any incoming requirements and changes to sustainability goals. The ability of an architecture to respond to change can also be modeled as a dimension for technical sustainability subject to the extent to which we can anticipate changes.

We now summarize the lessons learned:

1. The evaluation method largely depends on the expertise of the analysts and developers. From conducting the case study, a number of benefits of having a disciplined method to guide the evaluation were observed. By following the phases presented in the method, it was possible to provide a systematic and repeatable process of gathering relevant information, analyzing alternatives, and making decisions. However, it was also recognized that no matter how objective, there are still underlying values and subjectivity in the ways that people perceive, judge, and ultimately make decisions. Consequently, we consider that largely the success of the selection process depends on the experience and domain knowledge of the evaluation team.
2. The method to guide the evaluation needs to be customizable. It provided useful guidance to conduct the case studies. Some parts of the method offered valuable support, such as evaluating alternative architectures for value and debt. On the other hand, although approaches such as the obstacle analysis technique were beneficial to the project, their use was considered complex and time demanding. Even when using a different risk analysis approach, the evaluation exercise proved to be valid. In addition to that, it is obvious that there is no one-size-fits-all solution to developing software systems.
3. Good understanding of organizational requirements is vital for the success of selection process.
4. Sustainability debt is subjective. It can be affected by the following: (1) the decision makers and their expertise (e.g., stakeholders), (2) the elicited parameter values, (3) the importance of the requirements to the system design, and (4) the potential use and value of the system. The threshold of debt depends on the acceptance of stakeholders.
5. Value is subjective. It depends on the different variables. This can be modeled as a function of the utility of the system, the dependencies, and the acceptance level of the stakeholders.

16.7 Conclusion

We have motivated the need for architectural evaluation methods suitable for evaluating sustainability in architectures. In particular, we have presented an evaluation method based on the CBAM, portfolio theory, and technical debt for evaluating the sustainability goals in architectures. The method aims at identifying the portfolio(s) of architectural design decisions which are more promising for adding/delivering value while reducing risk on the sustainability dimensions. The proposed method hopes to identify more trade-offs and candidate ASs, which are risk and debt aware on dimensions related to sustainability. These can be technical, human, social, economic, and/or environmental. The ultimate goal is to develop an objective decision-support framework for reasoning about sustainability requirements in relation to architectural decisions under uncertainty. We have illustrated the approach with an Emergency Deployment System (EDS). The results show that the method can make the value, cost, and risks of architectural design decisions and sustainability requirements explicit.

References

1. Asundi J, Kazman R, Klein M (2001) Using economic considerations to choose among architecture design alternatives
2. World Commission on Environment and Development (1987) Report of the world commission on environment and development: our common future (the Brundtland report). Med Conflict Survival 4
3. Penzenstadler B (2013) Towards a definition of sustainability in and for software engineering
4. Brown N, Ozkaya I, Sangwan R, Seaman C, Sullivan K, Zazworka N, Cai Y, Guo Y, Kazman R, Kim M, Kruchten P, Lim E, MacCormack A, Nord R, Guo R, Ozkaya R (2010) Managing technical debt in software-reliant systems. In: Proceedings of the FSE/SDP workshop on Future of software engineering research, ser. FoSER '10, p 47
5. Ernst NA (2012) On the role of requirements in understanding and managing technical debt
6. Ojameruaye B, Bahsoon R (2014) Systematic elaboration of compliance requirements using compliance debt and portfolio theory. LNCS vol 8396
7. Chen L, Babar MA, Nuseibeh B (2013) Characterizing architecturally significant requirements. IEEE Softw 30
8. Penzenstadler B, Raturi A, Richardson D, Calero C, Femmer H, Franch X (2014) Systematic mapping study on software engineering for sustainability (se4s)
9. van Lamsweerde A, Letier E (1998) Integrating obstacles in goal-driven requirements engineering
10. Kazman R, Asundi J, Klein M (2001) Quantifying the costs and benefits of architectural decisions. In: Proceedings – International Conference on Software Engineering
11. Kazman R, Asundi J, Klein M (2002) Making architecture design decisions: an economic approach. Software Engineering Institute Technical Report CMU/SEI-2002-TR-035
12. Markowitz H (1957) Portfolio selection: efficient diversification of investments
13. Butler S, Chalasani P, Jha S, Raz O, Shaw M (1999) The potential of portfolio analysis in guiding software decisions. In: Proceedings of the First Workshop on Economics-Driven Software Engineering Research (EDSER1)

14. Sivzattian S, Nuseibeh B (2001) Linking the selection of requirements to market value: a portfolio-based approach. In: REFS 2001. pp 202–213
15. Khurum M, Gorschek T, Wilson M (2013) The software value map – an exhaustive collection of value aspects for the development of software intensive products. J Softw Evol Process 25
16. Clements P, Kazman R, Klein M (2001) Evaluating software architectures: methods and case studies. Addison-Wesley Professional
17. Kazman R, Bass L, Abowd G, Webb M (1994) Saam: a method for analyzing the properties of software architectures
18. Esfahani N, Malek S, Razavi K (2013) GuideArch: Guiding the exploration of architectural solution space under uncertainty
19. Letier E, Stefan D, Barr ET (2014) Uncertainty, risk, and information value in software requirements and architecture
20. Kazman R, Klein M, Clements P (2000) Atam: Method for architecture evaluation. Cmusei 4:83
21. Raturi A, Penzenstadler B, Tomlinson B, Richardson D (2014) Developing a sustainability non-functional requirements framework. In: Proceedings of the 3rd International Workshop on Green and Sustainable Software – GREENS 2014. pp 1–8
22. Van Lamsweerde A, Letier E (1998) B-Louvain la-neuve Belgium. Integrating obstacles in goal-driven requirements engineering
23. van Lamsweerde A (2001) Goal-oriented requirements engineering: a guided tour
24. Venters CC, Lau L, Griffiths MK, Holmes V, Ward RR, Jay C, Dibsdale CE, Xu J (2014) The blind men and the elephant: towards an empirical evaluation framework for software sustainability. J Open Res Softw 2
25. Van Lamsweerde A (2004) Elaborating security requirements by construction of intentional antimodels
26. Wohlin C, Runeson P, M Höst, Ohlsson MC, Regnell B, Wesslén A (2012) Experimentation in software engineering, vol 9783642290442
27. Babar MA, Zhu L, Jeffery R (2004) A framework for classifying and comparing software architecture evaluation methods, vol 2004

Printed in the United States
by Baker & Taylor Publisher Services